ELECTRON COOLING
and
NEW COOLING TECHNIQUES

Proceedings of the Workshop on

ELECTRON COOLING and NEW COOLING TECHNIQUES

Laboratori Nazionali di Legnaro, Legnaro, Padova – Italy
15 – 17 May, 1990

Editors

R. Calabrese
University of Ferrara, Italy

L. Tecchio
University of Torino, Italy

World Scientific
Singapore • New Jersey • London • Hong Kong

Published by

World Scientific Publishing Co Pte Ltd
5 Toh Tuck Link, Singapore 596224
USA office: 27 Warren Street, Suite 401-402, Hackensack, NJ 07601
UK office: 57 Shelton Street, Covent Garden, London WC2H 9HE

ELECTRON COOLING AND NEW COOLING TECHNIQUES

ISBN-13 9789810205119
ISBN-10 9810205112

INTERNATIONAL ADVISORY COMMITTEE

R. CALABRESE, *Univ. of Ferrara*
P. DALPIAZ, *Lab. Naz. of Legnaro*
E. JAESCHKE, *MPI Heidelberg*
D. MOEHL, *CERN, Geneva*
R. E. POLLOCK, *Univ. of Indiana*
A. N. SKRINSKY, *INP Novosibirsk*
T. TANABE, *Univ. of Tokyo*
L. TECCHIO, (Chairman) *Univ. of Torino*

LOCAL ORGANIZING COMMITTEE

R. CALABRESE, *Univ. of Ferrara*
G. CARUGNO, *Lab. Naz. of Legnaro*
P. DALPIAZ, *Lab. Naz. of Legnaro*
F. PETRUCCI, *Univ. of Ferrara*
P. SCHIAVON, *Lab. Naz. of Legnaro*
L. TECCHIO, *Univ. of Torino*
M. VENCO, *Lab. Naz. of Legnaro*

PREFACE

This workshop was organized to summarize the present knowledge on electron cooling experimental results and theory, and at the same time to discuss ideas for new cooling techniques and future applications.

There has been much progress since the first pioneer study of G. Budker in Novosibirsk. The major amount of work on electron cooling has been done in his laboratory, which constitutes a reference point for physicists working in this field.

Nowadays, several coolers are operating in different laboratories; the results show the very good performances and reliability of the electron cooling technique applied to the low and intermediate energies ion and/or proton/antiproton storage rings.

The study of new cooling techniques, like laser cooling, are in progress and the first results look very promising.

The workshop was divided mainly in three parts. The first day was devoted to the theoretical aspects of electron cooling technique and to the presentation of the experimental results from the several coolers actually in operation.

On the second day the technological problems connected with this subject (e-beam generation, energy recovery, vacuum, ...) were discussed.

All the activities about the new cooling techniques and future applications were discussed on the last day. In this section the first results about the laser cooling, the ultra-cold electron beam generation and the ion beam crystallization were presented.

We hope that this workshop has helped to promote the case of electron cooling and encouraged developments in this and other new cooling techniques.

It is the appropriate place here to thank our colleagues from the International Advisory Committee for the firm committment in preparing and carrying out this workshop. The help of the Local Organizing Committee, the secretaries of the Laboratori di Legnaro, Mrs. M. Venco, Mrs R. Nicolè and Mr. P. Schiavon is warmly acknowledged. In particular, we would like to thank the Laboratory Nazionali di Legnaro and the Istituto Nazionale di Fisica Nucleare for their generous support of this meeting.

R. Calabrese and L. Tecchio
Editors

CONTENTS

ELECTRON COOLING
and
NEW COOLING TECHNIQUES

CERN PS/90-43 (AR)

STATUS OF ELECTRON COOLING EXPERIMENTS AT LEAR

J. Bosser, M. Chanel, R. Ley, D. Möhl,
F. Ollenhauer, and G. Tranquille

European Organization for Nuclear Research (CERN)
CH-1211 Geneva 23

ABSTRACT

Since the initial cooling experiments performed with protons and antiprotons and reported in earlier papers[1,2], the LEAR electron cooler has undergone a series of improvements and has been used to successfully cool O^{8+}, O^{6+}, and H^- ions. This paper aims to present the major modifications made to the cooler and the LEAR environment since its installation in 1987, and also to review cooling results obtained with the different particles and ions. Future plans for the LEAR cooler will also be discussed.

1. INTRODUCTION

The electron cooling device for LEAR was developed by a collaboration between CERN and KfK Karlsruhe and was installed in LEAR in 1987. Initial cooling experiments were performed with protons in November 1987 and March 1988, and with antiprotons in March 1989[1,2]. In the period between the proton and $\bar{p}$ tests, cooling of H^-, O^{8+}, and O^{6+} ions was performed and in November 1989 O^{8+} ions was successfully accumulated in LEAR using electron cooling.

The LEAR cooler (Fig. 1) is situated at the center of the third straight section (SL3) of the LEAR machine (Fig. 2) and is designed to cool particles with an equivalent momentum up to 308.6 MeV/c. The electron beam is generated in an electron gun and is accelerated to the desired energy via four ring shaped anodes. It is then transported in a solenoidal field and bent into the cooling section by a 36° toroid magnet. After the cooling section, which is 1.5 m long, the electrons are deflected by another toroid to be recuperated by the collector. A detailed description of the cooler can be found in Ref. 3, and Table 1 gives the typical operational parameters.

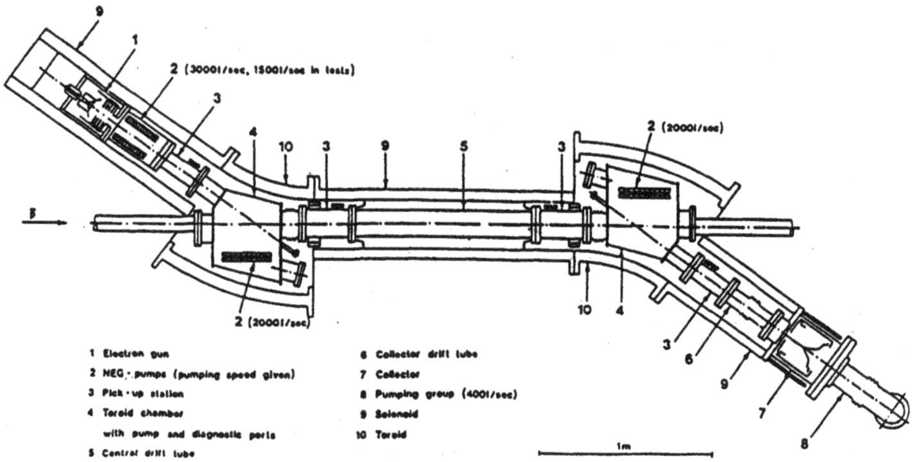

Figure 1 - The LEAR electron cooling device.

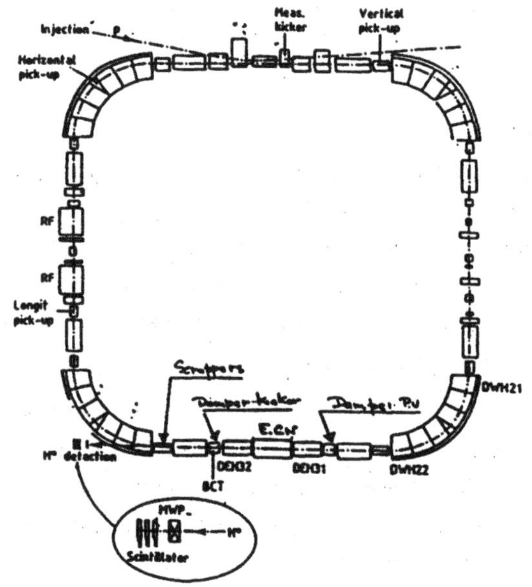

Figure 2 - Layout of the LEAR machine.

Table 1 - Operational parameters for the LEAR cooler.

Beam momentum	308.6	200.	105.0	147.4	MeV/c/n
e^- beam energy	26.93	11.48	3.24	6.23	keV
e^- current	2.5	0.7	0.088	0.24	A
Solenoid field	445.	300.	145.	268.	G

Apart from the obvious effect of cooling, the cooler also perturbs the circulating beam in a number of ways:

1. The hot cathode and electrons that are lost on the vacuum chamber cause an immediate increase in the mean pressure around the ring. Despite the use of NEG pumps in the collector region the pressure in SL3 has never been better than 6×10^{-12} Torr for 3‰ losses of a 3.2 keV electron beam.

2. The 36° toroids also produce a vertical magnetic field which kicks the beam by 10 mrad in the horizontal plane. Two horizontal deflecting magnets situated on either side of the cooler are used to correct the beam trajectory and to center it in the cooling region.

3. The longitudinal field of the main solenoid also causes the transverse betatron motions of the beam to couple. This could lead to beam loss if the machine is not properly adjusted and the working point is too near to a coupling resonance. Moreover a slight tune shift is observed when the cooler solenoid is switched on and also from the focusing effect of the electron beam. Table 2 shows the combined effect measured for various momenta.

Table 2 - Tune shifts due to electron beam and solenoid

Beam momentum	308.6	200.0	105.0
ΔQh (10^{-3})	-11.0	5.0	0.0
ΔQv (10^{-3})	-8.0	0.0	7.0

The control room for the cooler is situated some 30 m away from the apparatus itself in a building that houses all the necessary power supplies and the Faraday cage. At present the control system is completely stand alone and consists of an LSI 11/23 computer connected to a two crate CAMAC loop on a fibre optic link. All the power supplies are controlled via DACs of various resolutions and the important parameters are acquired via Fischer 12 bit ADCs. At present the LSI is being replaced by a workstation oriented control system similar to that used at LEAR. A detailed description of the control system can be found in Ref. 4. For routine operation at LEAR, planned for this year, it is envisaged to use the cooler in what we call the "pulsed mode" of operation. As the main perturbation induced by the cooler is the deformation of the horizontal orbit, the solenoid is controlled via a GFD (digital function generator) and the dipole bumps needed to correct the orbit are applied before any cooling is performed. In this manner the solenoid can be ramped synchronously with the LEAR machine and at all times the orbit is compensated. The correction coils are adjusted via a program on the ecool control system which sets the correct values for each energy. When the beam has been decelerated onto a standard machine "flat-top" the high voltage supply, which is also controlled by a GFD, is pulsed to the operational value in 120 ms and is kept on for 6 seconds. Figure 3 shows the principle

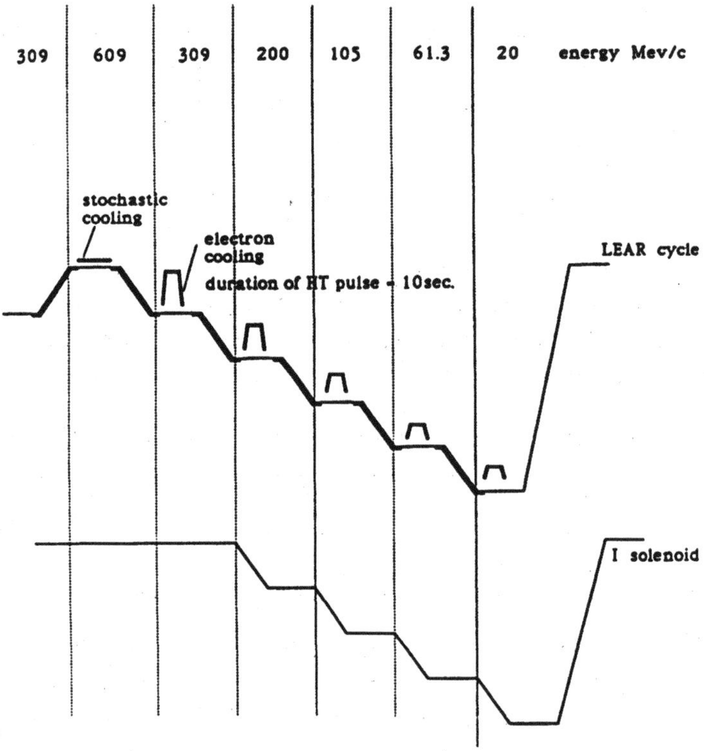

Figure 3 - Principle of the pulsed mode of operation

of this operation for the low energy (LE) cycle during which electron cooling can be used. In this way we have managed to decelerate proton beams to 105 MeV/c saving some 15 minutes in cooling time as compared to stochastic cooling.

The major hardware modifications made to the cooler since its installation have concerned its reliability and reproducibility in everyday operations. As explained previously GFDs have been introduced as a means of synchronizing the cooler with the LEAR machine for future operation. A great effort was made in increasing the high voltage stability which was the suspected cause of longitudinal instabilities in earlier experiments. A second, more stable, power supply was also added in parallel for electron energies of less than 3 keV and to allow even more flexibility, appropriate electronics were developed to enable a remotely controlled inversion of the uni-polar Delta power supplies used on some steering coils. For the automatic compensation of the cooler toroids and main solenoid, two tilted solenoids have been installed in place of the horizontal correctors DEH31 and DEH32. These solenoids have been calculated to completely eliminate the coupling effect of the main solenoid and are tilted by 15° to kick the beam by 10 mrad in the opposite direction to the cooler toroids. Their effect on the closed orbit can be seen in Fig. 4.

Software development was also very important during this time with a number of additional programs being written for demagnetizing the solenoid, reading and setting the electron cooler security synoptic and also to trigger the high voltage GFD on any of the eight tables available for use. Fibre optic cables were also laid down to transmit to the control room from the Faraday cage the current loss on the different components of the collector. It is envisaged to acquire these currents and status for an alarms/surveillance program currently under development. The hardware necessary for the remote acquisition of the electrostatic pick-ups has also been installed and will be tested during the year.

Apart from the new control system which is at present being implemented[4)], two major projects are under development and should be operational before next year. The first is the new collector which is being developed by a CERN/CAPT Lipetsk collaboration [5)] (CAPT is a department of INP Novosibirsk). It should greatly improve the recuperation efficiency of electrons and help to improve the reliability of this component. The present collector is too fragile to run with 27 keV electrons. High voltage beakdowns are too frequent to envisage its regular use at 308.6 MeV/c. It is for this reason why we will limit electron cooling operations to 200 MeV/c and under until the present collector can be replaced. The second project concerns the performance of LEAR and consists of the installation of a feedback system known as a damper acting on the circulating beam at

LEAR. In previous experiments it was found that a beam of about 10^9 (anti)protons became very unstable under the conditions (transverse emittances, momentum spread and particle intensity) reached with electron cooling. It was therefore decided to develop a feedback system to counteract this cherent instability The characteristics of this device will be explained in more detail in a later section.

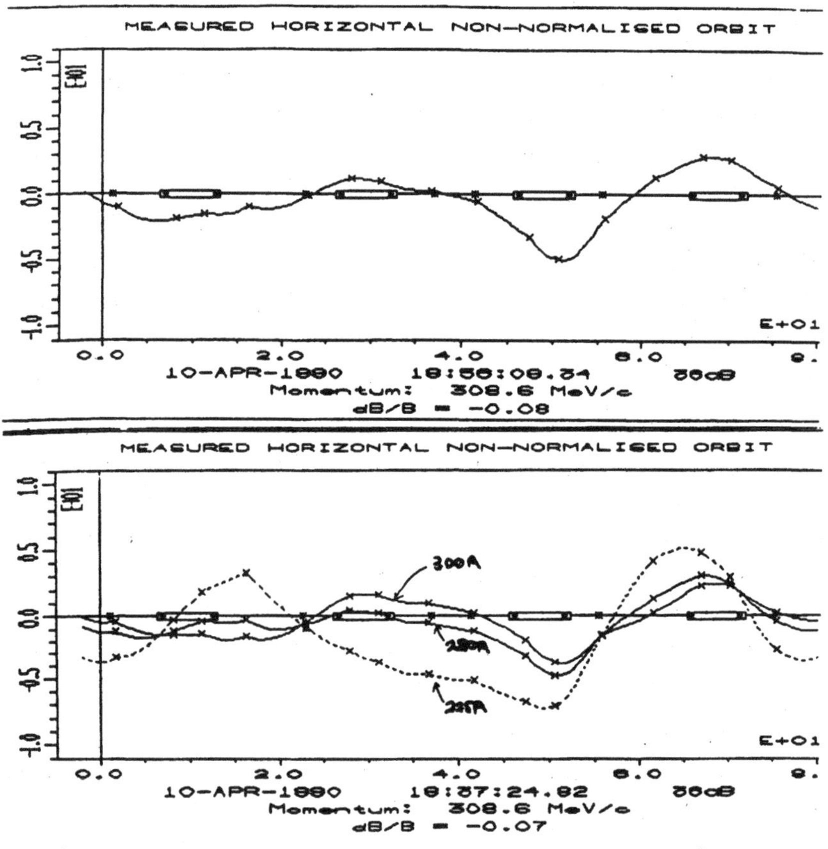

Figure 4 - Compensation of the cooler toroid deflection with two tilted solenoids. The first trace shows the normal machine closed orbit. The second trace shows three orbits with the ecool solenoid at 100 A and the compensation solenoids at 300 A, 280 A and 225 A respectively.

For operations in the future it would also seem desirable to replace the present electron gun with a variable perveance gun. This would allow us to increase the cooling efficiency at low energies.

2. INSTRUMENTATION

Electrostatic pick-up stations distributed around LEAR offer the possibility to measure the circulating beam position when bunched. In addition to these, the cooler has four such pick-up stations (Fig. 1) used for position monitoring of the electron beam. The two pick-ups in the drift section can also be used for machine closed orbit measurement when precise matching of the two orbits is required. At positions indicated in Fig. 2, Schottky pick-up electrodes are installed. The Schottky signal is the signal induced by the particles of a coasting beam and represents the density fluctuation. This signal can be observed on a spectrum analyzer where information on the momentum spread and emittance of the beam can be deduced. Great care is necessary in the interpretation of these signals when the beam is strongly cooled such that the random nature of the particle motion is lost. Another means of measuring the beam size is by the use of moveable scrapers installed a few meters upstream from the cooler which are used to sense the beams edges. This method is destructive as it interacts directly with the circulating beam and one should also be very careful in interpreting the results when the beam might be performing coherent oscillations. For protons with a momentum greater than 300 MeV/c the neutral hydrogen channel is a very highly effective way of measuring the cooled beam size and the alignment of the electron and coasting beams. The neutral hydrogen beam, which is created in the cooling section when an electron is captured by a proton, is observed in a detection system mounted at the extension of the vacuum tube of SL3. It consists of a multiwire proportional chamber (MWPC) with two planes and a stack of three scintillators. The MWPC covers an area of 10 cm × 10 cm and their thicknesses in the beam direction are 5 mm, 1 cm, and 1 cm, respectively.

3. EXPERIMENTAL RESULTS

Table 3 - Summary of results of electron cooling of protons and antiprotons

Beam momentum	308.6	200.0	105.0	MeV/c
$\Delta p/p$	0.7	0.4	0.4	‰
εh	1.3	3.0	5.0	π mm.mrad
εv	1.8	5.0	9.0	π mm.mrad
Cooling time ($\Delta p/p$)	0.6	0.3	0.3	seconds

further on. The types of measurements made have usually concentrated on the longitudinal plane and consist of measurements of the longitudinal frictional force and cooling times. Transverse emittances were measured using the scrapers or, when possible, using the neutral hydrogen channel.

3.1 Longitudinal cooling times

For this measurement two similar methods have been tested. The first method (method 1) consists in stepping away the electron beam energy via the high voltage power supply and then resetting it to the operational value. By observing the longitudinal spectral density in a narrow bandwidth around a harmonic of the revolution frequency on a spectrum analyzer which is triggered at the same time as the high voltage step, the cooling time can be determined. The second method (method 2) uses the spectrum analyzer in the same way, but instead of stepping away from the operational voltage, radio-frequency (RF) noise at a different harmonic of the revolution frequency is put on a longitudinal gap with bandwidth and power adjusted to blow the beam up to a momentum spread of about 10^{-3}. When this noise is switched off the spectrum analyzer is triggered and the spectral density evolution is observed. Examples of the signals observed are shown in Figs. 5 and 6.

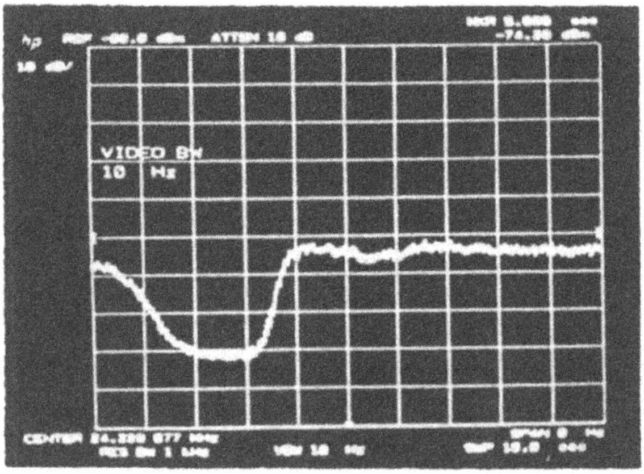

Figure 5 - Longitudinal cooling time measurement using method 1. The curve gives the noise density (which is proportional to the square root of the particle density) at the nominal momentum. Horizontal scale is 1 sec/div.

In order to obtain the velocity dependence of the frictional force F(v) from the equilibrium distribution ρ(v), one has to solve the one-dimensional Fokker-Planck equation for a frequency independent diffusion constant D:

$$\frac{\partial\rho}{\partial t}=\frac{\partial}{\partial v}\left(-F(v)\rho(v)+D\frac{\partial\rho}{\partial v}\right)$$

In the equilibrium case $\partial\rho/\partial t = 0$ the shape of the frictional force is determined by the normalized slope of the distribution function:

$$F(v)=D\frac{\partial\rho/\partial v}{\rho(v)}$$

The diffusion constant D is derived experimentally from an independent measurement (Ref. 2, p. 331). Figure 8 shows the result of two measurements made with 50 MeV $\bar{p}$s, one with aligned beams and the other with misaligned beams (angle of about 1 mrad).

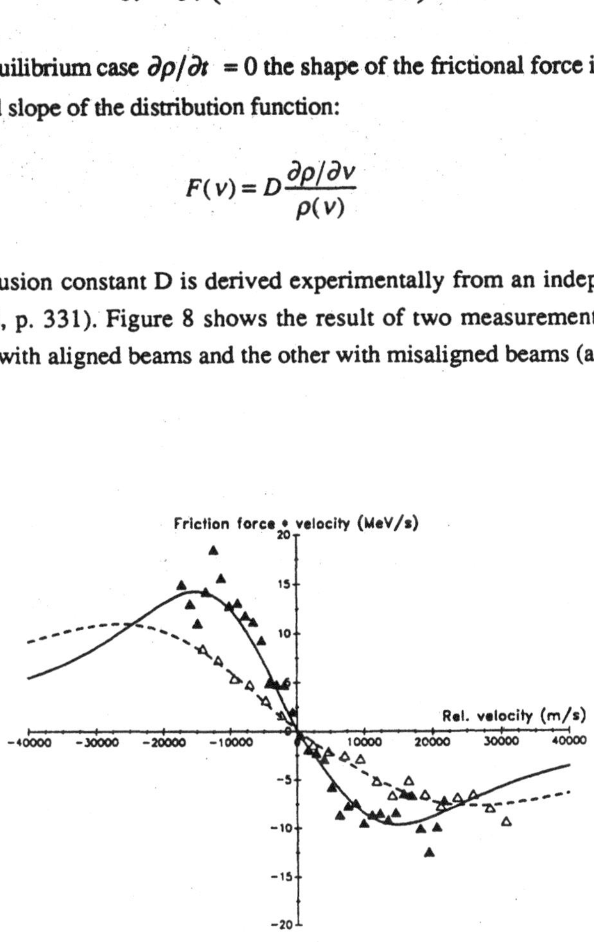

Figure 8 - The longitudinal frictional force F(v) as a function of v for 50 MeV $\bar{p}$s. The filled triangles correspond to the measurement made with aligned beams which the unfilled triangles to the measurement with misaligned beams.

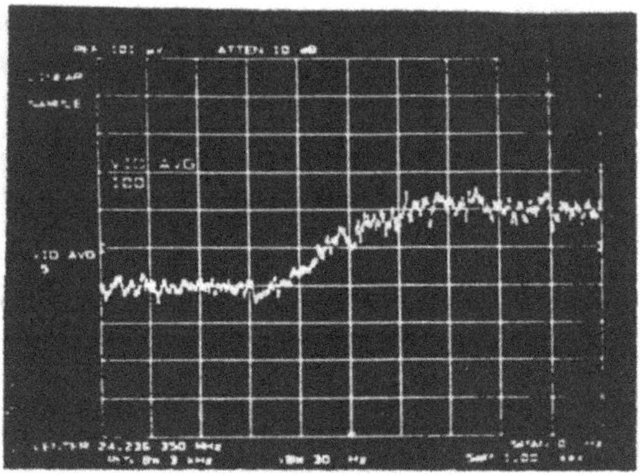

Figure 6 - Longitudinal cooling time measurement using method 2. The curve shows the increase in noise density as the heating noise is switched off. Horizontal scale is 0.1 sec/div

3.2 Longitudinal Frictional Force

In order to determine the frictional force at low relative velocities a method analyzing the distribution in equilibrium between a constant stochastic heating power and electron cooling was developed. The measurement was performed in the following way (Fig. 7). RF noise with a bandwidth of 10 kHz and a power of 5 μW/Hz was injected onto a cooled beam at 13.126 MHz on a RF gap. This produces a broader distribution of the momentum spread which can be increased or reduced by varying the attenuation on the noise.

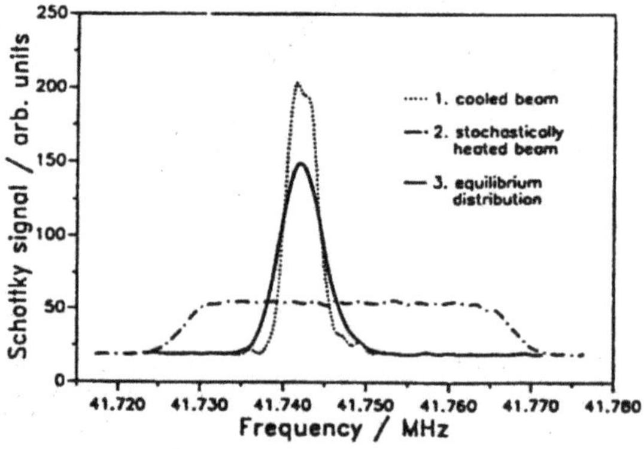

Figure 7 - Principle of the longitudinal frictional force measurement

3.3 Results with H^-, O^{6+}, and O^{8+} Ions

Very little time was given to electron cooling for tests with H^-, O^{6+}, and O^{8+} ions as the LEAR machine had other priorities. However, we managed to show that electron cooling could be used effectively to cool these ions with the exception of H^- where the cooled beam intensity decayed rapidly: probably due to mutual ("intra-beam") stripping of the H^- ions on the dense beam.

3.3.1 H^- ions

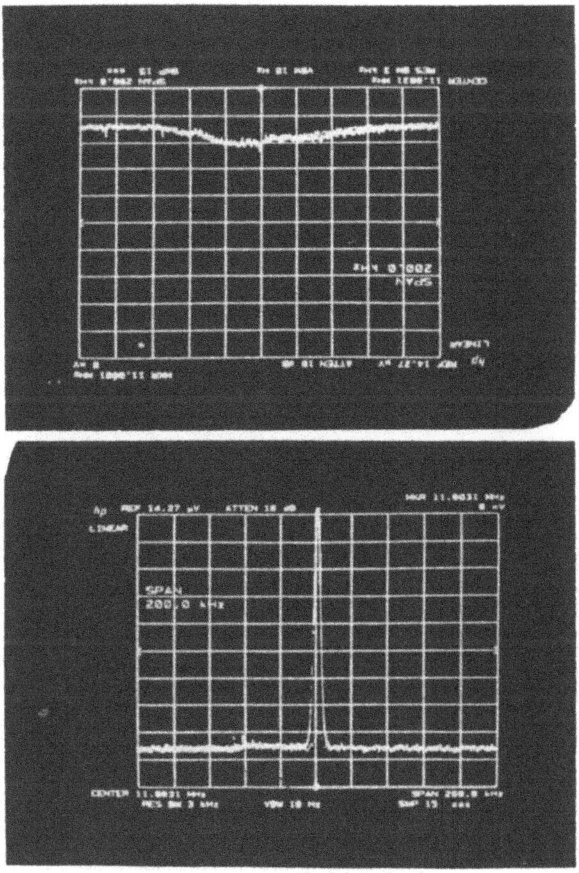

a)

b)

Figure 9 - (a) Longitudinal Schottky scan of H^- ions just after injection and (b) with electron cooling switched on. Horizontal scale: $\Delta p/p = 2 \times 10^{-4}$/div.

H^- ions (as for protons and other ions at CERN) are produced in the LINAC 1 where they are accelerated to 50 MeV kinetic energy. They are then transported via the 180° loop and the E2 transfer line normally used for proton transfers to LEAR where they can be injected in a number of different modes[6)]. The injected intensity was of the order of 10^8 ions per shot. For electron cooling tests the ions were injected via the normal channel with the LEAR machine in antiproton polarity. Electron cooling was kept on at 2 kV less than the operational and stepped up once a beam had been injected. Figure 9 shows the longitudinal distribution after injection and just after electron cooling has been stepped up to the operational value. If we look at the time evolution (Fig. 10) at the central frequency, we see a very sharp peak just after electron cooling is switched on which would seem to indicate an extremely fast cooling time. However, if one looks at the beam intensity evolution with and without electron cooling (Fig. 11) one sees that the lifetime of the H^- ions is greatly reduced when the cooler is on. This phenomenon is probably due to intra-beam stripping where the loosely bound electron is stripped off in a new collision with another H^- of the same beam forming a neutral hydrogen atom which is lost.

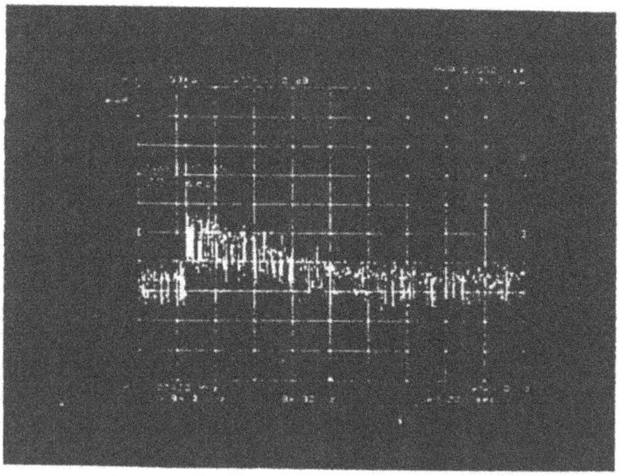

Figure 10 - Longitudinal cooling time for H^- ions.

3.3.2 O^{6+} and O^{8+} ions

The aim of the tests with O^{6+} and O^{8+} ions was to obtain some idea of the life-time of heavy ions in a good vacuum cooler ring and to test the stacking of ions[7)]. As will be seen later, electron cooling proved to be an invaluable tool in the stacking process and also to counteract the beam decay due to multiple scattering on the residual gas. O^{6+} ions were delivered at 115.6 MeV/c/n while O^{8+} ions had a momentum of 147.4 MeV/c/n. The measured momentum spread wad $\Delta p/p = \mp 3^0/oo$ and the number of particles was estimated to be about 4×10^6 per shot.

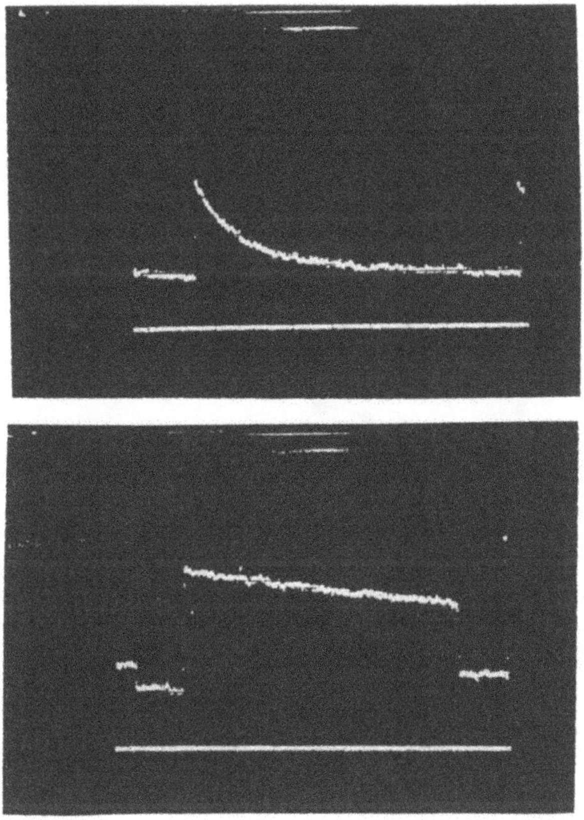

a)

b)

Figure 11 - Beam intensity evolution of H^- ions with (a) and without (b) electron cooling.

The multi-injection process of ions is performed in four consecutive phases. Figure 12 illustrates the principle of this injection mode where the two LEAR radio-frequency cavities (CRF41 and CRF42) are working respectively on the first and second harmonic of the revolution frequency. This effectively means that the circumference of the machine can be divided into two halves corresponding to what will be called the "injection bucket" and the "stack". The four phases are outlined as follows:

1. CRF41 is ramped to its operational voltage on harmonic 1 followed by CRF42 working on harmonic 2 100 ms later. The voltage on CRF41 is reduced adiabatically to zero and the injection kicker, which is synchronized on the first cavity and delayed in such a manner that the injected beam will fall into the injection bucket, is triggered.

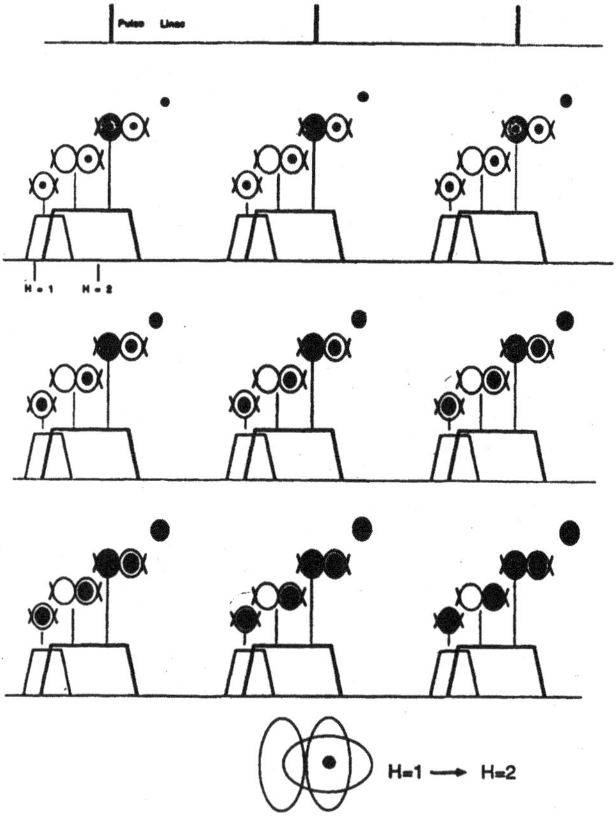

Figure 12 - Principle of the multi-injection process.

2. A few milliseconds after injection CRF42 is also reduced to zero and the beam is debunched, filling the whole circumference of the machine. Electron cooling, which is on during the whole process, reduces the momentum spread and transverse emittances of the coasting beam for a further 1.8 s.

3. CRF41 is then set on and bunches the beam on harmonic 1.

4. CRF42 is brought to its nominal value with the right phase so that the bunched beam is trapped within the "stack" and the empty "injection bucket" is filled as described under point 1 above.

The repetition rate for the above process is the LINAC 1 repetition rate of 2.4 s. This is ample time for electron cooling to reduce the beam characteristics to a $\Delta p/p$ of 4×10^{-4} and transverse emittances of the order of 3π mm.mrad. Longitudinal cooling time measurements in fact showed a reduction in momentum spread of a factor of 5 under 300 ms with emittances as low as 1π mm.mrad. Such small spreads in beam energy and size (Fig. 13) were only possible with the use of the damper which will be described in the next section.

4. THE TRANSVERSE FEEDBACK SYSTEM - THE DAMPER

4.1 Observations of instabilities

Transverse instabilities can be observed (Fig. 14) on both the Schottky and position pick-ups. If the spectrum analyzer is used to observe the variation in amplitude of an (n-q) sideband as a function of time, the instability manifests itself by an abrupt rise in the amplitude. This is followed by a cooling period of about 1 to 10 seconds before threshold is reached again. In the simple case of a dipole type instability where the beam oscillates like a string, the beam center is represented by a travelling wave type of oscillation with an amplitude $x_0(t)$ which grows exponentially in time (chromaticity is neglected):

$$x(s,t) = x_0 e^{i[(n \pm q)\omega_{rev} t \pm (ns/R)]}$$

Here n is the mode of the oscillation and q the non-integer part of the betatron tune Q. We can observe the signal on the spectrum analyzer at the lower frequencies where we find the most unstable modes. The mode frequency is given by $\varpi_n = (n \pm q)\varpi_{rev}$ and at LEAR it was found that the low order slow (minus) waves for $n \leq 3$ were the most unstable ones. Usually, when instabilities occur, beam is lost until the number of particles is about 10^9. At this intensity the blow-up occurs without noticeable particle loss. It should be noted that the same process occurs with stochastic cooling but for higher particle intensities.

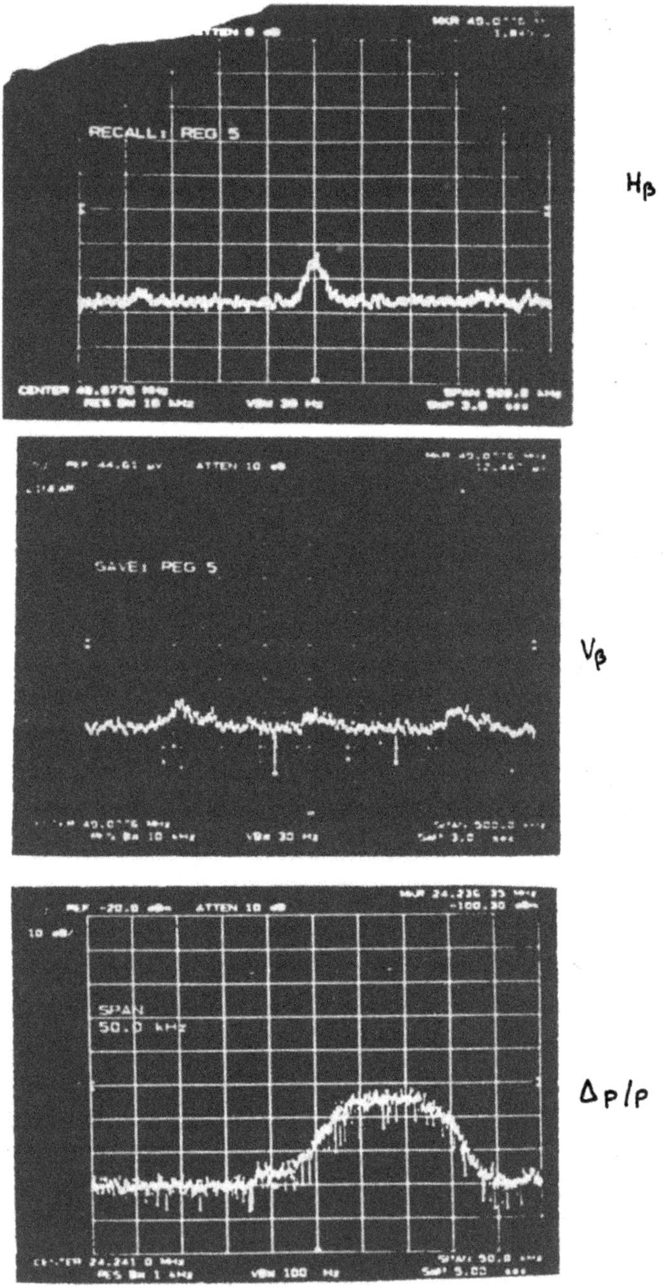

Figure 13 - Schottky scans of a cooled O^{8+} beam.

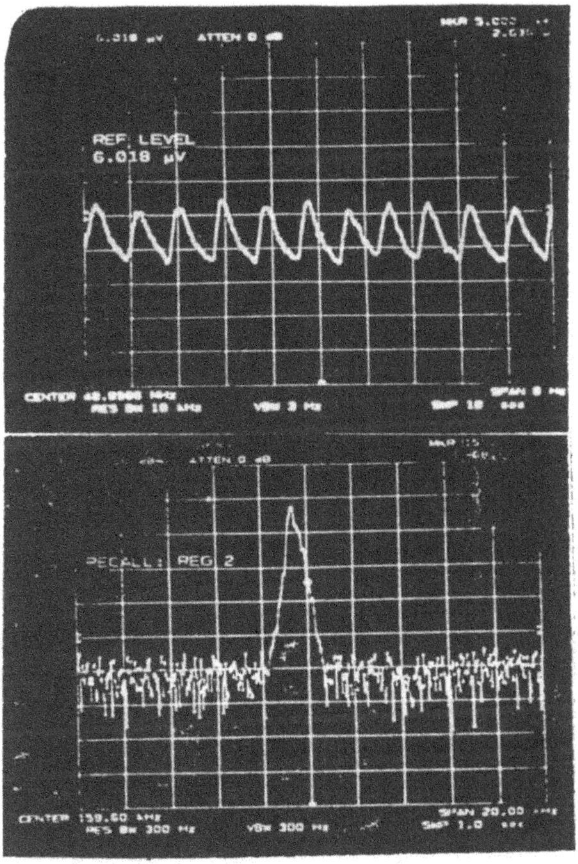

Figure 14 - Transverse instabilities observed on (a) the Schottky pick-up and (b) on the position pick-up. In Fig. a) the height of the transverse Schottky signal near 40 MHz is displayed. The horizontal scale is 1 s/div. The instability occurring at this frequency causes a large coherent oscillation which smears out and leads to an emittance growth during 0.2 s. This blow-up is then compensated during 0.8 s by the cooling before the next burst of instabilities occurs. In (b) the spectral density at low frequency near one of the bands where the instability occurs is displayed. The spike representing the beam oscillation jumps up with each burst of the instability.

4.2 The Causes of Instabilities

During the cooling process the beam density $N/\varepsilon_h\varepsilon_v\varepsilon_\ell$ increases and the beam can become unstable when this density is greater than a given threshold. An overview of instabilities in cooled beams can be found in Ref. 8. Under normal circumstances, a beam will resist coherent instabilities by virtue of Landau damping. A small difference in the oscillation frequency prevents the ensemble to respond coherently to the driving force exerted on it by beam induced fields. Cooled beams are susceptible to respond coherently for at least two reasons:

1) the tune spread due to non-linearities and the momentum spread decreases,
2) the induced fields such as direct space transverse charge fields increases as the beam cools down.

One can define a mode frequency shift: $\Delta\omega_n = \pm\Delta q\omega_{rev}$ and a mode frequency spread: $\delta\omega_n = (n-q)\delta\omega_{rev} - \omega_{rev}\delta Q$. If the mode frequency shift has an imaginary part, we see that the oscillation can self amplify. As a rule of thumb we can say that the stabilization of transverse instabilities by Landau damping requires that:

$$|\Delta\omega_n| \leq \frac{1}{\pi}|(n-q)\delta\omega_{rev} - \omega_{rev}\delta q|$$

An approximation of the threshold for both transverse and longitudinal instabilities is given in Table 4. We see that the loss of Landau damping for a beam of 10^9 particles occurs when the momentum spread $\Delta p/p$ is about 10^{-4}. This is also linked to the double peak structure observed on longitudinal Schottky scans of strongly cooled beams[9]. Many other phenomena may induce instabilities such as the direct space-charge field or the interaction of the beam with the surrounding structures or ions or electrons from the residual gas trapped in the proton beam.

Table 4 - Transverse (row 1) and longitudinal (row 2 instability thresholds

It is assumed that the coupling impedance is dominated by the space-charge contribution. The vertical beam dimension is assumed to be $\sqrt{2}$ times the horizontal dimension, and the number of particles is take to be 1×10^9. The formulae used can be found in Refs. 10 and 11.

Beam momentum	308.6	200.0	105.0	60.0	MeV/c
$(\Delta p/p)$ min	0.095	0.235	0.97	2.57	10^{-3}
$(\Delta p/p)$	0.8	1.26	2.56	4.17	10^{-4}

4.3 Damper Characteristics

The damper consists of a horizontal and vertical electrostatic pick-up and a horizontal and vertical kicker placed at an odd number of quarter wavelengths of the betatron oscillation away from the pick-up. The transverse position signal from the pick-ups is linearly amplified, delayed, and then applied to the kicker plates. The bandwidth is directly related to the number of modes to be corrected and the gain is determined by the growth rate of the instability and by the initial perturbation of the beam. The results obtained with our first assembly are very promising and show a drastic increase in the number of particles that can be cooled under stable conditions. A definite version will be installed this year and it will be designed to maintain the damping during ramping in energy. A closed orbit suppressor will be needed to reduce the strong signals which occur when the bunched beam is not perfectly centered at the position pick-up.

5. FUTURE PLANS

As mentioned earlier in the text, until the installation of the new collector we will bring the cooler into routine operation for momenta below 200 MeV/c. Once installed, it is hoped that the new collector will enable us to also work in the momentum range from 200 MeV/c to 380 MeV/c. A test bench for the reception tests of the new collector is at present being set up and will be used later on for testing new collector designs and the "variable perveance gun" that we are at present studying. The transverse feedback system also needs to be modified in order to be operational at all the energies at which electron cooling can be used. We also have plans to replace the electrostatic pick-up on the collector side with a pick-up that could be used for trapping ions that are extracted from the collector.

6. ACKNOWLEDGEMENTS

We would like to thank all the members of the LEAR and LINAC teams who participated actively in the ion runs. Many thanks also to D.J. Williams and J.C. Perrier for their help in the damper development, and to J.L. Vallet and M. Le Gras for their help in preparing the hardware used in the machine development sessions.

7. REFERENCES

1) Poth, H. et al., "First Results of Electron Cooling Experiments at LEAR", Zeitschrift für Physik A332 (1987).

Wolf, A. et al., "Results from Electron Cooling Experiments at LEAR", Proc. of EPAC 1988, Rome.

Bosser, J. "Electron Cooling Tests at CERN, CERN/PS/AR/Note 88-21 (1988)

2) Poth, H. et al., "Further Results and Evaluation of Electron Cooling Experiments at LEAR", NIM A287 (1990).

Bosser, J. et al., "Summary of the MD on Comparative Measurements of Electron Cooling with Protons and Antiprotons", CERN/PS/AR/Note 89-6 (1989).

3. Poth, H. et al., "The Electron Cooling Device Developed for LEAR, Proc. IVth Workshop on Physics at LEAR with Cooled Low Energy Antiprotons", Villars-sur-Ollon, 1987, and references within.

4) Tranquille, G. et al., "The New Control System for the LEAR Electron Cooler". This Workshop.

5) Bikovsky, V.et al., "The New Electron Beam Collector for LEAR". This Workshop.

6) Manglunki, D., "Measurements and Surprises with H^- Ions", Proc. IVth Workshop on Physics at LEAR with Cooled Low Energy Antiprotons", Villars-sur-Ollon, 1987, and references within.

7) Baird, S. et al., "Stochastic Cooling and Storage Tests with Oxygen Ions in LEAR", Proc. 1989 IEEE Particle Accelerator Conference.

8) Möhl, D., "Limitations and Instabilities in Cooled Beams, Proc. of the Workshop on Crystalline Ion Beams, Wertheim, FRG, October 1988.

9) Cocher, S. and Hofmann I., "On the Stability and Diagnosis of Heavy Ions in Storage Rings with High Phase Space Density", GSI 89-61 (1989).

10) Schnell, W. and Zotter, B., CERN/ISR-GS-RF/76-26 (1976).

11) Keil, E. and Schnell, W., CERN/ISR-TH-RF/69-49 (1969).

THE CRYRING ELECTRON COOLER

Håkan Danared
Manne Siegbahn Institute of Physics
S-10405 Stockholm
Sweden

ABSTRACT

The CRYRING electron cooler is designed to cool heavy, highly charged ions up to their maximum energy in CRYRING of 24 MeV per nucleon. It will also be used as an electron target for atomic-physics experiments at electron energies up to 20 keV. The design criteria and specifications of the cooler are reviewed and its status as of July 1990 mentioned.

1. INTRODUCTION

CRYRING is a facility for research in atomic, molecular and nuclear physics whose main components are an electron-beam ion source (EBIS) for heavy, highly charged ions, an RFQ linear accelerator and a synchrotron/storage ring[1]. The research program at the ring includes studies of photon–ion, electron–ion and atom– or ion–ion interactions. In order to obtain an ion beam of high intensity, low energy spread and long life time for such studies electron cooling will be applied to the stored beam. In addition the cooler can be used as an electron target for investigations of, primarily, electron–ion recombination processes.

2. GENERAL DESIGN

The storage ring has a circumference of 51.6 m. It has a sixfold symmetry with six straight sections 3.8 m long where quadrupole and sextupole magnets, etc, are located and six 3.5 m long straight sections without magnets. The electron cooler (see schematic diagram in figure 1) is positioned in one of the magnet-free sections together with two small dipole magnets, one on each side of the cooler, for correction of the horizontal deflection of the ions that is caused by the cooler toroids. The ion trajectory through the cooler can be kept fixed when the magnetic fields are varied by exciting backleg windings on the two ring dipoles adjacent to the cooler. Table 1 summarises the most important data for the cooler.

The ring has a maximum rigidity of 1.44 Tm, implying that protons can be accelerated up to 96 MeV and heavy, highly charged ions with a charge-to-mass ratio of 0.5 up to 24 MeV per nucleon. Since no physics with protons is foreseen for CRYRING, however, the maximum energy for cooling has been set to 24 MeV per nucleon which corresponds to an electron energy of 13 keV at equal ion and electron velocities. The lowest energy where cooling can be performed has to be determined experimentally; it may be possible to cool down to the injection energy of 300 keV per nucleon although the electron energy then becomes only 160 eV. When the cooler is used as an electron target the electron energy can be increased up to 20 keV giving a centre-of-mass energy $E_{\rm cm} \approx \Delta E/4E_0^2$ between ions and electrons.

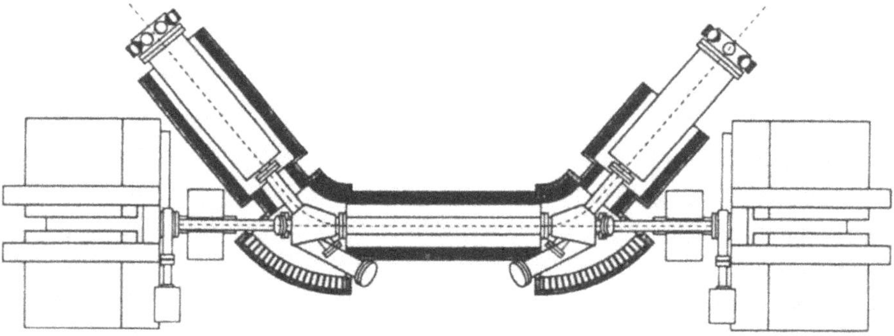

Figure 1. Schematic diagram of magnets and vacuum system, including correction dipoles and two ring dipoles.

There are plans for a second phase of the CRYRING project with an extracted beam for nuclear physics, demanding a high duty cycle of the ring. Fast acceleration is also necessary, for example, after injection of highly charged ions having high cross sections for recombination with the rest gas at low energies. Consequently, the ring has been built for fast ramping, allowing a full machine cycle in approximately 0.5 s. Also the electron cooler can be cycled at this rate—

the power supplies have rise and fall times of typically 100 ms within the ranges used during normal operation. The ability to switch parameters rapidly may be of interest in other contents as well, such as when changing between cooling and target modes when the cooler is used for atomic-physics experiments.

Table 1. Design parameters of the CRYRING electron cooler

Electron energy	≤ 20	keV
Gun perveance	$0.1 - 5$	$\mu A/V^{3/2}$
Electron current	≤ 3	A
Beam diameter	40	mm
Magnetic field	≤ 0.3	T
Ramping speed up to 0.1 T	≤ 0.8	T/s
above 0.1 T	≤ 0.3	T/s
Length of cooling solenoid	1.1	m

3. MAGNETS

The layout of the was to a large extent determined by the relatively small space available and led to a U-shaped design with toroid angles of 50° and where the length of the cooling solenoid is 1.1 m or 2.1% of the ring circumference. The length of the gun solenoid, 0.9 m, ensures that the electron gun is positioned in an almost homogenous field and gives space for non-evapourable-getter (NEG) pumps between the gun and the adjacent small solenoid. At the location of the electron gun the field has reached 98% of its nominal value and calculations have shown that the small and smooth increase of the field further inside the solenoid has no effect on the electron temperature. Also the 0.5 m long collector solenoid has space for NEG pumps.

The maximum field of the magnets is 0.3 T at a current of 600 A. All seven cooler magnets, plus the two correction dipoles, are connected in series to a single power supply stable to within 10^{-5} of full current (10^{-4} during ramping). The correction dipoles will in addition be provided with active shunts for individual regulation. Although the magnetic field in an electron cooler is rather weak the generation of eddy currents during the fast ramping must be considered in the case of the CRYRING cooler. For example the aluminium mandrels of the solenoids have a longitudinal slit filled with epoxy resin preventing the induction of currents around these mandrels.

The gun, cooling and collector solenoids have dipole coils attached to the inside of the aluminium mandrels, steering the electron beam in both planes so that it can be made to overlap properly with the ion beam and enter into the centre of the collector. Also the toroids have dipole coils to counteract heating and centrifugal drift of the electrons as they bend through the toroids. Because of the low electron energies in the CRYRING cooler small field imperfections that may exist, particularly at the interfaces between different magnets, do not cause any significant heating of the electrons; they behave adiabatically with respect to the changes in the field. Thus there are correction coils, ensuring a field which is as straight as possible, only in the cooling solenoid. Finally there is a solenoidal coil around the collector, connected in series with the main cooler magnets but generating a field pointing in the opposite direction, thus creating a rapid divergence of the field in the collector region.

4. ELECTRON GUN

The electron gun consists of a flat, 40 mm diameter, dispenser-type cathode, a Pierce anode made from copper and a set of eight acceleration anodes from stainless steel[2)]. The last anode and a mesh surrounding the beam further downstream are connected to ground. In the gun the electrons are accelerated adiabatically. The adiabatic condition is fulfilled if the distances along the symmetry axis of the gun over which the radial electrical fields from the acceleration anodes vary appreciably (on a scale set by the magnetic forces) are small compared to the electron gyro wavelength. This can be achieved for a wide range of perveances in this gun, at least between 0.1 and 5 $\mu A/V^{3/2}$, by selecting the voltages of the anodes properly. At perveances above 1 or 1.5 $\mu A/V^{3/2}$ the electron temperature is decreased if the gun is operated with positive voltages on some of the intermediate anodes. The whole perveance range can be used at all electron energies, the only limit being that of the electron current supply which can deliver maximally 3 A. The magnetic field can be varied independently of perveance and electron energy as long as the adiabaticity is maintained or, at least, as long as the transverse electron temperature does not significantly increase above the 0.1 eV resulting from the cathode temperature. At 20 keV electron energy the magnetic field thus required varies between 40 and 80 mT, the higher fields necessary for high perveances. At lower energies the field decreases with the square root of the energy.

Figure 2 shows the geometry of the gun together with equipotential lines and electron trajectories at a perveance of 0.2 $\mu A/V^{3/2}$, an electron energy of 20 keV and a magnetic field of 40 mT. At higher perveances a stronger electrical field has to be applied between the cathode and the first anode, generating stronger radial fields, and consequently the increase of the required magnetic field. Electron-trajectory calculations were performed using the Herrmannsfeldt code[3].

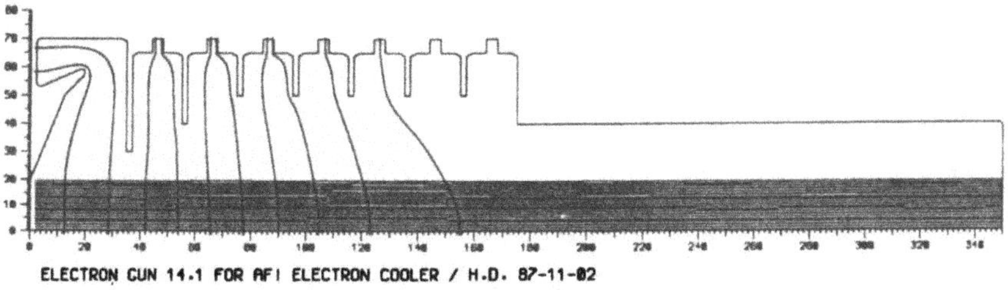

Figure 2. Electron gun with electron trajectories and equipotential lines at perveance 0.2 $\mu A/V^{3/2}$, electron energy 20 keV and magnetic field 40 mT.

5. COLLECTOR

In front of the collector there are a set of four deceleration anodes, similar to the acceleration anodes in the gun, out of which one is on ground potential. The collector itself consists of a conical part and a cylindrical part which, together

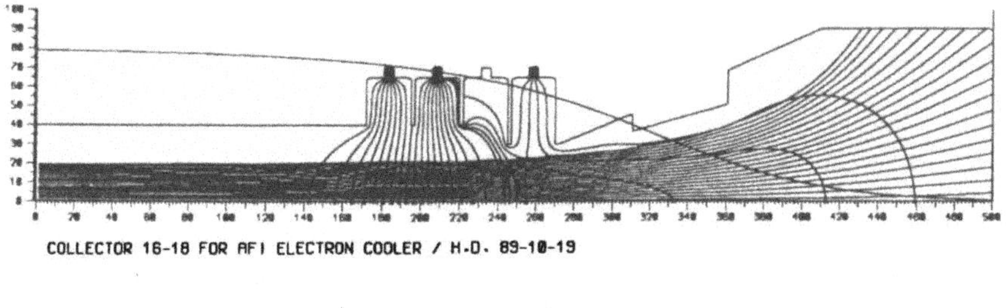

Figure 3. Electron collector with electron trajectories, equipotential lines and axial magnetic field at gun perveance 0.2 $\mu A/V^{3/2}$, electron energy 20 keV and magnetic field 40 mT inside the collector solenoid.

with the bottom plate, is water cooled. The entire collector is made from OFHC copper.

A schematic diagram of the collector is shown in figure 3 together with equipotential lines, electron trajectories and the longitudinal magnetic field on the symmetry axis. The perveance, electron energy and magnetic field are the same as in figure 2. The electrical potential has a minimum close to the entrance to the collector. At this point the magnetic field still is high, implying that secondary electrons returning from the collector surface become reflected both by electrostatic and magnetic forces. The collector perveance lies in the range between 15 and 18 $\mu A/V^{3/2}$, depending only slightly on the electron-beam parameters.

6. VACUUM SYSTEM

The aim for the vacuum in the ring is the low 10^{-12} torr region. In order to get as close as possible to this pressure also in the electron cooler the materials used for the vacuum system are, as far as it is possible, only 316 LN stainless steel, OFHC copper and aluminium oxide. The entire vacuum system will be baked to 300°C. In the gun and collector solenoids NEG pumps will be installed, probably also in the toroid chambers. The electron gun will be pumped from behind by a cryopump, the collector with an ion pump and the chambers of the correction dipoles by the same combination of ion and NEG pumps as is used around the rest of the ring. Although cryopumps not always are associated with these extremely low vacua tests have proven that a clean, mildly baked (to 60 – 70°C) cryopump can reach final pressures well into the 10^{-12} torr region.

The pressure will, as in the rest of the ring, be lowered by cooling the vacuum chambers with water holding a temperature of about 12°C. This cooling also counteracts the heating by eddy currents of, in particular, the flanges which occurs during fast ramping. The power dissipated in the vacuum system is very small but may, due to the efficient thermal insulation installed for use during bakeout, still cause a temperature rise of 5 – 10°C at maximum ramp rate without cooling.

The mechanical design of the vacuum system has been made as simple as possible, with gun and collector placed in cylindrical chambers of 250 mm diameter and mounted on the CF250 flanges on top of these chambers for easy assembly and disassembly. The toroid chambers are provided with vacuum pipes going through the toroid iron below the cooling solenoid and out of the cooler's symmetry plane, as shown in fig. 1. Through these pipes particle detectors for use in different atomic-physics experiments, crystals for X-ray spectroscopy, or other devices can be inserted into the toroid chambers.

7. PERFORMANCE EXPECTATIONS

Electron cooling of heavy highly charged ions is expected to be very rapid, a few hundredths of a second for an ion like Xe^{44+} at 10 MeV per nucleon. The other extreme is light, singly charged ions at a low energy of 300 keV per nucleon (the injection energy) requiring some minutes to cool down. The acceptance of CRYRING is 200π mm mrad horizontally and 100π mm mrad vertically and assuming an equilibrium between cooling and intra-beam scattering one may expect that the emittance can be reduced from these values by at least two orders of magnitude for typical heavy-ion beams.

The cooler will also have some adverse effects on the ion beam, the most important ones being the deflection in the toroids, a tune shift caused by the electron-beam space charge and excitation of the $Q_x - Q_y = 0$ resonance in the ring. Correction dipoles compensate for the deflection in the toroids; nevertheless the maximum magnetic field of 0.3 T can only be applied at the highest ion energies. This means that the field will have to be turned off, or at least strongly reduced, during injection. Since the tune shift from an electron beam of full power is around 1 at the injection energy also the electron current must be reduced during injection. The resonance $Q_x - Q_y = 0$ is driven by the solenoidal field of the cooler and causes coupling between ion motion in the horizontal and vertical planes. This resonance is easily excited in CRYRING since horizontal and vertical tunes are almost equal at the normal working point. To avoid negative effects from the coupling a compensating solenoid may have to be installed.

8. PRESENT STATUS

The magnets were delivered to MSI early this year. Also the magnet power supply and the high-voltage supplies are delivered. Presently field measurements with Hall probes and an electronic autocollimator are in progress and correction coils are being tested. Vacuum and electron-beam systems are designed and their manufacture will begin during the autumn. We plan to have the cooler installed in the ring and beginning to operate during 1991.

REFERENCES

1. Herrlander, C. J., Rensfelt, K.-G. and Starker, J., "CRYRING—a Heavy-Ion Storage and Synchrotron Ring", EPAC Rome 1988, 350 (1989).
2. Danared, H., "An Adiabatic Electron Gun for the CRYRING Electron Cooler", EPAC Rome 1988, 1033 (1989).
3. Herrmannsfeldt, W. B., "Electron Trajectory Program", SLAC – 226, Stanford, 1979.

ELECTRON COOLING AT IUCF --
0.3 MeV in 1988, 9 MeV in 1998?

Timothy JP Ellison

Indiana University Cyclotron Facility
2401 Milo Sampson Ln., Bloomington, IN 47405
USA

ABSTRACT

The operational status of the IUCF electron cooling system is reviewed along with measurements of cooling rates and equilibrium proton beam emittances. Some of the technical difficulties associated with extending this technology to higher energies for cooling 16 GeV protons in LISS are also discussed.

1 INTRODUCTION

Many of the design specifications for the IUCF electron cooling system can be found in an earlier reference[1]. In this paper a few of the operational aspects of the electron system are discussed along new measurements of the longitudinal drag rate and equilibrium proton beam properties. In the final section of this paper, I consider some of the technical aspects associated with the possible construction of a 9 MeV electron cooling system for LISS, the Light Ion Spin Synchrotron, an ≈16 GeV (proton) light ion accelerator/storage ring with cooling and spin manipulation insertion devices currently being discussed as the next accelerator for IUCF.

2 OPERATION

The IUCF 270 keV electron cooling system has now been in operation for over 2 years, since March 1988. Following a brief few-day commissioning period, the IUCF system has been routinely used by the accelerator operators as merely another insertion device in the ring. Although the highest proton beam energy achieved by the storage ring to date is 310 MeV (where the beam has been cooled) the electron cooling system has been operated at its design energy (270 kev), with electron beam currents in excess of 2.5 A, and collection efficiencies as high as 99.9999%. The system also ramps, cooling the beam both at the injection energy and after acceleration (the "cool-ramp-cool" mode of operation).

2.1 The Cool-Ramp-Cool Mode of Operation

We use the cool-ramp-cool mode of operation to accumulate successive batches of injected beam before acceleration, and to re-cool the beam after acceleration. The accumulation of beams using electron cooling is necessary to achieve higher beam currents than can be obtained with a single injection. Although we use cooling to inject successive batches of 45 MeV proton beams which have already been accumulated by using stripping injection of 90 MeV H_2^+, the cooling system is even more important for cooling accumulation of fully-stripped polarized proton beams which are kick-injected. Cooling at the higher energy is desirable for two reasons: firstly cooling allows operation with thicker targets and longer beam lifetimes (i.e. higher luminosity); secondly, cooling makes up for the severe rf longitudinal heating which occurs during the acceleration process[a].

We initially attempted the cool-ramp-cool mode of operation using the following sequence of events: gating off the electron current at the beginning of the ion beam acceleration cycle (ramping), resetting all the electron cooling system power supplies for cooling at the accelerated ion beam energy, and then gating the electron beam back on. We found, however, that in many instances ramps which worked fine while the electron beam operated quiescently at the settings for cooling the injected beam, would not efficiently accelerate beam when the electron beam

---

[a]The longitudinal heating is caused by the 512 Hz steps in the DDS (direct digital synthesizer) reference for the PLL in the low level rf system which operates open-loop. We are currently adding a beam phase feedback system to eliminate this problem.

was gated off.

Consequently, we decided it was necessary to keep the electron beam on during ramping of the ion and electron beam energies. This was accomplished in the following straightforward manner: the reference for the cathode potential power supply (HVPS) is merely stepped to the new value. The power supply ramps up linearly with a slew rate of about 100 kV/s, and decays exponentially with an ≈ 1 s time constant. An analog circuit which uses a voltage proportional to the measured cathode potential as an input and outputs a voltage proportional to the electron beam relativistic momentum is added to the references for the vertical steerers in the electron system which compensate for the electron beam centripetal drift in the toroids. Since this circuit is always in operation, the operators can arbitrarily change the electron beam energy without worrying about changes in the electron beam position.

2.2 Collection Efficiencies

On a day-to-day operating basis no effort is made to maximize collection efficiencies and we typically operate with losses on the order of 1 to a few times 10^{-4}. We have also demonstrated that by compensating the toroid ($\mathbf{R} \times \mathbf{B}$) centripetal drift with an electric field to provide an opposing ($\mathbf{E} \times \mathbf{B}$) drift, collection efficiencies of 99.9999% can be achieved. Although this technique has already been discussed in detail elsewhere[2], below I elaborate on a few points.

Firstly, whereas this technique would appear to remove one of the only escape mechanisms for trapped electrons, this is not necessarily the case. Low energy electrons can still be removed by a variety of means since the ($\mathbf{R} \times \mathbf{B}$) drift velocity depends upon the electron beam energy, whereas the ($\mathbf{E} \times \mathbf{B}$) drift velocity does not. Consequently, when the two drifts are set up to cancel a significant portion of the primary beam drift, they do not cancel for low energy electrons. The high energy electrons which escape the collector also should not create a significant problem: firstly, they constitute a very small current (on the order of 10^{-4} times the primary current) and secondly, they are not destined to forever oscillate within the system: on the order of 99.99% of electrons reflected from the collector will be captured in the collector after a single oscillation within the system. Consequently, one does not need to *exactly* compensate the (**R** × **B**) drift with an (**E** × **B**) drift, merely compensate well enough so that an electron can oscillation once or twice through the system without being lost.

Secondly, using an (**E** × **B**) drift in the main (cooling) solenoid region to compensate for the (**R** × **B**) drift in the toroids does not preclude operation with the trajectories of both the electron and ion beams along the axis of this solenoid providing one also uses a magnetic dipole in this region. One can easily verify that, although the transverse electric field and superimposed transverse magnetic dipole field cause deflections of the proton and electron beams in different planes, if the fields are chosen such that electron beam will travel parallel to the axis of the solenoid, the electric and magnetic forces also balance for a proton beam traveling parallel to the axis of the solenoid.

Finally, I would like to point out that when doing calculations of collector efficiencies, the important parameter to optimize is *not* how many electrons in the primary electron beam are able to travel to a surface on the collector (something which can be rather easily accomplished), but rather how many of the high energy electrons backscattered from the electron surface are able to escape the immediate vicinity of the collector.

3 PERFORMANCE

Below, I briefly summarize new measurements of the longitudinal drag rate and equilibrium cooled beam properties.

3.1 Measured Longitudinal Drag Rate

The various techniques used to measure the longitudinal drag rate along with results obtained using these techniques have been previously described[3]. Since these measurements were discussed, we have invested two additional shifts of beamtime investigating the longitudinal energy spread of the electron beam. Initial measurements showed that the longitudinal drag rate increased with electron beam energy error until the 0.45 A, 24 keV electron energy was "shifted" by about 10 eV. This apparently indicates an electron beam effective energy spread of ± 10 eV, a value far in excess of what should result from the known cathode potential voltage fluctuations (± 1 eV) and somewhat larger than what we expect from intrabeam scattering[4] (± ≈ 3.5 eV).

The technique[3] of balancing the longitudinal drag rate with an emf produced by a large transformer (betatron accelerator) was used as a tool to minimize the effective electron beam longitudinal energy spread. It was found that very small changes in the angular alignment between the two beams significantly affected the longitudinal drag rate. This is demonstrated in Figure 1 where the proton beam

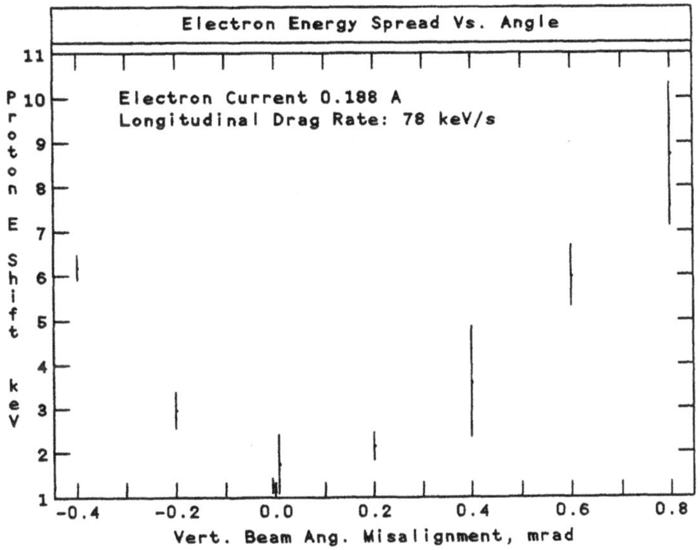

Figure 1. Shift in the equilibrium energy of a 45 MeV proton beam necessary for the electron cooling system to provide a longitudinal drag rate of 78 keV/s to compensate for the betatron accelerator emf as a function of the angular misalignment of the electron and proton beams.

energy shift for a given emf is shown as a function of the angular misalignment between the proton and electron beams. Presumably this large increase in the effective electron beam longitudinal energy spread with very small angular misalignments results from the combination of a corresponding large increase in the equilibrium cooled proton beam transverse size and the unneutralized electron beam space charge. When the effective electron beam energy spread was minimized, to value of about ± 2.5 eV for a 0.188 A electron beam, we observed longitudinal drag rates about four times greater than were previously reported[3] for small velocity differences (< 10^4 m/s). The measured rate for large velocity differences, however, was unchanged and still in good agreement with the nonmagnetized theory.

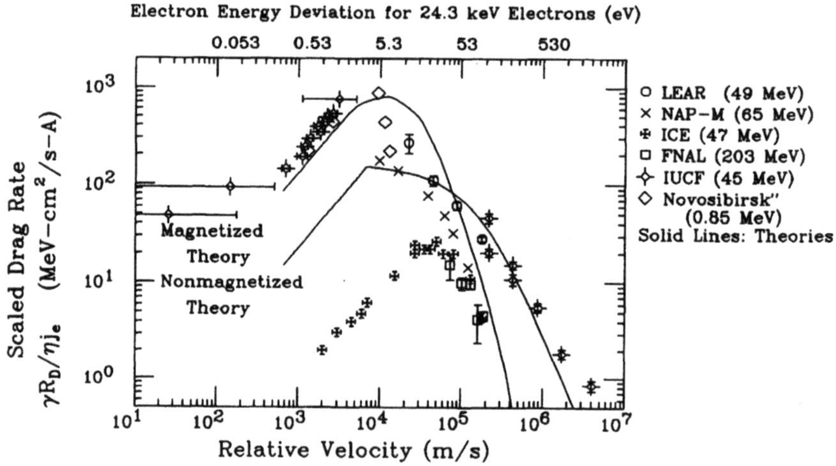

Figure 2. Scaled longitudinal drag rate measurements from various laboratories along with the nonmagnetized theory (solid line).

Scaled longitudinal drag rates from Novosibirsk (NAP-M)[5], ICE[6] (P/2 gun), FNAL[7], LEAR[8], IUCF and the single pass cooling experiments at Novosibirsk[9] are displayed in Figure 2. The data has been scaled as follows: the measured longitudinal drag rate, R_D (MeV/s), is multiplied by γ (E/Mc^2) and divided by η (the length of the cooling region divided by the ring circumference) and j_e (the electron density, A/cm^2). Consequently, the units are [(MeV/s)/(A/cm^2)]. The solid line is the prediction of the simple nonmagnetized theory assuming an electron beam transverse temperature of 1050 °C.

The IUCF data for small longitudinal velocity differences between the proton and electron beams (< 10^4 m/s) were taken while operating with a 0.2 A (j_e = 0.04 A/cm^2) electron beam. It should be noted that our experience has been that the in this regime the longitudinal drag rate does necessarily scale linearly with electron beam current. Apparently, even higher longitudinal drag rates for small velocity differences have recently by the LEAR group[10].

There is also another indirect measurement of the longitudinal drag rate which can be made: a measurement of the frequency deviation in the frequency-modulated longitudinal Schottky signals caused by ripple in the cathode power supply. For example, using data from Figure 2, we expect a to see a maximum longitudinal drag rate of about 800 keV/s for a 45 MeV proton beam cooled by a 0.2 A (2.54 cm diameter) electron beam at IUCF where η = 0.032. Consequently,

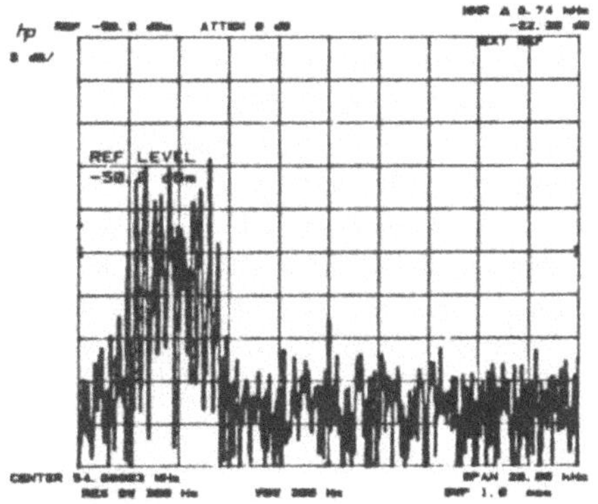

Figure 3. Schottky signal from a 45 MeV coasting proton beam cooled by a 0.2 A electron beam (CF 54.699 MHz, h = 53; RBW = VBW = 300 Hz; Vertical: 5 dB/div; Horizontal: 2 KHz (0.1 s)/div).

the proton beam should be able to track electron beam energy fluctuations with slew rates less than 800 keV/s × (m/M) = 435 eV/s, or a 60 Hz cathode potential ripple amplitude of about 1.8 V resulting in a proton beam energy modulation of about ± 3.3 keV. A Schottky signal spectrum resulting from such behavior is shown in Figure 3. The spikes which occur every 1/60th of a second in sweep time, the U.S. line voltage period, are characteristic of a pure tone frequency modulated by a 60 Hz triangular waveform. The width of the signal is a direct measurement of the frequency deviation.

My interpretation of this signal is that the proton beam energy is being coherently modulated by:

$$\pm\, 2.9 \text{ keV} = \pm\, \tfrac{1}{2}\beta^2\gamma Mc^2 \cdot (3.1 \text{ kHz}/54.6 \text{ MHz}) \cdot (\eta^{-1} = 1.15)$$

where here η refers to the frequency dispersion in the ring (i.e. $\eta = \gamma^{-2} - \gamma_t^{-2}$, where $\gamma_t Mc^2$ is the ring transition energy). The longitudinal drag rate necessary in order for this to be true is about 700 keV/s (= 6 keV • 120 s^{-1}), a value consistent with our estimate using the maximum drag rate shown in Figure 2. The instantaneous proton beam energy spread is considerably smaller than the range over which it is modulated; however, the conventional swept-frequency spectrum analysis technique is not able to measure this quantity.

3.2 Measured Proton Beam Equilibria

3.2.1 Measured longitudinal equilibrium energy spread

Numerous longitudinal Schottky signal spectra were recorded for 45 MeV proton beams with intensities ranging from about 0.02 to > 1,000 μA. Restricting the measurements to beam currents of less than 5 μA, where the Schottky signal splitting is negligible, measurements of the relative beam momentum spread (FWHM) ranged from 0.5 to 7 × 10^{-5}. Generally we have observed that the relative proton beam momentum spread increases with increasing electron beam current, presumably due to the effect mentioned above.

We have also measured the equilibrium proton beam energy spread by recording the time spread of rf-bunched beams. A summary of these measurements, along with results from Fermilab[11] and LEAR[8,12], are show in Table I. One expects the bunched beam energy and time spread to increase with beam current, as with coasting beams, due to intrabeam scattering; we have not, however, studied this in a systematic manner.

Table I. Comparison of measured bunch lengths for electron cooled beams for beams at FNAL, LEAR and IUCF.

rf/Ring Parameters					Beam Properties			
$(\gamma-1)Mc^2$ (MeV)	V (kV)	h	η $(\Delta f/f)(p/\Delta p)$	C (m)	I (μA)	I_{peak} (mA)	Δt (ns)	ΔE (keV)
Fermilab								
114	0.04	7	0.72	135	0.08			6.4
IUCF								
45	2.05	9	0.87	86.83	38	2	1.7-2	6-7
45	5.3	9	0.87	86.83	35	2	2	11.4
148	2.3	9	0.7	86.83	18.3	1	1.3*	< 9.6
LEAR								
50	0.1	1	0.95	83.25	75	1	60.4	19.2
50	5	1	0.95	83.25	75	3	24.2	54

*Resolution limit of oscilloscope.

The Fermilab measurement is based upon an analysis of the bunched beam frequency spectrum. The IUCF data is a direct measurement of the beam time spread from viewing the bunched beam in the time domain using a high bandwidth longitudinal pickup. The LEAR measurement is actually a measurement of the neutral beam (H) time spread.

3.2.2 Measured proton beam transverse equilibrium emittance

The equilibrium transverse emittance was measured by sweeping a few μA, 45 MeV proton beam across a 7 μm diameter carbon fiber and measuring the secondary emission current from the fiber as shown in Figure 4 and Figure 5. The β and η (dispersion) function values are 1.4 m and 4.15 m at the profile monitor.

The measurement shown in Figure 4 is consistent with a transverse emittance (1 σ) of 0.023 $\pi\mu$m, or an energy spread (FWHM) of 9 keV. The 9 keV energy spread is consistent with the measured beam energy spreads shown above in Table I; consequently, we can only use this measurement as an upper limit on the equilibrium beam emittance. This value of the emittance corresponds to a proton beam transverse temperature of about 1 eV in the cooling region where the beta functions are larger by a factor of 1.7. Although this upper limit for the proton beam transverse temperature is significantly larger than that measured at Novosibirsk, the IUCF measurement was made with rf bunched beam with peak currents significantly higher than at Novosibirsk. Figure 5 demonstrates the large beam size increase resulting from a small angular misalignment of the beams.

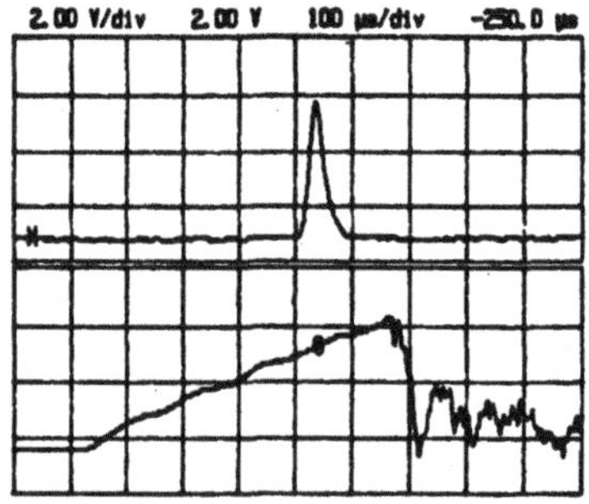

Figure 4. Transverse profile of a 45 MeV proton beam cooled with a 0.2 A electron beam. Secondary emission current (beam profile, upper trace, 1.56 mm/div horizontally) and beam position monitor output (lower trace, 3.67 mm/div vertically) vs. time (100 µs/div).

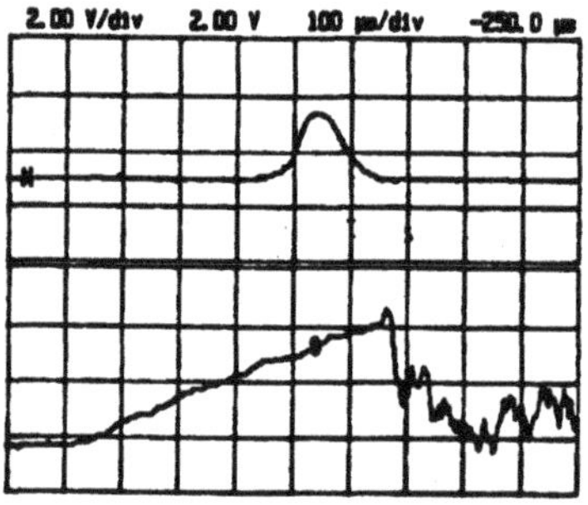

Figure 5. Same signals as in previous Figure; However, in this case the proton and electron beams are misaligned in the cooling region by 0.47 mrad.

4 INTERMEDIATE ENERGY ELECTRON COOLING IN LISS

4.1 Cooling rates

In the literature, many expressions for the electron cooling time contain a severe relativistic scaling law. For example, excerpting reference 13, page 326,

> The cooling rate depends on the magnetic and velocity regimes. For the unmagnetized case, we find
>
> $$\lambda = \frac{2\pi n Z^2 e^4 c^3 \Lambda_c}{m m_p \beta^3 \gamma^5} \frac{L_C}{2\pi R}$$
>
> It is the strong inverse dependence of λ on γ that is the source of the greatest difficulty in electron cooling. Cooling cannot be done this way at high energy and it is necessary to decelerate particles in order to cool them.

Keeping in mind that cooling times are on the order of ½ second for $\beta\gamma \approx$ ½, we can thus expect cooling times on order of 10^5 s ($\approx$ 1 day) for ten GeV protons ($\beta\gamma \approx 10$). This not very encouraging.

There are assumptions made that lead to this $\beta^3\gamma^5$ scaling. They are:

1). The proton beam transverse angles exceed the electron beam transverse angles, and

2). The proton beam angles are kept constant as we scale up in energy.

There is another assumption which can be made which makes electron cooling both appear to be attractive at higher energy and points out very clearly the strong points and limitations of electron cooling. This assumption is: the proton beam rest frame velocity spread is less than the electron beam transverse velocity spread. When this assumption is made, the expression[14] for the maximum longitudinal drag rate has only a factor of γ in the denominator[b]; transversely[15] in this regime, the exponential damping rate falls off as γ^{-2}.

---

[b]Of course, if we assume $\Delta E/E$ to remain constant with energy, and define the cooling time as $\Delta E/R_D$, then the cooling time would scale as γ^{-2} as does the transverse cooling time.

4.2 Numerical example

By way of example, I look at the cooling of 11 GeV/c protons, a momentum typical for the proposed LISS (Light Ion Spin Synchrotron) at IUCF, and close to the momentum of the Fermilab stochastically-cooled antiproton accumulator with which we can make comparisons. Using predictions of the scaled nonmagnetized theory of electron cooling shown Figure 2 and a value for η (the ratio the cooler length to ring circumference) of 0.03, we expect a maximum longitudinal drag rate of about > 250 (keV/s)/(A/cm^2). Using the parameters for the cooling device listed in Table II, longitudinally we expect a maximum longitudinal drag rate in excess of 80 keV/s, and transversely we expect the cooling times to be 100 s.

Table II. Parameters for a 9 MeV electron cooling system. The values shown in **bold** text are used for this numerical example.

Beam Energy	Beam Current	Beam Radius	Cooling Length
(MeV)	(A)	(cm)	(m)
2 - 9 (**6**)	1 - 10 (**1**)	**1**	**15**

In order for our assumption that the rest frame proton velocity be less than the electron beam transverse velocity spread be true, we impose the following constraints:

1). The maximum proton beam transverse emittance must be less than 0.2 $\pi\mu$m ($\approx 25\pi\gamma^{-2}$ μm) assuming horizontal and vertical beta functions of 100 m.

2). The proton beam energy spread must be less than $\pm$ 6 MeV ($\approx \pm\ 0.5\gamma$ MeV).

We thus see that the time to cool the beam from a $\pm$ 6 MeV ($\Delta p/p \approx \pm$ 0.05%) energy spread to equilibrium is about 100 s (varying inversely with the initial beam energy spread), the same as the transverse cooling time constant.

Although this "precooled" beam quality sounds quite high (especially compared to the emittance of antiprotons coming off a production target), the accelerator would be filled at low energy (≈ 0.5 GeV) by a machine such as the IUCF Cooler, where the beam quality should meet these requirements; in addition, we would expect the transverse emittance to decrease as $1/\beta\gamma$ during acceleration.

4.3 Comparison with stochastic cooling at the FNAL Antiproton Source

By comparing these cooling parameters with those observed at the FNAL antiproton source[16], we see that electron and stochastic cooling are very complementary rather than competing technologies as illustrated by Dieter Möhl in the previous E-COOL conference[17]. For high intensity beams, stack core densities of 3×10^{10}/eV-s with 2.5×10^{11} particles, corresponding to an energy spread of about 5 MeV, are achieved. Such an equilibrium energy spread is ideally suited for further cooling, or "freezing" by an electron cooling system. Transversely, equilibrium emittances of about 2π μm are achieved, an emittance which is still about 10 times too large for effective electron cooling.

4.4. Technical considerations for high energy cooling

There are three major technical requirements for cooling at high energy, none of which appears to be out of the grasp of present technology.

4.4.1 Electron beam transverse temperature

The maximum longitudinal drag rate varies inversely with the electron beam transverse temperature. Thus, we need to be able to accelerate and transport a high intensity electron beam from the cathode to the cooling section without increasing its normalized emittance[c]. For magnetically confined beams, we would

---

[c]Adiabatic acceleration keeps the product of the transverse area and temperature constant; here it is assumed that the transverse beam size is kept constant.

wish to scale the magnetic field strength as $\beta\gamma \cdot 1$ kG. The Novosibirsk[18] group has built and operated a 1 MeV 1 A electron recirculation system using a relatively low field of 900 Gauss which has achieved collection efficiencies of up to 99.9%. The low field for this system required the use of resonant focussing optics (i.e. the system would only operate at certain discreet setting of the magnetic field); presumably, the electron beam did not adiabatically traverse the system.

Above about $\beta\gamma \approx 3$ - 5, a system with magnetically-confined beams will become quite massive and inelegant. Fortunately, the space charge forces fall off as γ^{-2}, so that at higher energies it is not necessary to use magnetically confined beams. The near-adiabatic acceleration of 2 and 2.5 MeV nonmagnetically confined Ampere-intensity electron beams has already been demonstrated[19] using the test facility at the National Electrostatics Corp.

4.4.2 Cathode potential regulation

One would like to have the electron beam energy slow drift be less than about 60 eV at an electron energy of 6 MeV ($\Delta E/E \approx \Delta p/p \approx 1 \times 10^{-5}$). This would keep the average cooled proton beam momentum stable to the same degree (110 keV at an energy of 11 GeV). Since the longitudinal drag rate is relatively low, the ripple will have the effect of reducing the strength of the drag rate for proton energy deviations less than (eM/m) times the ripple voltage amplitude but will not cause any significant amount of coherent energy shifting. For example, ripple at 60 Hz would at most cause a ± 0.5 keV (= ± 60 keV/s · 120^{-1} s) coherent energy shifting in the cooled proton beam. The present state of the art for voltage stability at the National Electrostatics Corporation[20] (NEC) is about 100 V at 6 MV, and the company is presently working on increasing the relative regulation by a factor of about two. (The stability goal for a heavy ion machine currently under construction is 20V/2MV, 10 ppm). This degree stability is obtained using feedback from a slit system for heavy ions, something which is more difficult when using electrons. Consequently, some amount of development is needed for optimizing the feedback system. Since the electron beam will be so intense, one might consider modulating a small portion of the electron current and using noninterceptive beam position monitors, or simultaneously operating a weak (few μA) H^- beam and use a conventional slit feedback system.

4.4.3 Electron collection efficiencies

At high energy, where the electron cathode potential power supply can only source current on the order of a few hundred μA, collection efficiencies in excess of 99.99% are essential. A collaboration of FNAL, U. of Wisconsin, and NEC[21] has already achieved collection efficiencies of close to 99.99% recirculating a 120 mA electron beam at 2 MeV. Since the collection efficiencies should remain constant with energy, this achievement comes to within a factor of 10 of the minimum required. It should be noted that the operation was not limited to 120 mA due to lack of charging current: the system would become unstable well before this limit was reached. Consequently, there is not a trivial solution to this problem.

By replacing the magnetic steerers in this design with electrostatic steerers, the optics could be made such that the beam reflected from the collector would oscillate once through the system and be returned to the collector, as discussed in Section 2.2. The electric fields required to bend the electrons with a 2 m radius of curvature, $\approx$ 3 MV/m, can be reached with conventional technology. Alternately, this oscillation of the backscattered electrons can be achieved using a tandem Pelletron such as in the compact free electron laser system being developed in Santa Barbara[22]. However, it is not clear that increasing the collection efficiencies by a few orders of magnitude would solve the instabilities, the causes of which are unknown.

4.4.4 Conclusions

I do not think there are any insurmountable problems in achieving the required collection efficiencies and power supply regulation, though further development will be necessary. At IUCF we are investigating continuing the work started by the Santa Barbara/NEC/U. of Wisc./FNAL collaboration. The cost of putting an MeV electron recirculation system into operation for testing, with necessary modifications, will be about 350 k$ if done at NEC and about 150 k$ if done in Santa Barbara on a positively charge tandem machine. The cost of a complete multi-MeV electron cooling system would be about 1 M$ + 1 M$/MeV.

REFERENCES

1. Otto C. Dermois, Timothy J.P. Ellison, and D.L. Friesel, "The Indiana University Cyclotron Facility electron cooling system", *Proc. 11th Int. Conf. on Cyclotrons and their Applications*, edited by M. Sekiguchi, Y. Yano, K. Hatanaka (Tokyo, Japan, 1986) 153-156.

2. Timothy JP Ellison, Robert J. Brown, and Brian D. DeVries, "The IUCF electron cooling system collector performance", Nucl. Instr. and Meth., **B40/41** (1989) 864-869.

3. Timothy JP Ellison, Dennis L. Friesel, and Robert J. Brown, "Status and performance of the IUCF 270 keV electron cooling system, *Proc. of the 1989 IEEE Particle Acc. Conf., Acc. Eng. and Tech.*, (Chicago, IL, 20-23 March 1989).

4. N.S. Dikansky, V.I. Kudelainen, V.A. Lebedev, I.N. Meshkov, V.V. Parkhomchuk, A.A. Sery, A.N. Skrinsky, V.N. Sukhina, "Ultimate Possibilities of Electron Cooling", preprint 88-61, Institute of Nuclear Physics, 630090, Novosibirsk 90, USSR.

5. N.S. Dikansky, V.I. Kudelainen, I.M. Meshkov, V.V. Parkhomchuk, D.V. Pestrikov, A.N. Skrinsky, B.N. Sukhina, "The study of fast electron cooling", Institute of Nuclear Physics, 630090 Novosibirsk, preprint 79-56.

6. Mary Bell, John Chaney, Heiner Herr, Frank Krienen, Poul Møller-Petersen, Guido Petrucci, "Electron Cooling in ICE at CERN", N.I.M. **190**, 237-255 (1981).

7. T. Ellison, W. Kells, V. Kerner, F. Mills, R. Peters, T. Rathbun, D. Young, and P.M. McIntyre, "Electron cooling and accumulation of 200-MeV protons at Fermilab", IEEE Trans. Nucl. Sci., NS-30, No. 4, 2636–2638 (1983). This data was derived from the data contained in the original FNAL electron cooling logbooks by the author.

8. H. Poth, W. Schwab, B. Seligmann, M. Wörtge, A. Wolf, S. Baird, M. Chanel, H. Haseroth, S.E. Hill, R. Ley, D. Manglunki, G. Tranquille, J.L. Vallet, P. Dittner, "First results of electron cooling experiments at LEAR", CERN-EP/88-98, 9 August 1988.

9. From reference [4], page 36, Figure 12 (H^+ beam).

10. Helmut Poth, "Electron cooling: theory, experiment, application", CERN-EP/90-04 (10 January 1990, to appear in Physics Reports).

11. R. Forester et al., IEEE Trans. Nucl. Sci., **NS-28**, No. 3, 2386 (1981).

12. A. Wolf et al., "*Results for electron cooling experiments at LEAR*", *in Proceedings of the European Particle Accelerator Conference*", Vol. 1, edited by S. Tazarri (7-11 June 1988, Rome) (World Scientific, Singapore) pp. 206-206.

13. F.T. Cole and F.E. Mills, "Increasing the phase-space density of high energy particle beams", Ann. Rev. Nucl. Part. Sci., **31**, 295-235 (1981).

14. Tim Ellison, "Electron Cooling", Ph.D. thesis, Ind. Univ. (1990).

15. Allan H. Sørensen and Ejvind Bonderup, "Electron cooling", N.I.M. **215**, 27-54 (1983).

16. W. Kells, J. Marriner, R. Pasquinelli, S. Mtingwa, "Performance of Tevatron I core stochastic cooling systems", *Proc. of the 1987 IEEE Particle Accelerator Conference*, Vol. 2, edited by Eric R. Lindstrom and Louise S. Taylor (IEEE, Piscataway, NJ) pp. 1090-1091.

17. D. Möhl, "A comparison between electron cooling and stochastic cooling", in ECOOL 1984 -- Proc. of the Workshop on Electron Cooling and Related Applications (edited by H. Poth, 1984) KfK 3846, pp. 293 -301.

18. M.E. Veis, B.M. Korabelnikov, N.K. Kuksanov, P.I. Nemytov, R.A. Salimov, and A.N. Skrinsky, "Electron beam with an energy 1 MeV in the recuperation regime", in *Proceedings of the European Accelerator Conference* (7-11 June 1988, Rome, World Scientific) edited by S. Tazzari, pp. 1361-1363.

19. D.B. Cline, D.J. Larson, W. Kells, F. Mills, J. Adney, J. Ferry, and M. Sundquist, "Intermediate energy electron cooling for antiproton sources using a Pelletron accelerator", IEEE Trans. Nucl. Sci., Vol. NS-**30**, No. 4, 2370 -2372 (August 1983).

20. Jim Ferry, National Electrostatics Corp. (private phone communication, 3 August 1989).

21. J.R. Adney, M.L. Sundquist, D.R. anderson, D.J. Larson, and F.E. Mills, "Successful DC recirculation of a 2 MeV electron beam at currents more than 0.1 A", *Proceedings of the 1989 IEEE Particle Accelerator Conference*, Vol. 1, edited by Floyd Bennett and Joyce Kopta (IEEE, Piscataway, NJ) p. 348.

22. G. Ramian and L Elias, "The new UCSB compact far-infrared FEL", NIM A **272** (1988) 81–88.

MEASUREMENT OF $p+p \to p+p+\pi^0$ WITH AN ELECTRON-COOLED BEAM

H.O. Meyer

Department of Physics, Indiana University and
Indiana University Cyclotron Facility, Bloomington, IN 47405

ABSTRACT

The total cross section for the reaction $p+p \to p+p+\pi^0$ was measured at nine center-of-mass energies from 1.5 MeV to 23 MeV above threshold. This experiment was carried out in the Indiana Cooler, a storage ring constructed recently. The experimental advantages of an electron-cooled proton beam were utilized. The data cover an energy range where only the lowest possible angular momentum state contributes to the exit channel. There is evidence that the measured energy dependence of the total cross section is not compatible with that predicted by current models based on pion-s-wave production and rescattering.

1. INTRODUCTION

Intermediate-energy light ion storage rings with electron cooling are new tools in the hands of the nuclear physicist and a number of facilities are presently commissioning their machines for nuclear research with stored beams and internal targets. At the Indiana University Cyclotron Facility (IUCF) we have recently completed the first nuclear physics experiment with such a device. This paper is a report on this experiment. The participants included A. Berdoz, F. Dohrmann, J.E. Goodwin, M.G. Minty, H. Nann, P.V. Pancella, S.F. Pate, R.E. Pollock, B. v. Przewoski, T. Rinckel, M.A. Ross, F. Sperisen,and the author, all from IUCF.

The experiment described here addresses the production of π mesons in nucleon-nucleon (NN) collisions. This inelastic part of the NN interaction plays an important role in meson-exchange models and must be treated coherently with the elastic interaction. Information on pion production in the NN system is also needed as a basis in models of pion production or absorption in nuclei.

2. THE $pp \to pp\pi^0$ REACTION NEAR THRESHOLD

Customarily, the partial waves in the exit channel of $NN \to NN\pi$ reactions are labelled[1] by $L\ell$, where L is the angular momentum of the nucleon pair and ℓ is the angular momentum of the pion with respect to the nucleon pair. Within 100 MeV of threshold, because of the short range of the strong interaction, only Ss, Sp, Ps, and Pp final states contribute significantly; within 25 MeV, the Ss configuration should dominate. This prediction is strengthened for the reaction $pp \to pp\pi^0$ by the fact that the Sp final state is forbidden in reactions with isospin T=1 in initial and final NN states.

In the $pp \to pp\pi^0$ reaction, the normally dominant pion production via an intermediate ΔN state, containing an excited nucleon, is suppressed by the fact that the formation of the N and the Δ in a relative S state is forbidden. The rescattering contribution is also small since the dominant isovector component of low-energy πN scattering cannot contribute. Thus, near threshold, the $pp \to pp\pi^0$ reaction is expected to be strongly dominated by the direct-production Born term.

The energy dependence of the cross section is customarily expressed in terms of η, the largest possible center-of-mass pion momentum (with nucleons at rest relative to each other) divided by the pion mass. Based on phase space arguments, the assumption of $\eta << 1$, and a simple treatment of the final state interaction, one expects[2] for the total cross section close to threshold

$$\sigma_{tot} = B_0\,\eta^2 + B_1\,\eta^6 + B_2\,\eta^8 \qquad (1)$$

where the three terms correspond to the Ss, Ps, and Pp final states, respectively.

Clearly, if the bombarding energy is sufficiently close to threshold, only the final state of type Ss contributes to the process $pp \to pp\pi^o$. In this case, the two final nucleons are in a ^1S_o state, and the entrance channel is limited to a single partial wave ($^3P_0$). Since this simplification is expected to facilitate the theoretical interpretation of the data, there is special interest in this near-threshold energy region. Previous experiments, however, were hampered by the fact that the cross sections are small (because resonant production is suppressed in $pp \to pp\pi^o$) and by the fact that the process of interest is in competition with background pion production from nuclei other than hydrogen in the beam (e.g., the windows of target cells). The experiment described here is the first to cover an energy range low enough that one might expect only the Ss final state to contribute.

3. THE EXPERIMENT

3.1. Overview

This experiment has been carried out with the Indiana Cooler[3] a newly constructed storage ring which has been equipped with an intense, co-moving electron beam for phase space cooling of the stored proton beam.[4] The result consists of a measurement of the total cross section for the process $pp \to pp\pi^o$ at a number of bombarding energies close to threshold. Due to the properties of the novel experimental environment, it becomes feasible to measure the four-momenta of both outgoing protons, making it possible to completely reconstruct the three-body final state.

3.2. The Experimental Setup

The hexagonal lattice of the storage ring is shown in Fig.1; three straight sections are for injection, electron cooling and beam manipulation, while the others are provided for experiments. The setup for this experiment has been mounted in the center of the section preceding the electron cooling device. Fig.2 shows the multi-purpose target chamber and the support frame for the detectors. Aperture functions at the target are $\beta_x=0.51$ m, $\beta_y=1.03$ m, and the dispersion is 4.1 m.

A close-up of the target chamber which houses an internal hydrogen target is shown in Fig.3. A gas jet, emerging from a 0.11 mm diameter nozzle (N), cooled to 40 K, crosses the stored beam. Half of the gas flow (typically $6{\cdot}10^{19}$ molecules per s) is pumped away through a catcher opposite the jet (C). The remainder of the gas is removed in a three-stage differential pumping arrangement (1-3 in Fig.3). This results in a hydrogen target of $2{\cdot}10^{15}$ to $5{\cdot}10^{15}$ atoms/cm^2 total thickness.

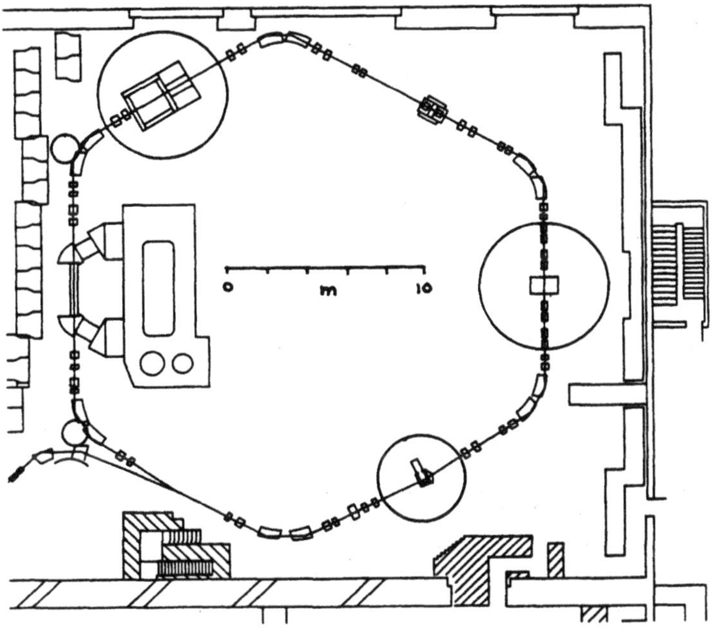

Fig.1. Floor plan of the Indiana Cooler. The circles show experimental areas. The present experiment was carried out in the straight section preceding the electron cooler.

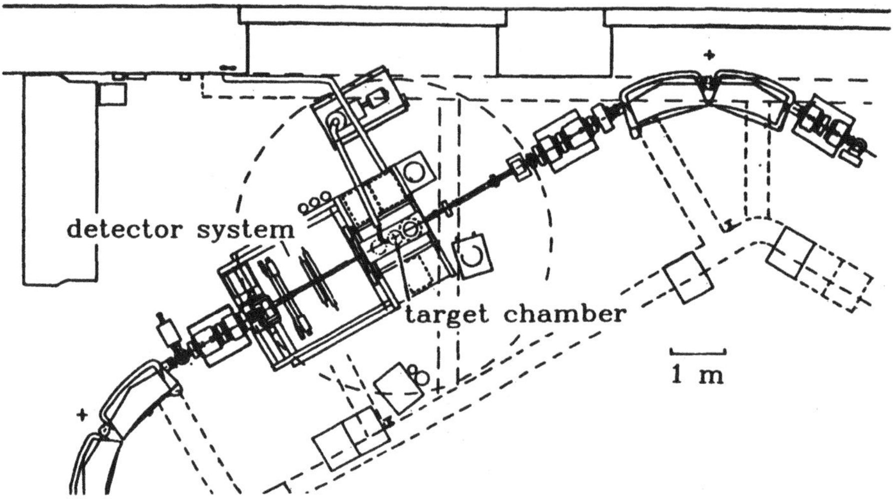

Fig.2. View of the experimental straight section preceding the electron cooler.

About 80% of the target thickness is concentrated within ± 1 cm of the jet; the jet itself, measured at beam height, is 3 mm (FWHM) wide.[5] At the target, the unobstructed area through which the beam passes is 12 mm high and 24 mm wide. The pumping apertures on the downstream side are holes in 24 μm thick aluminum foils and the end of the target box consists of a 127 μm thick steel foil. This minimizes the amount of material the exiting charged particles have to traverse before being detected.

The detector arrangement (see Fig.4) downstream of the target chamber has cylindrical symmetry around the beam axis. A 1.5 mm thick scintillator (F), segmented into quadrants, is followed by two pairs of orthogonal wire planes, and by a 102 mm thick (E) and a 6.4 mm thick (V) scintillator, both segmented into octants. The wire chambers feature a central hole for the beam pipe to pass through

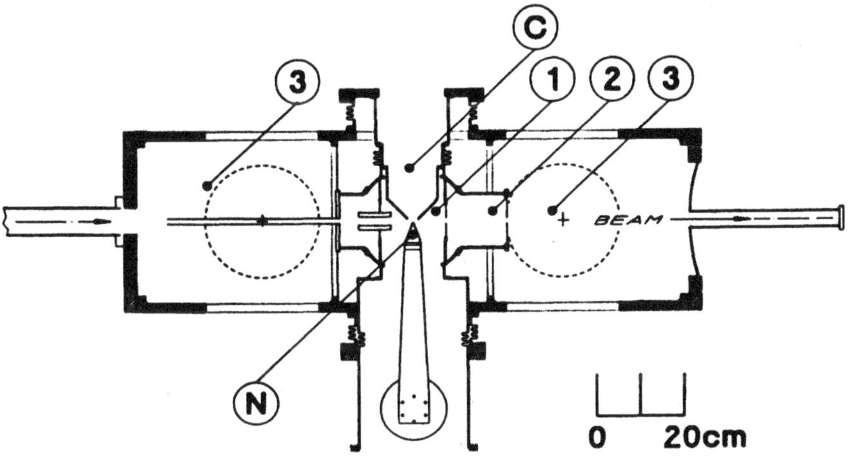

Fig.3. Side view of the target chamber and the internal hydrogen gas jet target. Note the thin exit foil on the downstream end. The labels are described in the text.

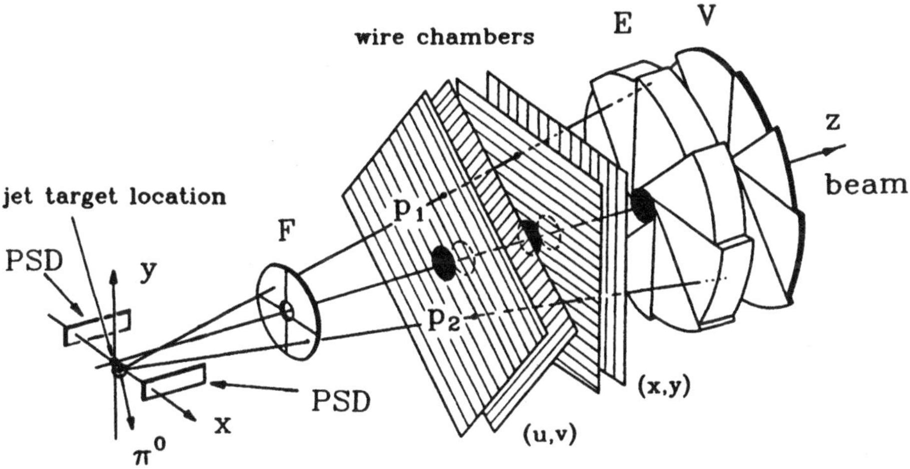

Fig.4. Schematic view of the detector arrangement. The components are described in the text.

and have been specially designed[6)] for this experiment. The direction of each outgoing charged particle is determined by the wire chambers; two pairs of planes, oriented as shown in Fig.4, are needed to resolve ambiguities when more than one charged particle is detected.

Near threshold, the laboratory angle of protons from the $pp \to pp\pi^0$ reaction is limited: the detector was built to completely cover this angular region with the exception of the central hole for the beam pipe. As the bombarding energy is changed, the positions of the detector elements are adjusted in order to accommodate the corresponding changes in the kinematically allowed cone. In addition, for the measurements above 290 MeV, the angular acceptance of the detector is increased by relocating the jet target from the middle position in the target box to a position closer to the exit foil. The angular acceptance of the detector system is from 4° to 20° in the most upstream position, and 2° to 10° in the most downstream position. The thickness of the E detector is chosen to stop up to 120 MeV protons. The impact position of the protons on the E detector is deduced from the wire chamber information and used to correct for a position dependence of the light collection efficiency in the elements of the E detector.

The trigger for a $pp\pi^0$ event is a coincidence between E and F, but no pulse in V, and, in addition, a response from more than one segment of the E detector, signalling a multi-particle final state. In a separate event stream, for which the trigger is a coincidence between all three scintillator planes and no condition on the multiplicity, elastically scattered protons are acquired (see Sect. 3.3). In order to evaluate the importance of accidental coincidences, the experiment is also triggered at random times during data taking. By this method, the accidental rate was found to be negligible.

3.3. Determining the Integrated Luminosity

The absolute normalization of the cross section was obtained from p+p elastic scattering, observed concurrently with pion production. Together with the tabulated[7)] value of the p+p elastic scattering cross section, this leads to a determination of the luminosity. This procedure is necessary, since in a storage ring environment it is difficult to accurately measure the beam current or the target thickness individually.

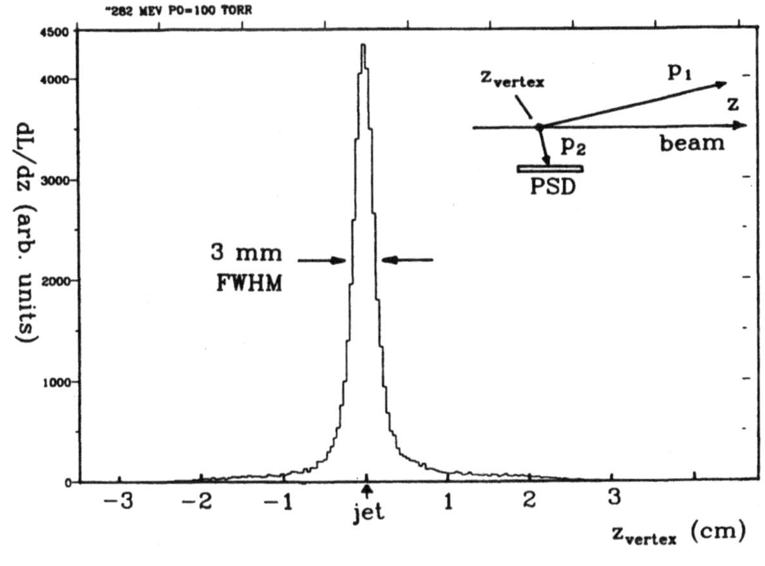

Fig.5. Differential luminosity along the beam in the target region. Since the beam size is smaller than the jet diameter, this is proportional to the gas density.

When detecting elastically scattered protons at forward angles, one finds that a large fraction of the observed particles are beam protons outside the ring acceptance which have been scattered from material close to the beam. In order to reject such events, low-energy recoil protons from p+p elastic scattering are observed in coincidence by two 7 mm by 45 mm position-sensitive silicon detectors ('PSD' in Fig.4). They are mounted 11 cm from the beam in a horizontal plane on either side of the target with their position-sensitive direction along the beam axis. Practically all such coincidences satisfy the kinematics of p+p elastic scattering. The position of the recoils in the silicon detector and the angle of the forward protons determine the location of each scattering center, and allow the reconstruction of the luminosity along the beam. This yields the density profile of the gas jet (see Fig.5). In addition, information on the uniform gas density outside the jet region is obtained; this is needed since the luminous volume active for pion production is somewhat larger than for coincident elastic scattering.

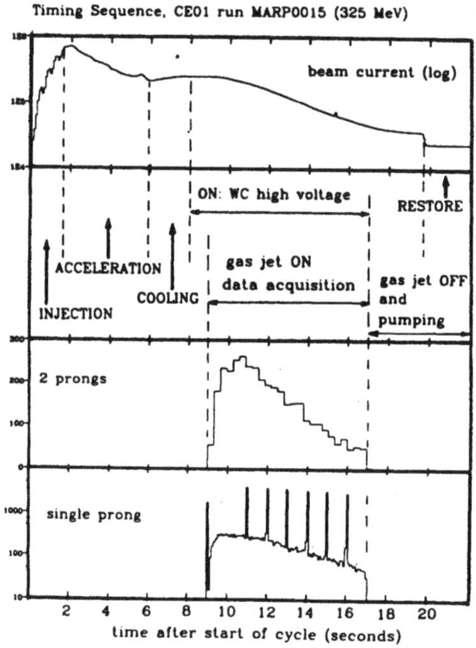

Fig.6. Time structure of a single cycle. The top panel shows the stored beam current, the two bottom panels show the count rate for two-prong and single-prong events.

3.4. Measurements

Data are acquired in so-called cycles. The time structure of such a cycle is shown in Fig.6. Each cycle consists of injection of 45 MeV protons from dissociating a 90 MeV H_2^+ beam on a thin carbon stripper foil (2 s), acceleration to the final energy (4 s), data acquisition while cooling (10 s), and restoration of initial conditions (4 s). Also shown in Fig.6 are the rates for two-prong events (candidates for pion production) and single prong events. The spikes in the single-prong rate are due to events for which no coincidence with a recoil was required; those events

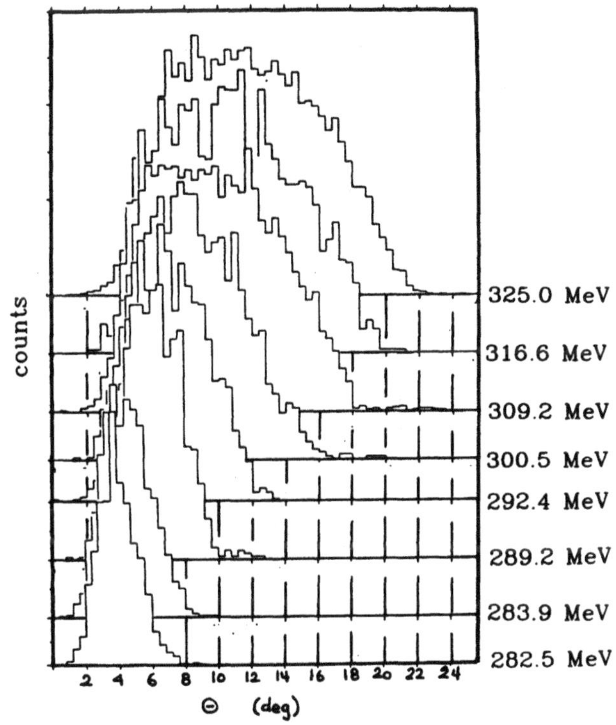

Fig.7. Histograms of angular distributions of the protons from the $pp \to pp\pi^0$ reaction, illustrating the increase of the kinematically allowed cone as the bombarding energy is increased.

were acquired for diagnostic purposes. During data taking, the RF cavity was run at a constant frequency; the harmonic number was 6. Stored beam currents ranged from 5 to 25 μA, resulting in a typical average luminosity of $5 \cdot 10^{28}$ $cm^{-2}s^{-1}$. During a total time for production runs of 110 h an integrated luminosity of about 16 nb^{-1} was achieved.

Measurements have been carried out at nine bombarding energies ranging from 282.5 MeV to 325 MeV. At energies below 290 MeV, the absolute value of the bombarding energy is determined with a precision of ± 200 keV from the strongly energy-dependent maximum laboratory reaction angle of the $pp\pi^0$ protons (illustrated in Fig.7). From the known RF frequencies, a ring circumference of 86.77±0.01 m is deduced from these data. At energies above 290 MeV, where the maximum proton angle varies less rapidly with energy, the bombarding energies were then determined from the circumference and the respective RF frequencies.

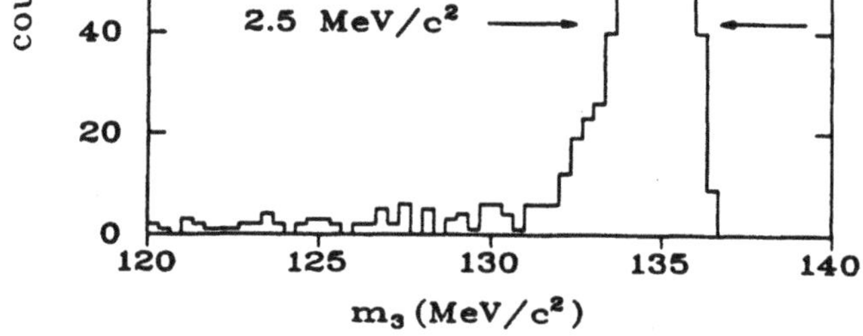

Fig.8. Reconstructed mass m_3 of the unobserved particle. The bombarding energy in this case is 282.5 MeV.

4. DATA ANALYSIS

4.1. Definition of $pp\pi^0$ Events

Charged particles in the E detector are identified as protons by a condition on the relation between the time of flight from the F to the E detector and the energy deposited in the E detector. The particle tracks reconstructed from the wire chambers are required to be consistent with the individual elements of the F and E detector which have registered a hit. From the energy in the E detector and the direction obtained from the wire chambers the mass m_3 of the unobserved particle is reconstructed: as can be seen from Fig.8, a clean peak at the π^0 mass is observed. The area under this peak determines the number N_π of observed $pp\pi^0$ events.

A significant number of two-prong events deviate from the expected pattern of two hits in all of the four wire chamber planes. Spurious extra hits are eliminated by the condition that the real event is the one for which the distance between the two reconstructed tracks is the smallest. Missing hits are either due to the two prongs lining up along the wire direction in one of the planes or due to a chamber inefficiency which is determined independently from single-prong events. Thus, events with a hit missing in one of the planes, or with one additional hit in up to two of the planes, are also included in the data sample.

4.2. Corrections

A Monte Carlo program is used to simulate the response of the detector system, including the effects of multiple scattering, the wire separation, energy and time resolution of the scintillators, and the finite target volume. Excellent agreement with the measured spectra is obtained. The three-body final state is taken to be phase-space distributed with the option to include the S-state final state interaction between the two outgoing protons via an effective range expansion.[8)] At bombarding energies beyond 300 MeV, the final state interaction has a noticeable effect. This is demonstrated in Fig.9 where simulations with and without the final state interaction are compared to the data. The accepted value of -7.82 fm.[9] was used when calculating the solid curve in Fig.9. In order to explain the angular distributions, the effective range has to be chosen about four times smaller than expected. This means that the final state interaction is important in determining the kinematical distribution of this reaction, but that the method of using an effective range expansion to describe the NN interaction is an unacceptable approximation.

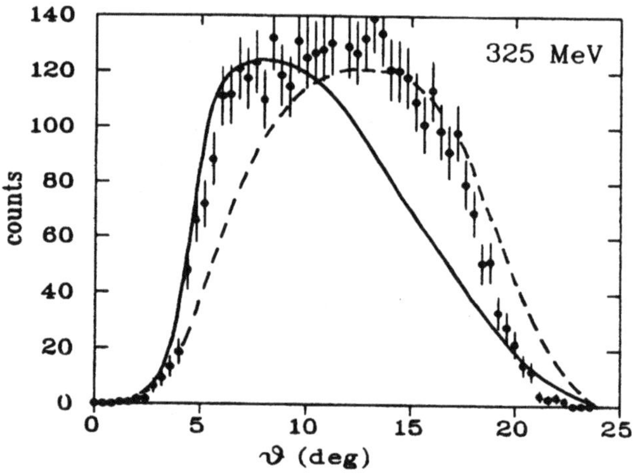

Fig.9. Histogram of the angular distribution of the protons from $pp \rightarrow pp\pi^0$ at 325 MeV. The dashed curve is a simulation assuming phase space distribution of the three outgoing particles. The solid curve represents a calculation where the effective range expansion has been used to take into account the final state interaction between the two protons. The normalization of both curves is arbitrary.

Special care is taken to determine the fraction of $pp\pi^0$ events that are not seen due to the dead zone in the center of the detector. The phenomenological extrapolation into the dead region is based on the shape of the angular distribution predicted by the Monte Carlo simulation. The final state interaction is included but has very little effect very close to threshold. The measured number of $pp\pi^0$ events is then corrected accordingly. Fig.10 illustrates the size of this correction for the case of 286.5 MeV bombarding energy; in Fig.10, the data (dots) are compared to a full simulation (a) and one without events with at least one prong in the central hole (b). The correction factor due to loss in the central hole ranges from 1.69 at the lowest, to 1.24 at the highest energy.

Additional corrections arise from the efficiencies of the scintillators and the wire chambers. The scintillator efficiencies are estimated from measured total reaction cross sections and total (p,n) cross sections. The wire chamber performance characteristics, i.e., the role of accidental hits, missing wires, the distribution of wire cluster sizes, and the fusion of individual clusters into one, has been simulated by a Monte Carlo calculation. The calculation is adjusted to reproduce the experimental distribution of multiplicities and then used to determine the wire chamber efficiency

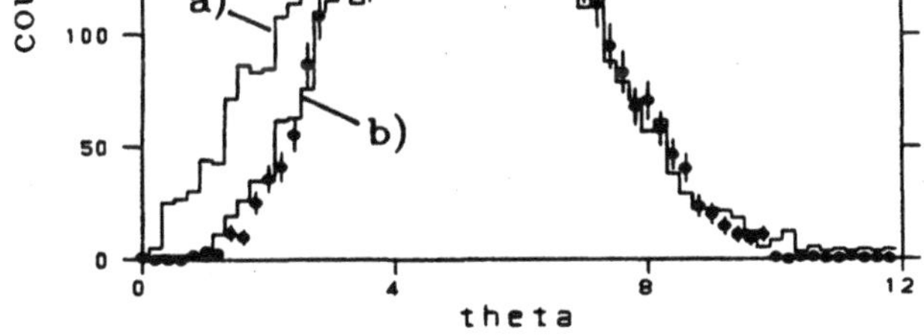

Fig.10. Angular distribution of outgoing $pp\pi^0$ protons at 286.5 MeV bombarding energy. Histogram (b) is a Monte Carlo simulation of the experiment. In calculation (a) the effect of the hole is ignored.

for two-prong events. The simulation is also used to determine the fraction of events with both protons in the same segment of the E detector (which would fail to trigger the experiment). The combined correction due to these three effects is at most 5%.

4.3. Tests

A number of test runs were carried out with the intention to check the experimental procedure. Measurements at the same energy but with different detector and target positions, and separated in time by two months were found to yield compatible values for the cross section. This supports the validity of the hole correction and the reproducibility of the procedure. At most energies, runs were also taken with gas bleeding into the target region instead of entering through the cold nozzle. This results in a uniform gas density, defined by the differential pumping impedances. Compatible results verify the procedure used to determine the absolute normalization.

5. RESULTS AND DISCUSSION

The final results for the total cross section σ_{tot} are computed from the observed number of pions, N_π, the absolute, integrated luminosity L deduced from p+p elastic scattering, and a correction factor g. This factor is mainly due to the loss of events in the dead zone of the central hole; smaller contributions include the measured wire chamber efficiency and the scintillator efficiency, as described above. The results are shown as solid dots in Fig.11, together with previous measurements.[10-13] The errors shown contain the contribution from counting statistics as well as estimated uncertainties in the luminosity L and the correction factor g. The errors do not include an estimated 5% error of the overall normalization of the data which is due to the uncertainty of the p+p elastic scattering cross section obtained from the literature.[7]

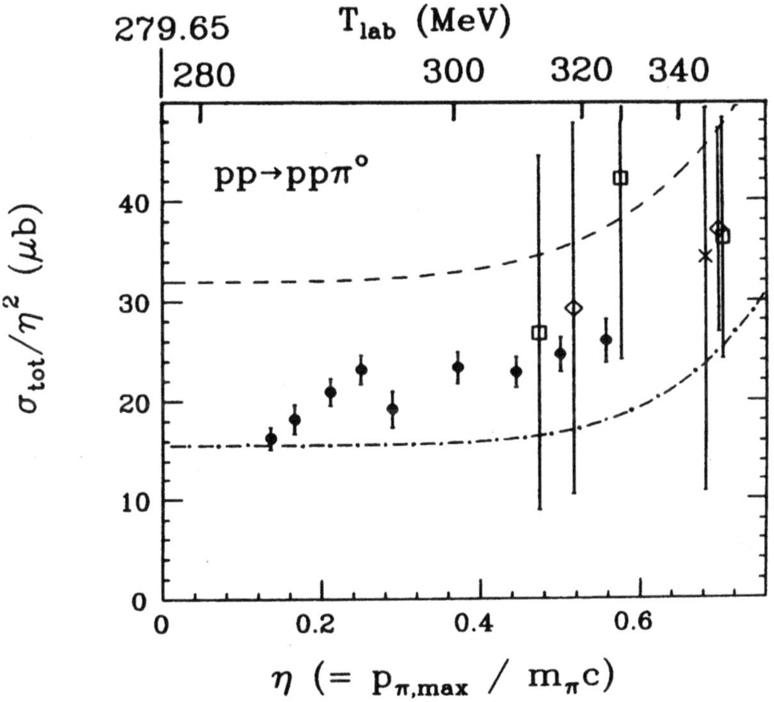

Fig.11. The total cross section for $pp \rightarrow pp\pi^0$ versus the dimensionless pion momentum. The data are from present work (solid dots), Ref.10 (cross), Ref.11 (squares), and Ref.13 (diamonds). The curves represent Eq.1 and are discussed in the text.

Also shown are two previously published, polynomial fits according to Eq. 1 from Refs. 11 and 13. These fits (dashed line[11]), dot-dashed line[13])) yield the coefficients $B_0=32$ μb and $B_0=15.6$ μb for the contribution of the lowest angular momentum state. This should be compared with theoretical values for B_0 ranging from 11 to 18 μb.[14,15]) As can be seen from Fig.10, our measurement is not compatible with a near-threshold energy dependence of σ_{tot} proportional to η^2, implied by Eq.1. Since this energy dependence follows from simple calculations of s-wave production and rescattering[14]), the present measurement provides a strong incentive for theoretical efforts to reevaluate the energy dependence of the $NN \rightarrow NN\pi$ process close to threshold. The simple energy dependence is based on phase space arguments and an effective range expansion for the final state interaction. The validity of these assumptions is also questioned by the failure to account for the angular distributions at the higher energies within this frame work. The theoretical work should be facilitated by the fact that only one partial wave contributes and that the transition matrix element is dominated by the Born term.

6. OUTLOOK

The work reported here is the first nuclear physics experiment carried out with a storage ring with electron cooling. It clearly illustrates the potential of the new technology: aside from the fact that synchrotron acceleration gives IUCF access to the $NN\pi$ threshold, this experiment has benefited from the small interaction energy spread, the low background due to the windowless target, the accessibility of small angles, and the observability of low-energy recoils.

A promising future application of this technology involves polarized beams and polarized targets. At present, considerable experience with storing and manipulating polarized proton beams has already been gained at IUCF, and the target thickness that can be achieved with a polarized atomic beam source injecting into a buffer cell[16)] is reaching the order needed for internal target experiments. A measurement of spin correlation coefficients for the $pp \rightarrow pp\pi^0$ reaction, for example, would yield additional information on individual angular momentum states in the final state.

7. ACKNOWLEDGEMENTS

We are grateful to the operating team of the new machine, in particular to D. Friesel, T. Sloan, and V. Derenchuk for their tireless effort in bringing the new machine on line. The work has been supported in part by the Deutscher Akademischer Austauschdienst and by the US National Science Foundation under grant NSF PHY 87-14406.

REFERENCES

1. A.H. Rosenfeld, Phys.Rev. 96, 139 (1954).

2. M. Gell-Mann and K.M. Watson, Ann.Rev.Nucl.Sci. 4, 219 (1954).

3. R.E. Pollock, Proc. IEEE Particle Accelerator Conf., Chicago 1989, ed. F. Bennett and J. Kopta, IEEE Conf. 1989, p. 17.

4. F.T. Cole and F.E. Mills, Ann.Rev.Nucl.Part.Sci. 31, 295 (1981), and references mentioned therein.

5. F. Sperisen, A. Berdoz, H.O. Meyer, R.S. Barbieri, and R.A. Bonham, Nucl. Instr. Meth. A274, 604 (1989).

6. K. Solberg, A. Eads, J.Goodwin, P. Pancella, H.O. Meyer, T. Rinckel, and A. Ross, Nucl. Instr. Meth. A281, 283 (1989).

7. R.A. Arndt, L.D. Roper, R.A. Bryan, R.B. Clark, B.J. VerWest, and P. Signell, Phys.Rev. D28, 97 (1983), and the data base SAID mentioned therein.

8. K.M. Watson, Phys. Rev. 88, 1163 (1952).

9. H.P. Noyes, Ann. Rev. Nucl. Sci. 22, 465 (1972).

10. R.A. Stallwood, R.B. Sutton, P.H. Fields, J.G. Fox, and J.A. Kane, Phys.Rev 109, 1716 (1958).

11. A.F. Dunaitsev and Yu.D. Prokoshkin, Zh.Eksp.Teor.Fiz. 36, 1656 (1959) (Sov.Phys.JETP 9, 1179 (1959).

12. F. Shimizu, Y. Kubota, H. Koiso, F. Sai, S. Sakamoto and S.S. Yamamoto, Nucl.Phys. A386, 571 (1982).

13. S. Stanislaus, D. Horvath, D.F. Measday, and A.J. Noble, Phys. Rev. C41, R1913 (1990).

14. D.S. Koltun and A. Reitan, Phys.Rev. 141, 1413 (1966).

15. V.P. Efrosinin, D.A. Zaikin, and I.I. Osipchuk, Z.Phys. A322, 573 (1985).

16. W. Haeberli, Proc.Int.Workshop on Polarized Sources and Targets, Helv.Phys.Acta 59, 597 (1986).

ELECTRON COOLING OF HEAVY IONS AT THE HEIDELBERG COOLER STORAGE RING TSR

M. Steck, M. Blum, M. Grieser, E. Jaeschke,
C.M. Kleffner, D. Krämer, W. Ott, R. Repnow

Max-Planck-Institut für Kernphysik
D-6900 Heidelberg
Postfach 103980

ABSTRACT

An overview of the electron cooling experiments with heavy ions at the Heidelberg cooler storage ring TSR is given. They prove the increased cooling rate for highly charged ions as well as the beam emittance reduction to values below 1 πmmmrad. Emphasis has been put to the observation of longitudinal noise signals of the cooled beam which reveal strong coherent beam behaviour. New beam accumulation techniques with electron cooling have been developped even allowing dual beam operation.

1. INTRODUCTION

The Heidelberg heavy ion cooler ring TSR[1]) is the first operational storage ring of a new generation of heavy ion storage rings worldwide. A large variety of ions ranging from protons up to uranium can be injected from the Tandem-postaccelerator combination[2]). The relatively low beam intensity from a negative ion source is compensated in the storage ring by beam accumulation through multiturn injection and rf stacking[3]. The maximum magnetic rigidity $B\rho$= 1.5 Tm allows storage of heavy ions with energies up to 30 MeV/u for completely stripped ions with A/Z= 2. For phase space compression of the stored beams the storage ring TSR is equipped with an electron cooler[4] covering all the range of energies of heavy ions which can be stored.

2. ELECTRON COOLER

The electron cooling device is located in one of the four straight sections of the storage ring. The magnetic system consists of five solenoid and two toroid magnets with a bending angle of 45 degrees (Fig. 1) to guide the electron beam and includes correction dipoles for the ion beam as well as steerer coils in all solenoid magnets for the electron beam.

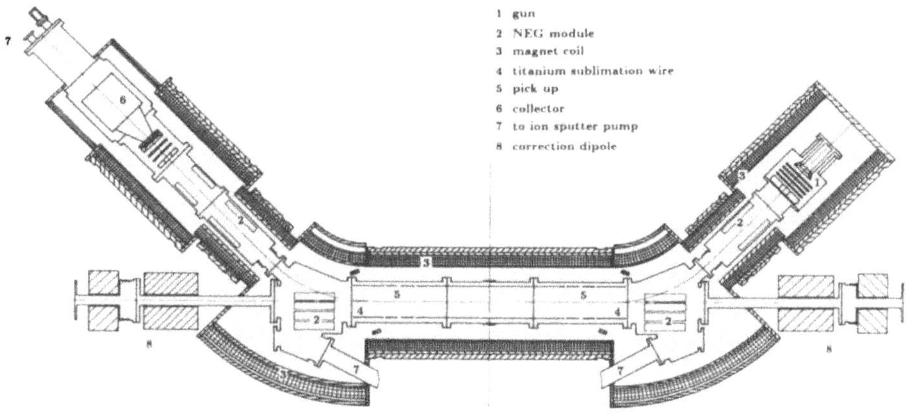

Fig. 1 Cutaway drawing of the TSR electron cooler

A low temperature electron beam of 5 cm diameter is extracted from a thermal cathode and accelerated in a five electrode system to the final energy. The good collection efficiency of the electron beam collector and the high pumping speed which is provided for the electron cooler facilitates electron beam operation in the 10^{-11}mbar range. The main parameters of the electron cooler are listed in Table 1.

Electron Beam	
kinetic energy	0-16 keV
electron current	0-3 A
current collection efficiency	0.9995-0.9998
beam diameter	50.8 mm
cathode temperature	1200 K
Magnetic Guiding Field	
maximum field	3 kG
variations of longitudinal field	$\pm 2 \cdot 10^{-2}$
field parallelity in cooling section	$\leq 5 \cdot 10^{-5}$
effective length of cooling section	1.2 m
Operation in Resonant Optics Mode	
gun perveance	0.4-6.6 μP
magnetic guiding field	0.2-0.6 kG
cyclotron motional energy	<0.1 eV
total transverse energy	0.2-1.0 eV
Vacuum	
pumping speed for H_2 (NEG)	18000 l/s
sputter ion pumping speed	860 l/s
basic pressure	$2 \cdot 10^{-11}$ mb
basic pressure with hot cathode	$<3 \cdot 10^{-11}$ mb
degassing rate with electron beam (per 1 mA of loss current)	$1 \cdot 10^{-7}$ mb·l/s

Table 1 Parameters of TSR electron cooler

Operational experience over more than 1000 hours of electron beam operation proved a very reliable performance of the system the parameters of which are fully computer controlled.

3. ADJUSTMENT OF COOLING PARAMETERS

The electron energy and the corresponding cathode potential can be easily calculated from the ion energy and the gun properties which are known from computer calculations. However the knowledge of the quantitative amount of space charge compensation by residual gas ions is indispensable. As the operational experience has shown that stable cooling conditions can only be achieved for uncompensated electron beam clearing electrodes in the collector side toroid are used to extract slow ions from the electron beam.
The electron beam adjustment in order to maximize the cooling rate includes the transverse parallel displacement and the angular alignment of the electron beam. For parallel displacement the entrance position of the electron beam to the cooling section is varied by means of the gun solenoid steerer coils. The revolution frequency of a coasting beam as a function of the transverse displacement resembles the radial space charge variation of the electron energy and the minimum revolution frequency corresponds to correct matching of electron and ion beam center (Fig. 2a). The angular alignment of the electron beam is monitored by observation of the temporal evolution of a higher harmonic of the bunching frequency after beam injection and adiabatic capture into a bucket. The amplitude growth with time shows a pronounced maximum over a range of typically ± 1 mrad (Fig. 2b) indicating the correct alignment of electron and ion beam direction. These optimization steps preceed all cooling experiments that will be reported.

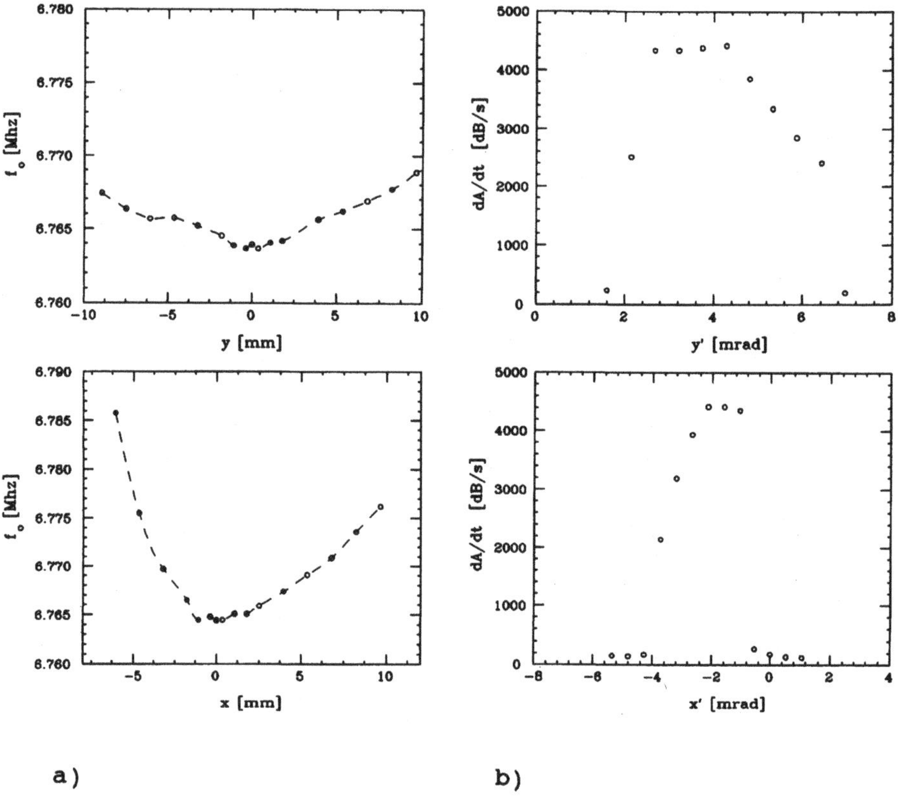

Fig. 2 Optimization of electron beam for cooling
a) revolution frequency (11th harmonic) versus parallel displacement, b) cooling rate for bunched beam (2nd harmonic of bunching frequency) versus angle

4. COOLING FORCE

First measurements of the drag force which the electrons exert on the ion beam have been executed. For an ion beam after multiturn injection the drag force was determined by detuning the electron velocity in discrete steps with respect to the velocity of the newly injected ions. The time for dragging the ions to the same velocity was monitored with the longitudinal Schottky pick up.
A maximum longitudinal drag force of 1.4 eV/m for carbon ions ($^{12}C^{6+}$, 73 MeV) was measured (Fig. 3) when cooled by an electron beam of 0.38 A ($n_{el}=3.5\cdot10^7 cm^{-3}$). Nevertheless higher cooling forces can be expected if the transversely hot ion beam ($T_{ion} \sim$ 500 eV) is precooled. Measurements of cooled ion beams by stepping the cathode potential are difficult for heavy ions as the high stability high voltage power supply responds too slowly to steps of the control voltage. Additional problems exist for fast scans with the standard spectrum analyzer used for Schottky noise detection. Therefore only the determination of the maximum cooling force was possible and the values are listed in Table 2.

Ion	n_{el} [$10^7 cm^{-3}$]	F_{max} [eV/m]
d	3.5	0.23
$^{12}C^{6+}$	3.5	5.5
$^{28}Si^{14+}$	2.3	7.8
$^{32}S^{16+}$	3.5	>40

Table 2 Maximum longitudinal cooling force for cold ion beams

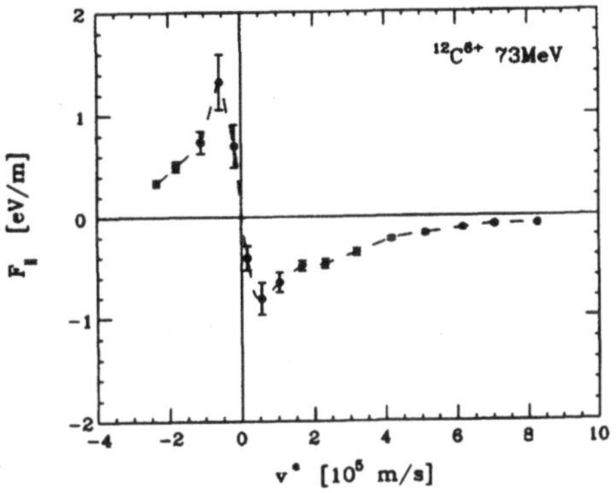

Fig. 3 Cooling force for hot ion beam after multiturn injection

5. TRANSVERSE COOLING

Information on the transverse size of the stored beam mainly comes from measurements of the beam profile of the ions which are lost in the first dipole magnet behind the cooler after capturing an electron in the cooling section. A measurement of the beam profile for $^{16}O^{6+}$ ions[5)] which originate from a $^{16}O^{7+}$ 142 MeV stored and cooled beam demonstrates the beam size reduction by electron cooling (Fig. 4).

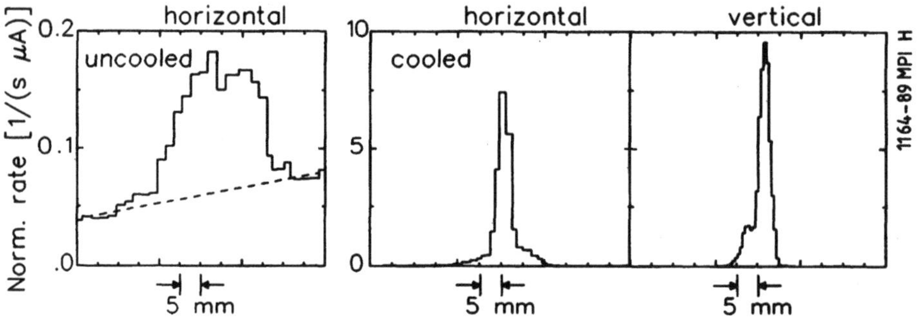

Fig. 4 Profile of ions after electron capture
horizontal without cooling (left)
horizontal and vertical with cooling (right)

Due to the limited detector resolution of 2 mm this profile measurement gives an upper limit of the beam emittance $\epsilon = 0.5\pi$mmmrad in horizontal and vertical direction as well. Destructive beam diagnosis by scraper measurements and acceleration with beam loss at the ring acceptance confirmed an emittance value of $\epsilon < 1$ πmmmrad.
Additional information provides the analysis of the longitudinal noise spectra of the cooled beam. For high intensity ion beams the beam impedance can be determined

from the separation of the plasma frequency f_n from the harmonic n of the revolution frequency f_o via the relation

$$\frac{ImZ_n}{n} = - \frac{\beta^2 E}{Ze\eta} \quad \frac{f_n^2}{n^2 f_o^2} \; I_{Ion}$$

Assuming an impedance which is dominated by the inductive space charge impedance

$$\frac{ImZ_n}{n} = - \frac{Z_o}{\beta\gamma^2} \; (0.5 + \ln \frac{b}{a})$$

the average beam emittance was calculated resulting in the values of the beam emittance listed in Tab. 3. The exceptionally high value for silicon beam is likely to be caused by insufficient optimization of cooling parameters. In general this analysis is in fair agreement with a reduction of the cooled beam emittance to values on the order of 10^{-1} πmmmrad.

Ion	Velocity [v/c]	Impedance [kΩ]	Beam Emittance [πmmmrad]
P	0.208	8.37	0.50
d	0.114	18.80	0.07
$^{12}C^{4+}$	0.113	23.41	0.005
$^{12}C^{6+}$	0.114	17.42	0.15
$^{12}C^{6+}$	0.157	12.51	0.15
$^{16}O^{8+}$	0.114	13.83	1.26
$^{28}Si^{13+}$	0.087	14.58	6.41
$^{28}Si^{14+}$	0.094	13.86	5.78
$^{32}S^{16+}$	0.114	15.32	0.51

Table 3 Impedances and emittances of cooled ion beams

6. LONGITUDINAL BEAM NOISE

The detection of longitudinal noise signals originating from the circulating ions have been the main object of investigations of cooled heavy ion beams at the TSR. Good agreement has been found with the main characteristics predicted by theoretical calculations for cooled beams[6]. The main aspect of the longitudinal noise signal is the observation of two coherent signal maxima attributed to plasma waves which are co- and counterpropagating to the beam. These plasma waves cause coherent signals which can be observed for sufficiently high beam intensity (Fig. 5) exceeding a certain threshold intensity which depends on the attainable beam temperature.

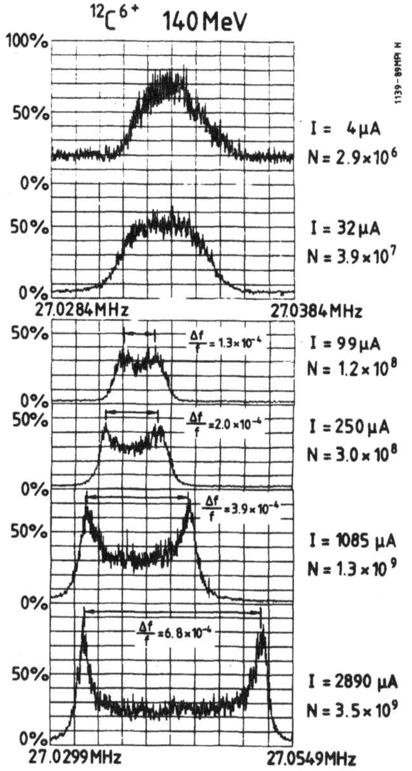

Fig. 5 Longitudinal noise signal for various beam intensities

Another aspect of the noise spectrum of a cooled beam has been clearly demonstrated in the measurements, the suppression of the integral power of the noise signal (Fig. 6). The integrated power of the noise from a cooled beam is not dependent on the particle number as for normal Schottky noise, but it is proportional to the beam temperature. The noise power of the signal of a cooled beam is significantly reduced compared to the noise power of the same intensity for an uncooled beam.

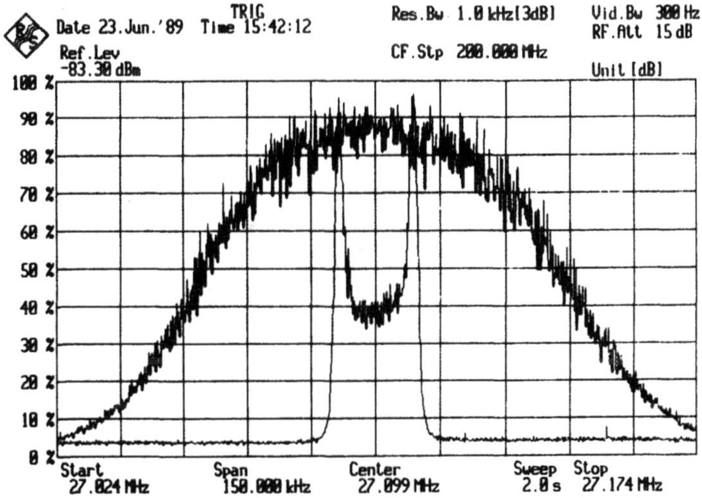

Fig. 6 Noise spectrum of uncooled and cooled ion beam ($^{12}C^{6+}$ 140 MeV)

Systematic measurements of the intensity dependence of the parameters of the spectrum on the beam intensity have been performed. As a typical example a direct comparison of three ion species (d 12.2 MeV, $^{16}O^{8+}$ 98 MeV, $^{32}S^{16+}$ 196 MeV) which were stored and cooled under identical settings of cooler and storage ring is shown in Fig. 7. The coherent frequency was determined from the separation of the two coherent signal maxima. The momentum spread of the beam is according to the theoretical description revealed by the analysis of the spectral distribution. The noise power is above the critical intensity proportional to the beam temperature and below the critical intensity proportional to the particle number.

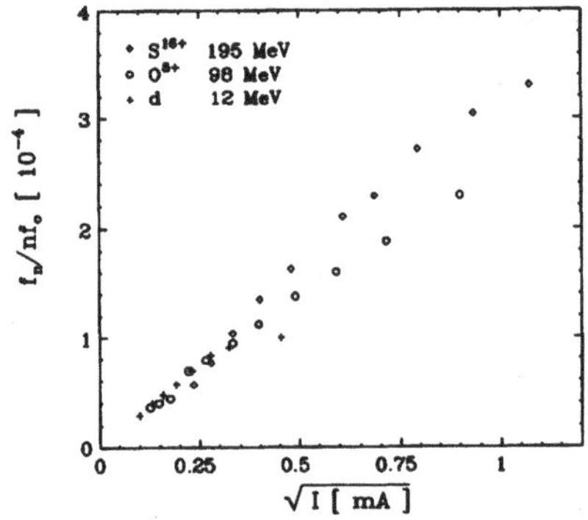

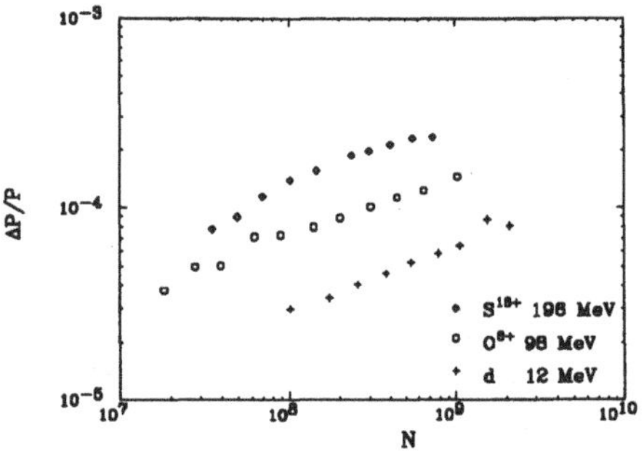

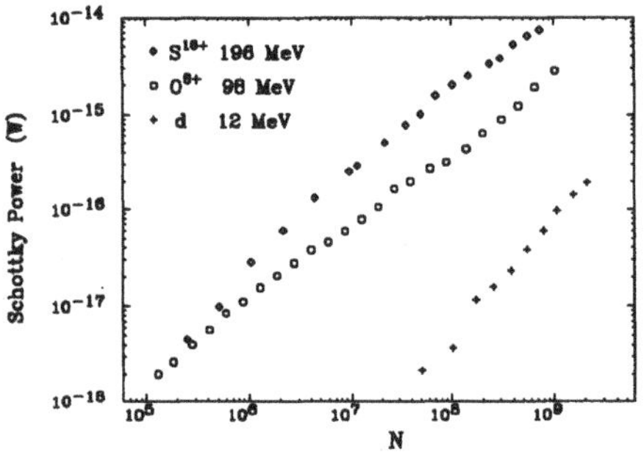

Fig. 7 Comparison of spectrum parameters for cooled ion beams (d 12.2 MeV, $^{16}O^{8+}$ 98 MeV, $^{32}S^{16+}$ 196 MeV)
a) coherent frequency,
b) momentum width,
c) noise power.

The coherent frequency increases proportional to the square root of the intensity in agreement with the theoretical prediction. The momentum width shows an increase proportional to $N^{1/3}$ which can be explained as a consequence of the equilibrium between electron cooling and heating due to intrabeam scattering. The noise power increases with the particle number N^a with an exponent a= 0.7-1.0. No deviation was observed from this almost linear behaviour from the lowest particle number that was

detectable up to the maximum number that was limited by transverse instabilities.
The scaling of the spectrum parameters with the ion charge is also well understood. For the coherent frequency relative to the center frequency a gradient which is solely dependent on the ion velocity is predicted. The experimental results show minor differences for the different ion species which might be caused by the transverse beam size, but are still within the experimental error.
The momentum spread increases with the charge of the ion. For constant particle number the ratio was determined as 1 : 2.4 : 4.4 (d : O : S) with a $N^{1/3}$ dependence on the particle number. From this a ratio of the longitudinal beam temperature of 1 : 47 : 310 follows. Above the critical particle number which in the present measurements is some 10^7 the noise power is proportional to the longitudinal beam temperature. By comparison of the noise power at constant particle number a ratio of of 1 : 35 : 190 for the beam temperatures was deduced. Both methods therefore confirm within the experimental accuracy a Z^2-dependence of the beam temperature in agreement with estimates for a beam of constant emittance.
This has also been confirmed for a similar experiment comparing deuteron and carbon beam stored under the same experimental conditions[7)].

7. BEAM ACCUMULATION

Electron cooling has been applied for the accumulation and successive storage of high intensity ion beams. This novel scheme uses the reduction of the transverse phase space volume by cooling to store the beam at a stack position which is not affected during consecutive multiturn injection cycles. This method requires proper setting of the fast ramping injection components and the momentum of the circulating ions in such a way that the stacked ions can circulate in spite of the new injection process. A deceleration of the ions after injection by detuning of the cathode potential by some permille proved to be useful but not necessary.

A Schottky spectrum of the accumulation process is shown in Fig. 8. The successively injected multiturn batches are decelerated by the electron beam to the stack position at a momentum 6 $^{o}/oo$ below the injection momentum. As the phase space volume of the stored and newly injected beam is continouosly reduced this scheme faces no limit opposed by considerations of phase space conservation. Intensity limits originate either from the onset of coherent oszillations or the balance of the finite beam

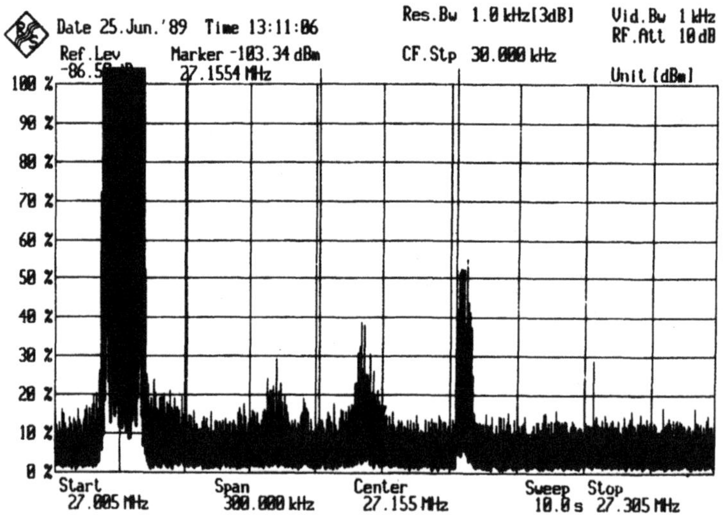

Fig. 8 Noise spectrum illustrating the process of beam accumulation of multiturn batches by deceleration and cooling with the electron beam.

lifetime and the injection rate. Therefore even low intensity beams which have a long lifetime and preferably a short cooling time can be accumulated up to the instability limit.

With the combination of multiturn injection and beam accumulation by electron cooling maximum intensities of typically some 10^9 particles have been stored stably. Intensities up to $3 \cdot 10^{10}$ ($^{12}C^{6+}$ 73 MeV) have been achieved by combination of multiturn injection, rf stacking and electron cooling[8]. The modulated frequency of the rf cavity decelerates the ions in this case filling the longitudinal phase space and bringing the ions closer to the detuned electron velocity. The reason for achieving much higher stored beam intensity compared to the case without rf stacking is not fully understood. One possible explanation is the overcoming of a certain intensity range which corresponds to an unstable area of the beam impedance by the fast filling injection process when appliing rf stacking. Careful adjustment of chromaticity corrections in order to facilitate rf stacking might also increase the range of stable ion currents. Measurements of the coherent frequency for large numbers of stored ions indicate an increase of the transverse beam size which seems to be necessary to stabilize the storage of high beam currents.

A recent progress based on the accumulation scheme employing the electron cooler is the simultaneous storage of two different ions. This was realized by successive injection of two ions of almost equal magnetic rigidity and velocity. This way neither the cooler parameters nor any of the ring and injection components had to be changed. In the first part about $1 \cdot 10^9$ deuterons (12.2 MeV) were accumulated and stored. Because of the lifetime of 6 hours there was enough time to change the ion source to oxygen beam ($^{16}O^{8+}$ 96 MeV) which was injected with the same ring setting. Due to missing fine adjustments of the injection line only a small fraction of the primary oxygen current was actually stored but by accumulation with the electron cooler up to $1 \cdot 10^8$ oxygen ions were collected in the ring. Because of the difference in the nuclear binding energy of the two ions they were observed as two well separated bands in the noise spectrum (Fig. 9). This dual beam storage technique might find ample application for experiments using one beam component as a calibration source or reference for reactions of the second component.

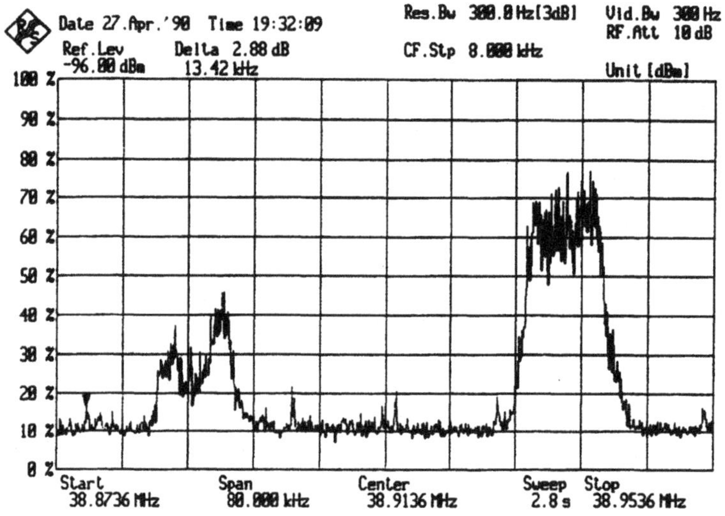

Fig. 9 Noise spectrum of two ion beams (d 12.2 MeV left, $^{16}O^{8+}$ 96 MeV right) stored simultaneously after accumulation by electron cooling.

8. REFERENCES

1. P. Baumann, M. Blum, A. Friedrich, C. Geyer, M. Grieser, B. Holzer, E. Jaeschke, D. Krämer, C. Martin, K. Matl, R. Mayer, W. Ott, B. Povh, R. Repnow, M. Steck, E. Steffens, W. Arnold, Nucl. Instr. and Meth. A268 (1988) 531

2. M. Grieser, R. Hammel, B. Huck, H. Ingwersen, E. Jaeschke and R. Repnow, IEEE Trans. Nucl. Sci. NS-30, No.1 (1983) 1095

3. G. Bisoffi, M. Grieser, E. Jaeschke, D. Krämer and A. Noda, Nucl. Instr. and Meth. A287 (1990) 320

4. M. Steck, M. Blum, A. Friedrich, E. Jaeschke, M. Jung, V. Kössler, D. Krämer and K. Matl, Proc. 1st Europ. Part. Acc. Conf., World Scientific (1989) 1172

5. A. Wolf et al., First Atomic Physics Experiments with Cooled Stored Ion Beams at the Heidelberg Heavy Ion Storage Ring TSR, XVI Int. Conf. on the Physics of Electronic and Atomic Collisions, New York, July 26-Aug. 1, 1989

6. V.V. Parkhomchuk, D.V. Pestrikov, Sov. Phys. Tech. Phys. 25 (1980) 818

7. M. Grieser, M. Blum, D. Habs, G. Huber, M. Jung, E. Jaeschke, C.M. Kleffner, D. Krämer, W. Petrich, R. Repnow, S. Schröder, M. Steck, Cold Heavy Ion Beams at the Heidelberg TSR, 2nd Europ. Part. Acc. Conf., Nice, June 12-16, 1990

8. M. Steck, G. Bisoffi, M. Blum, A. Friedrich, C. Geyer, M. Grieser, B. Holzer, E. Jaeschke, M. Jung, D. Krämer, K. Matl, W. Ott and R. Repnow, Nucl. Inst. and Meth. A287 (1990) 324

STATUS REPORT ON THE ELECTRON COOLING STUDIES IN THE TARN II

Tetsumi Tanabe

Institute for Nuclear Study, University of Tokyo
Tanashi, Tokyo, 188 JAPAN

ABSTRACT

Electron cooling experiments have been made in a synchrotron storage ring TARN II since September 1989. After a brief summary of the ring and the cooling device, the results of the cooling experiments on a proton beam are presented. Electron cooling studies on D^+ and H_2^+ beams are also discussed.

1. INTRODUCTION

Strong phase space compression of ion beams achieved by the electron cooling technique is now opening up new possibilities for the accelerator technology and the physics research. In 1983, we proposed to apply the electron cooling to heavy ion beams[1] as well as a proton beam which had already been demonstrated at that time. Construction of the electron cooling device has started in 1984. A program[2] to double the size of the storage ring TARNI has begun at the same time. The electron cooling device produced its first off-line electron beam at the end of 1988 and it was then installed in the newly commissioned ring TARN II in summer 1989. The first electron cooling experiment was achieved in September 1989[3].

First, the outline of the TARN II is briefly described. Secondly, the main parameters of the electron cooling system are presented. Finally, results of the electron cooling experiments are discussed.

2. TARNⅡ RING

Figure 1 represents the layout of the TARNⅡ ring and the injector sector focusing cyclotron. The cyclotron has a K-number of 68. The maximum injection energies, however, are limited by the maximum magnetic rigidity of the magnets in the beam transport line from the cyclotron to the ring and by the maximum voltage of the electrostatic inflector of the injection system. They are 20 MeV for proton and 40 MeV for $^4He^{++}$. The cyclotron can also supply with heavy ions up to Ne.

The ring has a diameter of 24 m and a six-fold symmetry. In the long straight sections, an injection system, an rf cavity, an electron cooling system and beam diagnostic elements have been installed. In near future, a slow beam extraction system will be installed. Vacuum pressure of the ring is about 10^{-9} Torr, as the beam pipes are still unbaked. The main parameters of the ring are listed in Table 1. The first beam acceleration has recently been performed in March 1990: 20-MeV proton beam was accelerated up to 100 MeV in 3.5 sec. The electron cooling has been studied mainly at the injection energies due to the problem for the stability of power supplies which excite dipole and quadrupole magnets of the ring.

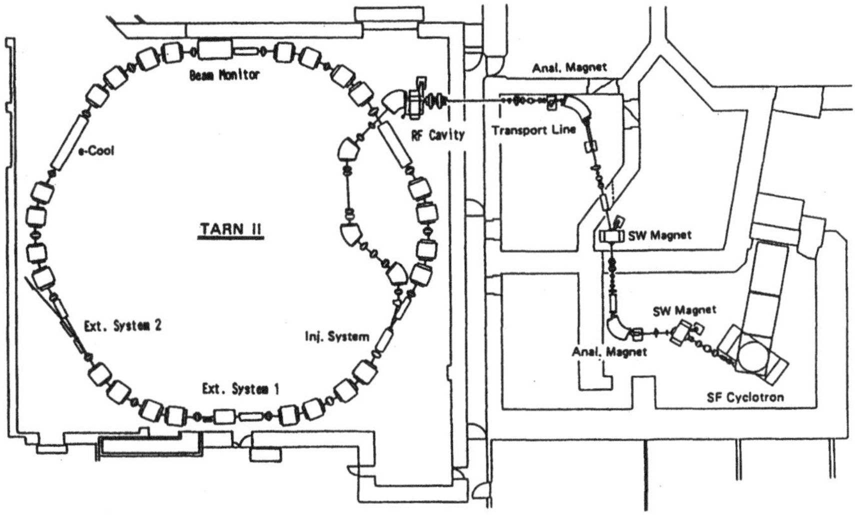

Fig. 1. Layout of the TARNⅡ ring and the injector cyclotron.

Table 1 Main parameters of TARN II ring.

Maximum magnetic rigidity	6.1 T·m	
Maximum beam energy proton	1.1 GeV	
ion with q/A=1/2	370 MeV/u	
Circumference	77.76 m	
Average radius	12.376 m	
Radius of curvature	4.045 m	
Length of long straight section	4.2 m	
Superperiodicity	6	
Rising time of magnet excitation	3.5 sec	
Repetition rate	0.1 Hz	
RF frequency	0.55 - 8.7 MHz	
Harmonic number	2	
	A)	B)
Horizontal β in cooling section	10.2	11.3 m
Vertical β in cooling section	3.7	2.3 m
Momentum dispersion in cooling section	4.7	4.6 m
Horizontal betatron tune	1.71	1.78
Vertical betatron tune	1.80	2.17

3. ELECTRON COOLING SYSTEM

The maximum design energy of the cooling device is 110 keV, which is equivalent to 200 MeV/u for ion beams. Figure 2 represents the layout of the cooling device. Electrons are emitted from a flat dispenser cathode with a diameter of 5 cm and accelerated to final energy by an acceleration column. Electrons then enter into a drift region and are bent by 45 degree to the cooling region which is 1.5 m in length. At the end of the cooling section, electrons are again bent by 45 degree upwards and are collected after decceleration. The whole system is immersed in a uniform solenoid field. So far we have tested up to 70 keV and 2 A. Typical relative current loss of the electron beam is in the range of 10^{-4}, which depends on the electron energy and current. Details of the design and the results of the off-line electron beam test are described in refs. 4-7. The orbit deformation of ion beams due to toroids is corrected by steering magnets, but there is no element to cancel the solenoid field.

4. ELECTRON COOLING EXPERIMENTS

4.1. Proton Beam Cooling

Electron cooling experiments have been performed primarily on

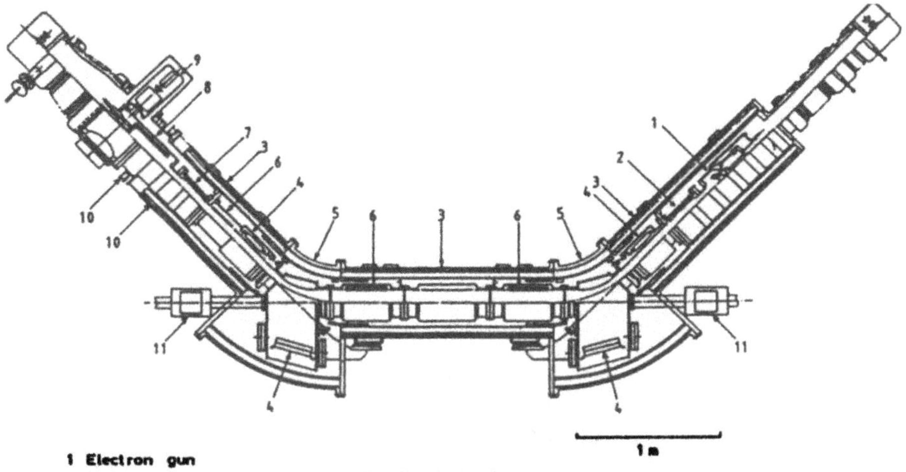

Fig. 2. Layout of the electron cooling device.

the proton beam with an energy of 20 MeV which corresponds to an electron energy of 11 keV. A typical electron current is 0.4 A and the solenoid field to confine the electron beam is 600 G.

The electron cooling can be observed by the increase of the beam lifetime, since the transverse phase space compression due to the cooling suppresses the divergence of the beam induced by multiple scattering with residual gas. The rate of the lifetime increase depends, however, strongly on the tune values. The lifetime of the cooled beam is 20 sec for the tune values, Q_H=1.71 and Q_V=1.80 (parameter A in Table 1), which are close to a linear coupling resonance. On the other hand, it increases to 500 sec for the tune values, Q_H=1.78 and Q_V=2.17 (parameter B in Table 1). Such a dependence can originate from the coupling between horizontal and vertical motions due to the cooler solenoid field. Figure 3 shows the increase of lifetime with cooling for the latter tune values.

The rapid decrease of beam emittance with cooling also allows to

a)

b)

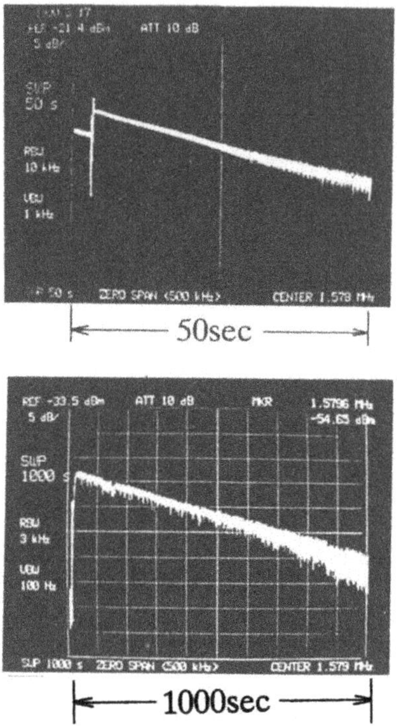

Fig. 3. Lifetime of a) uncooled and b) cooled proton beam.

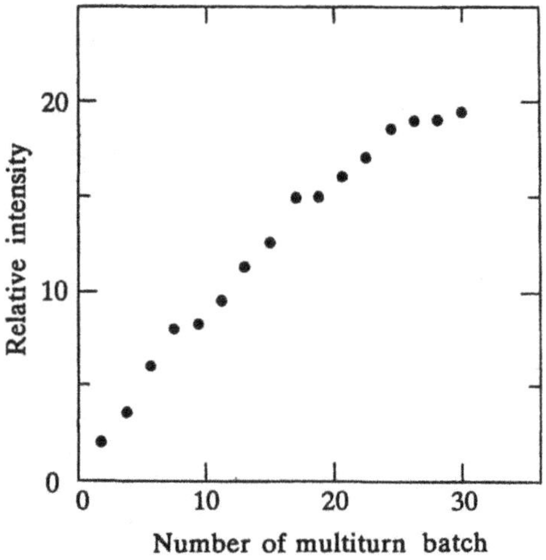

Fig. 4. Intensity increase during cooled stacking as function of the number of multiturn batches.

stack the beam in the transverse phase space. First, the horizontal phase space area is fully occupied by the multiturn injected beam. The area is then compressed with cooling in a few seconds. The beam is again multiturn injected without kicking out the cooled beam. The same procedure is repeated many times, resulting in the increase of the beam intensity. Figure 4 shows the increase of the neutral beam intensity which is proportional to the number of cooled beam. An equilibrium is reached after 30 stackings in 2 minutes and the integrated multiplication factor of 20 is obtained.

The change of momentum resolution with cooling can be observed by measuring the frequency spread of the higher harmonics of Schottky signals for a coasting beam. Figure 5 represents the 38th harmonic of the

a)

b)

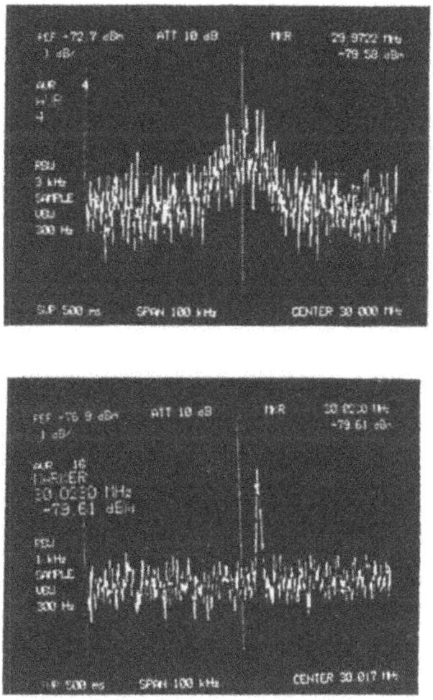

Fig. 5. Longitudinal Schottky signal spectra for a) uncooled and b) cooled proton beam.

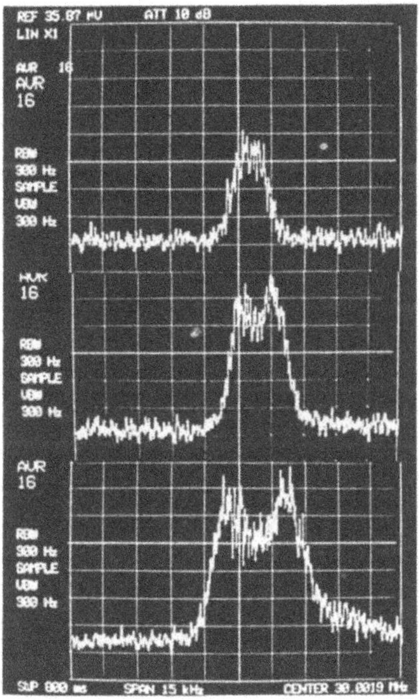

Fig. 6. Splitting of Schottky signal spectra with increase of stored beam intensity. From the top, numbers of proton beam are about 3×10^{7}, 10^{8} and 3×10^{8}.

revolution frequency for uncooled and cooled beams. Momentum spread ($\Delta p/p$) decreases from the order of 10^{-3} to the order of 10^{-5} within a few seconds. When scanned with a high resolution of a spectrum analyzer, splitting in the frequency spectrum of the Schottky signal is also observed as shown in Fig. 6. The splitting becomes conspicuous with the number of circulating beam, which is originated from two plasma waves travelling with and against the beam direction[8)]. The momentum width can be deduced by comparing the spectra with theoretical predictions taking account of the collective effects, which is $3\sim5\times10^{-5}$.

When circulating protons pass through the interaction region, they capture cooling electrons at a certain probability. Detection of such an electron-captured neutral hydrogen brings a lot of important

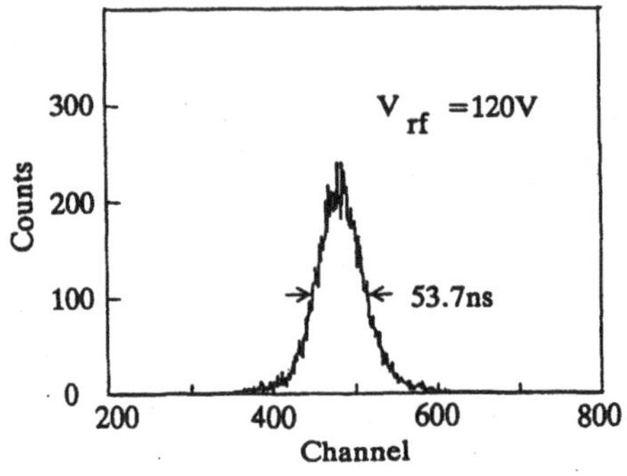

Fig. 7. Time distribution of hydrogen atoms with respect to the rf phase for a bunched beam at an rf voltage of 120 V.

information for the electron and the proton beam quality. The size of the neutral beam is measured with a position sensitive detector (PSD) with a sensitive area of 47×8 mm^2 which is installed at the exit of a thin window situated at 4.5 m downstream of the cooling section. The horizontal and vertical beam sizes (FWHM) are 6 mm and 8 mm, respectively. From the known beta functions, one can deduce the emittance and the beam radius of the circulating beam at the cooling section: The horizontal and vertical emittance are 1 π mmmrad and 2.6 π mmmrad, respectively, which contrast with the initial values of $\varepsilon_H \sim 100$ π mmmrad and $\varepsilon_V \sim 25$ π mmmrad. Proton beam radii at the cooling section are about 2.3 mm and 1.7 mm in horizontal and vertical directions, respectively.

For the bunched beam cooling, the bunch width decreases as the result of longitudinal cooling and synchrotron oscillation. The bunch width can be measured by observing the distribution of the arrival time of neutral hydrogen relative to the rf phase. A signal from a plastic scintillator started a TAC, which was stopped by a signal from the rf. Figure 7 shows the result of the bunch width measurements. The width (FWHM) of 53 ns at an rf voltage of 120 V corresponds to the bunch length of 3.3 m, which contrasts with the initial value of about 0.6 μ sec. The bunch width decreases with the increase of rf voltage. It reaches down to 15 nsec at 400 V,

a)

b)

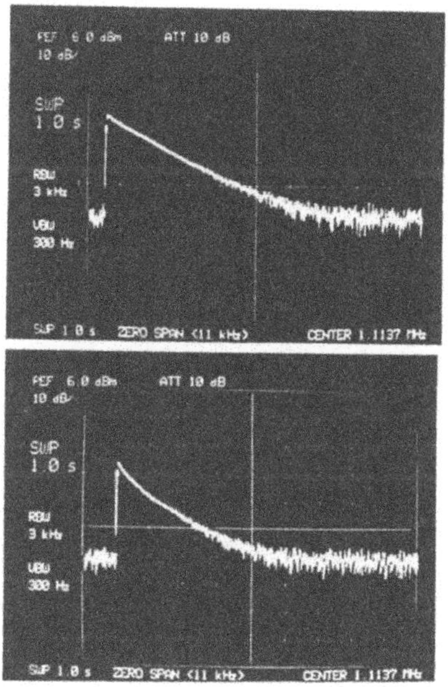

Fig. 8. Lifetime of H_2^+ beam a) without and b) with electron beam of 0.4 A.

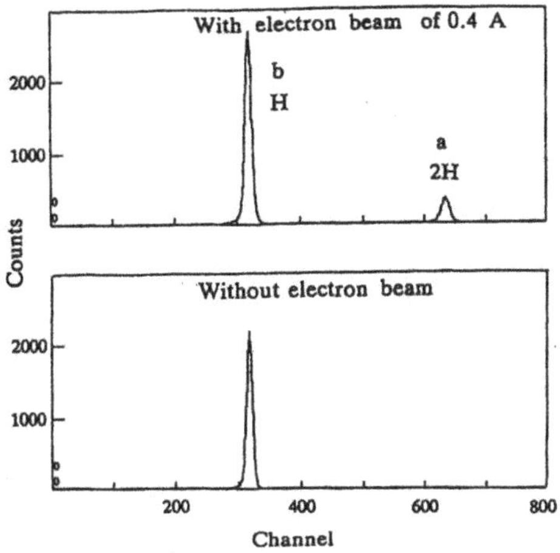

Fig. 9. Energy spectra of neutral hydrogen atoms from the H_2^+ beam with and without electron beam. Peaks a and b correspond mainly to dissociative recombination and dissociative excitation processes, respectively.

4.2. Deuteron and H_2^+ Beam Cooling

The cooling of molecular ion beams is very attractive as it has not yet been observed experimentally and there is no theoretical prediction. First, a deuteron beam with an energy of 20 MeV was cooled and the lifetime increase by a factor of 10 was observed. Then, the circulating beam was replaced by the H_2^+ beam with the same velocity, keeping the cooler and the ring parameters constant. The results of lifetime measurements are shown in Fig. 8 for the H_2^+ beam with and without electron beam. The lifetime is too short for the cooling to be established. It also clearly decreases when switching on the velocity-matched electron beam with a current of 0.4 A.

In order to study the beam loss mechanism, neutral hydrogen atoms are detected with the PSD. Figure 9 shows the energy spectra of neutral hydrogen with and without the electron beam. One can clearly observe H-atom pairs depositing full energy (20 MeV) with electron beam (peak a), which arise from a dissociative recombination process ($H_2^+ + e \rightarrow H + H^*$). On the other hand, the peak b corresponds to single H atoms mainly from a dissociative excitation process ($H_2^+ + e \rightarrow H^+ + H + e$), which deposit only half the energy (10 MeV). The measurements were also performed changing the electron energy in small steps. The double and single H atom production processes have a maximum and a minimum cross sections at the electron velocity which is equal to the H_2^+ velocity, respectively. This confirms the view that they correspond to the dissociative recombination and the dissociative excitation processes. Detailed studies on such a merged beam process will also contribute to the research of the atomic physics.

The ion source of the injector cyclotron is a PIG type and higher vibrational excited states of H_2^+ ions can already be populated at the ion source. These highly excited H_2^+ ions seem to be easily dissociated via electron impact in comparison with low-lying states. If one can store only the low-lying H_2^+ ions, the lifetime is expected to become longer. Such an ion source has already been realized, which chemically reacts away the higher vibrational states using a gas mixture of H_2 and rare gases (He and Ne) in the ion source[9]. In order to make the beam lifetime longer, the ion source can be important in addition to the improvement of vacuum.

ACKNOWLEDGEMENTS

The author would like to thank members of the TARN II group who contributed to the electron cooling research. The electron cooling and the electron capture studies were supported by Grant in Aid for Scientific Research (A) and (B) from the Ministry of Education, Science and Culture, respectively.

REFERENCES

1. Tanabe,T., Sekiguchi,M., Sato,K., Noda,A., Kodaira,M., Takanaka,M., Tanaka,J., Tsujikawa,H., Honma,T., Katayama,T., Mizobuchi,A. and Hirao,Y., "Electron Cooling Project at INS", Proc. of Workshop on Electron Cooling and Related Applications (ECOOL1984), Karlsruhe, pp. 127-135 (1984).

2. Katayama,T., "Heavy Ion Accelerator and Cooler -TARN2-", Proc. of 11th Int. Conf. on Cyclotrons and their Applications, Tokyo, pp. 128-133 (1986).
3. Tanabe,T., Chida,K., Honma,T., Hattori,T., Kanazawa,M., Katayama,T., Mizobuchi,A., Nakai,M., Noda,A., Noda,K., Sekiguchi,M., Soga,F., Ueda,N., Watanabe,S., Watanabe,T., Yoshizawa,M. and Ando,A., "Initial Electron Cooling Studies in the TARN II", Proc. of 7th Symposium on Accelerator Science and Technology, Osaka, pp. 201-203 (1989).
4. Tanabe,T., Sekiguchi,M., Sato,K., Kodaira,M., Noda,A., Tanaka,J., Tsujikawa,H., Honma,T., Takanaka,M. and Hirao,Y., "Electron Cooling Device for TARN II", IEEE Trans. NS-32, 2412 (1985).
5. Tanabe,T. and Watanabe,T., Nucl. Instr. and Meth. A278, 241 (1989).
6. Tanabe,T., Watanabe,T. and Honma,T., "Electron Cooling at INS", Proc. of the 1989 Particle Accelerator Conference, Chicago, pp. 651-653 (1989).
7. Tanabe,T., Particle Accelerators 32, 113 (1990).
8. Poth,H., Schwab,W., Seligmann,B., Woertge,M., Wolf,A., Baird,S., Chanel,M., Haseroth,H., Hill,C.E., Ley,R., Manglunki,D., Tranquille,G., Vallet,J.L. and Ditter,P.F., Z. Phys. A332, 171 (1989).
9. Sen,A., McGowan,J.W. and Mitchell,J.B.A., J. Phys. B20, 1509 (1987).

ELECTRON COOLING SYSTEM PROJECT OF THE STORAGE RING AT THE INSTITUTE FOR NUCLEAR RESEARCH (THE UKRAINIAN ACADEMY OF SCIENCES).

A.E. Valkov, A.V. Dolinsky, A.V. Zhmendak, S.N. Pavlov, A.I. Papash
(Institut for Nuclear Research of the Ukrainian Academy of sciences).
N.S. Dikansky, V.I. Kudelainen, D.V. Pestrikov, B.N. Sukhina
(Institute of Nuclear Physics, Novosibirsk)

1. INTRODUCTION.

During recent years the methods of charged particle beam cooling have been developed extensively (1,2), opening the unique possibilities of investigations in the field of elementary particle physics and in nuclear physics (3).

In accordance with this trend the heavy ion storage ring with electron cooling is planned to build at the Institute for Nuclear Research (Kiev). At the present report the general characteristics of projecting storage ring are given. The physical aspect of electron cooling system construction are considered in more details.

2. GENERAL RING CHARACTERISTIC.

The ion storage ring is a zero gradient synchrophasotron which includes 8 45° - beam-bending magnets and 26 quadrupole lenses. It is supplied by the injection system, two RF-cavity resonators and two superthin internal targets. The electron cooling system is located at one of the straight sections with 6 m lenght. The storage ring has the following parameters: ion energy range $E_i / A_i \cong 5{,}5 - 200$ MeV/u, magnetic rigidity $B \cdot R \cong 4{,}5$ T· m, circumference C=56m, beta-functions: average $\beta \cong 9$ m, at electron cooling section $\beta_e \cong 4$ m, at the main target $\beta_t \cong 0{,}9$ m, dispersion (ψ – function): at the cooling section $\Psi_e = 0$, near the injection system $\psi_{in} \cong 6$ m, Acceptance $\varepsilon_A = 120\,\pi$ mm.mrad, mean vacuum $\sim 10^{-10}$ torr . The more detailed description of the ring has been presented in Ref. (4). Isochronous cyclotron U-240 is used as an injector. Now it produces the ion beams of elements from H^+ to Ne. The ions are grouped in bunches with time pulse with $\sim$ 5 ns ; following with repetition time $\sim$ 100 ns. The particles number in one bunche is N_b 10^5 - 10^7 (see table 1).

Table 1

Ion	A_i / Z_i	E_i / A_i Mev/n	N_e particles
P	1	50	2,1 . 10^7
d	2	25	1,6 . 10^7
$^3He^{2+}$	1,5	50	5,2 . 10^6
$^4He^{2+}$	2	25	7,8 . 10^6
$^{12}C^{3+}$	4	5,5	1,2 . 10^6
$^{16}O^{4+}$	4	10	4,7 . 10^5
$^{14}N^{4+}$	3,5	5,5	3,3 . 10^5
$^{20}N_2^{5+}$	4	5,5	2,5 . 10^5

The specific energy of ions E_i/A_i ranges from 5,5 to 50 MeV/u (A_i - ion atomic weight). Beam emittance is $\varepsilon \cong 30\ \pi$ mm.mrad and energy spread is $\Delta E_i / E_i \cong 0,5\%$.

3. STORAGE CYCLE.

Since the ring under consideration shall be used for nuclear studies, the problem of high luminosities L is of paramount importance. The corresponding calculations show (4) that the total number of stored ions N should exceeds 10^9 particles,i.e. it is required to inject into the storage from 10^2 to 10^4 microbunches. As about 10 particle bunches are packed at one ring period and the emittance value is great, it is necessary to use the following injection procedure. When one storage revolution is filled the inflector is turned out. Ions are decelerated by the additional RF - cavity losing about 1% of energy and pass into the internal orbit. Then the beam is cooled and so the phase becomes free for next particle portion. The electron cooling limits this process in time.

The efficiency of electron cooling system is determined to a considerable extent by a choice of the following parameters: magnetic field strenght H, its inhomogenity $\Delta H_\perp/H$ and stability $\tilde{H}/H$ cooling section lenght l_e, transversal velocity spread $\Delta V_{e\perp}/V_{e||}$, energy stability of electron beam and also current density j_e and diameter d_e. As it was shown in experiments made on NAP-M the values $\Delta V_{e\perp}/V_{e\parallel}$ and $\tilde{E}_e/E_e$ must be less than 10^{-3} and 10^{-4} respectively (5). Let's consider the rest parameters. It should be noted that the beam injected from U-240 has longitudinal and transverse velocity spread $\Delta\vartheta_{i\perp} \cong \Delta\vartheta_{i\parallel} \cong \Delta\vartheta_{e\perp}$. In the range of lower velocities the sufficiently large values of ion friction force may be realized in a regime of adiabatic collisins only (6). This requires a satisfaction of ineguality $\rho_e < n^{-1/3}$ where ρ_e is Larmor radius and n_e is electron density. Using the typical values n_e 10^8 - 10^9 cm^{-3} and $T_{e\perp} \cong mV^2_{e\perp}/2 \cong 0{,}1$ eV, we obtain $H \cong 1$ кOe. As it follows from the experiments made on "Solenoid model" (7), the further increase of magnetic field is inexpedient at least for positive ions, if we have no other reasons to grow H. Let us estimate the cooling time τ_e for beam injected from U-240. In the mentioned velocity range, the friction force for adiabatic collisions takes the form (6)

$$F \cong \frac{4\pi n_e e^4 Z_i^2 \Lambda_{ie}}{m\upsilon_p^2}$$

where eZ_i is ion charge, Λ_{ie} is Coulomb logarithm for ion-electron collisions, m is electron mass and υ_p is ion velocity in the coordinate system, moving together with the electron beam. Then the damping rate of betatron oscillations may be presented in the form

$$\lambda \cong \frac{4\pi n_e e^4 Z_i^2 \Lambda_{ie}}{m\upsilon_s^3 M_p A_i(\theta_x^2 + \theta_z^2 + (\Delta P_{i\parallel}/P_i)^2)^{3/2}} \cdot \eta$$

where M_p A_i and υ_s are ion mass and velocity, respectively, in laboratory coordinate system, $\theta_x^2 + \theta_z^2$ is ion angular spread, $\Delta P_{i\parallel}/P_i$ is beam momentum longitudinal spread and $\eta = l_e/C$ is ratio of cooling section lenght to storage ring circumference. We see, that the damping rate should be linearly increased with increasing of n_e. However electron beam space charge leads to the rise of transverse gradient of longitudinal electron velocities. That is equivalent to an appearance of additional momentum spread $\Delta P_{i\parallel}/P_i$. Taking this effect into account we obtain the following expression for betatron oscillation damping rate,

$$\lambda \cong \frac{4Z_i^2 r_p \Lambda_{ie} c^2}{A_i \upsilon_s \varepsilon} \cdot \frac{u}{(\varepsilon/\beta_e + u^2)^{3/2}}; \quad u \cong \frac{\pi r_e n_e \varepsilon \beta_e c^2}{v_s^2}$$

where r_p and r_e are classical proton and electron radii, c is velocity of ligh

This function $\lambda(u)$ has maximun at

$$u = \sqrt{\frac{\varepsilon}{2\beta_e}} \quad .$$

This allows us to find the optimun value of electron beam current

$$I_{opt}\,(A) = \frac{\upsilon_s\, UC}{c \cdot 15} \sqrt{\frac{\varepsilon}{2\beta_e}} -$$

where U is electron gun potential. Then the maximun value of damping rate is .

$$\Lambda_{max} \cong \tau_e^{-1} \cong 1,5 \frac{Z_i^2\, \eta\, r_p\, \Lambda_{ie} c^2}{A_i\, \upsilon_s\, \varepsilon^2}$$

The results of calculations carried out by using of above formulae in supposition and $\Lambda_{ie} \cong 10$ and $l_e \cong 3$m are given in table 2. One can see, that the values of electron beam current density j_e change within the range 11mA/cm^2 - 270 mA/cm^2 and $\tau_e \cong 0,2$ -1 sec. In the case of ion stripping up to nucleus state the cooling time τ_e^* may be about 4 times smaller. Let's estimate the luminosities L, which may be achieved at these values of τ_e^* . As it is know $L = f_s N t_z$, where f_s is ion repetition frequency in the storage ring and t_z is target thickness, which equals atom density n_t in the target multiplied by target lenght l_t. Let's express t_z over the lifetime of ion beam at the target

$$t_z \cong \frac{1}{n_t \sigma_e \upsilon_s} \frac{c}{l_t} \quad \text{i.e.} \quad t_z \cong \frac{1}{\tau_t\, \sigma_e f_s}$$

Here σ_e is cross-section of the processes, leading to ion beam loss (8,9).The ion number in storage ring may be presented in the form $N = N_{mb} \cdot mn$, where m is number of bunches packed at one ring circumference and $n = \tau_{in} / \tau_e$ is the storage cycle number during the injection time τ_{in}. In other to increase the luminosity averaged in time, let's require $\tau_{in} \leq \tau_t$. Then the expression for L may be reduced to the form

$$L \cong \frac{1.8\, N_1}{\beta_{in}\, \tau_e \sigma_e}$$

where β_{in} is relativistic factor at the injection energy. The results of calculations of luminosities for the energy 200 MeV/u and target atomic number $Z_A \cong 90$ are given in table 3. One can see, that the luminosities are sufficiently high (of order of 10^{29} - 10^{30} cm^{-2} sec^{-1}). The above considerations were used as a basis of the original project (4). It was not taken into account, however, that for the enhacement of research possibilities of acceleratin- storage complex it is necessary to secure the high luminosities within the whole energy range from 200 MeV/u up to injection energy.

Table 2

ion	E_e keV	β_{in}	J_e A/cm^2	I A	τ_e sec	τ_{e*} m sec
P	26	0,31	0,27	1	0,74	-
d	14	0,23	0,11	0,4	1,1	-
$^3He^{2+}$	26	0,31	0,27	1	0,55	-
$^4He^{2+}$	14	0,23	0,11	0,4	0,55	-
$^{12}C^{3+}$	3	0,11	1,1 10^{-2}	4.10^{-2}	0,35	87
$^{14}N^{4+}$	5,4	0,15	2,7 10^{-2}	0,1	0,3	100
$^{16}O^{4+}$	3	0,11	1,1 10^{-2}	4.10^{-2}	0,26	66
$^{20}Ne^{5+}$	3	0,11	1,1 10^{-2}	4.10^{-2}	0,2	52

Table 3

ion	P	$^4He^{2+}$	$^{12}C^{6+}$	$^{16}O^{8+}$	$^{20}Ne^{10+}$
L(cm^{-2}sec^{-1})	5.10^{29}	1.10^{30}	2.10^{30}	6.10^{29}	2.10^{29}

At the same time the cross-section of losses increase rapidly when E_i decreases. This effect leads to diminishing of L (see table 4). Moreover the situation is complicated by the fact, that in experiment the friction force are often lower than values predicted by the theory . For istance for NAP-M the semiempirical formula was obtained (5)

Table 4

Luminosity magnetudes at some ion beam energy values for Ne $Z_A \cong 90$

E_i / A_i MeV/u	5,5	10	50	100	200
L $cm^{-2}\ sec^{-1}$	1.10^{25}	3.10^{25}	1.10^{27}	8.10^{27}	2.10^{29}

$$\lambda \cong \frac{66\, r_p r_e n_e c^4 \eta}{[(\alpha \upsilon_s)^2 + \Delta\upsilon_{i\perp}^2 + k\Delta_{i\parallel}^2]\sqrt{\upsilon_{e\perp}^2 + \Delta\upsilon_{i\parallel}^2 + \Delta\upsilon_{i\perp}^2}}$$

where α ia a coefficient of magnetic field inhomogenity. The value of k was obtained equal to 11. This has been connected with a presence of dispertion at the cooling section. In order to extrapolate this formula for our case, it is necessary to multiple λ by Z_i^2 / A (which permits us to take into account the presence of multicharged ions). Moreover k should be supposed equal to 1, since in the project under consideration we have ψ_e=0 at the cooling section. Assuming $\alpha \cong 5.10^{-5}$, we obtain the following cooling times :

H^+ - $\tau_e \cong 2{,}1$ sec; $^4He^{2+}$ - $\tau_e \cong 1{,}7$ sec; $^{20}Ne^{10+}$ - $\tau^*_e \cong 0{,}27$ sec

that is considerably higher than the values presented in the table 2. It follows that the decreasing of cooler lenght over 3 m leads to dangerously low values of luminosity. An increase of l_e above 3m is not possible because of chosen storage ring geometry.The luminosity of device may be increased by a number of ways. Firstly we can use multiturn exchange injection of heavy ions.At typical for our conditions energies $E_i \cong 100$ MeV that gives the advantage by L approximately 3 times (this possibility is foreseen in the project). Secondly, may be decreased by mean of emittance and beam energy spread reducing. Such experiments are making at U-240 now, but it is unlikely to expect the values of ε essentially lower than 10π mm mrad. At last the using of cooling sistem with compensated by charge the electron beam is possible (10). In this case the terms containing n_e are absent in the denominator of expression of betatron oscillation damping rate and current density j_e may be increased up to theresold value for ion-drift in stability

$$j_e \cong \frac{\upsilon_s^2 H}{3,8\, l_e}$$

At H ≅ 1kOe, current density may be enlarged by an order of magnitude. But the quality of electron beam should be essentially damaged due to transverse-longitudinal relaxation process.At H=3kOe the values $j_e \cong 0,3 - 2,5 A/cm^2$ may be achieved allawing to obtain electron cooling time belonging to to millisecond range (see table 5).We limit ourselfs by the range $j_e \cong 0,2 - 1 A/cm^2$, since the magnetic compression of electron beam is required in order to obtain $j_e > 1 A/cm^2$. It doesn't contribute the essential

Table 5

ion	j_e A/cm^2	I_e A	τ_e msec	τ_e^* msec
P	2,5	9,5	42	-
d	1,4	5,3	46	-
$^3He^{2+}$	2,5	9,5	32	-
$^4He^{2+}$	1,4	5,3	23	-
$^{12}C^{3+}$	0,3	1,1	7,5	1,9
$^{14}N^{4+}$	0,6	2,1	7,4	2,4
$^{16}O^{4+}$	0,3	1,1	5,6	1,4
$^{20}Ne^{5+}$	0,3	1,1	4,5	1,1

changes in storage ring operation because σ_e increases more slowly for light particles than for heavy ions when E_i decreases. The calculations made on the base of semiempirical formula for transverse - longitudinal relaxation of electron beam (11), show that this process would be suppressed to a great extent at H 3kOe and values of j_e given in table 5. It is necessary also to take into account that the particles stored in the ring would ba scattered by ions compensating the electron beam. However the corresponding analysis shows, that this effect is negligible. It should be pointed out that operation with superthin targets is characterized by $t_z \cong 10^{12} - 10^{14}$ cm^{-2}, $Z_A \cong 90$ whereas for compensating ions the production of its concentration and the cooling section lenght is $t_e < 10^{11}$ cm^{-2} at mean atomic number $\overline{Z_i} \leq 7$. Thus the system of electron cooling with charge compensated electron beam allows to provide the acceptable luminosity of the device in all anvestigated ion energy range for any superthin target.

4. THE OPERATION WITH SUPERTHIN TARGET.

Besides above mentioned problems the system of electron cooling must provide the suppression of multiple ion beam scattering on superthin target. It is know (12) that the condition of such suppression has a form

$$t_z < \frac{F_m \eta \upsilon_s^2 R_s}{mc^4 Z_A Z_i^2 r_e^2 \Lambda_t}$$

where F_m is maximun value of friction force, R_s is mean storage ring radius and Λ_t is Coulomb logarithm for fast particle scattering on the target. Using $F_m \cong e^2 n^{2/3} Z^2_i$ (11) we have $t_z < 10^{16}$ cm^{-2} for $Z_A \cong 90$ and Ei / Ai 200 MeV/u.This condition is rather soft, as for $t_z > 10^{14}$ cm^{-2} the beam lifetime on the target becomes smaller than a second. The processes of multiple scattering would result in the increasing of ion beam energy spread. This effect must be compensated by the action of electron cooling. Let's estimate the value of energetics pread. It is know that $\Delta E_i / E_i \cong 2\ \Delta P_i / P_i$ if the momentum spread is small. Since time-average process of multiple scattering has a diffuse character we must write

$$\Delta E_i / E_i \cong 2 \sqrt{<\Delta p^2>} / p_i$$

The average momentum spread by one order magnitude is $<\Delta p^2> \cong \tau_e\, d\Delta p^2 / dt$ In order to calculate the rate of increasing of root-mean-square deviation we use the well-know relation (3)

$$\frac{d\Delta p^2}{dt} \cong \frac{8\pi e^4 Z_A Z_i^2 \Lambda\, n_m}{\upsilon_s} \frac{l_t}{C}$$

In the region of low velocities $\Delta\upsilon_i < \Delta\upsilon_{e||}$ the friction force is $F \cong F_m \upsilon_p / \upsilon_F$ and

$$\lambda \cong \tau_e^{-1} \cong \frac{F_m}{\upsilon_F M_p A_i} \eta$$

where $\upsilon_F \cong 2{,}2 \sqrt{\dfrac{2e^2 n^{1/3}}{m}}$.

Then the steady state energetic spread is

$$\frac{\Delta E_i}{E_i} \cong \sqrt{\frac{16\pi t_z Z_A Z_i^2 e^4 \Lambda}{M_p A_i \upsilon_s^3}} \sqrt{\frac{\upsilon_F}{Fl_e}}$$

For Ne^{10+} we obtain $\Delta E_i / E_i \cong 8 \cdot 10^{-6}$ at $E_i / A_i \cong 200$MeV/u, $Z_A \cong 90$, $t_z \cong 10^{14}$ cm^{-2} and $j_e \cong 1A/cm^2$.We see from above formula, that energy spread is determined mainly by F_m / V_f ratio. In the experiment this ratio may be decreased and respectively the value $\Delta E_i / E_i$ increases, if we don't exclude the effects leading to the increasing of longitudinal electron temperature $T_{e\parallel}$, as $F_m \sim T^{-1}{}_{e\parallel}$ and $V \sim \sqrt{T_{e\parallel}}$. Transverse - longitudinal relaxation and inhomogenity of magnetic field in solenoid may show such effect. As it was mentioned above, the former effect will be suppressed at $H \cong 3$kOe for current densities presented in table 5. In order to eliminate the influence of magnetic field inhomogenity it is necessary to satisfy condition

$$\frac{\Delta H_\perp}{H} < \left(\frac{e^2 n_e^{1/3}}{E_e}\right)^{1/2}$$

where E_e is electron beam energy. For $Ei / Ai \cong 200$ MeV/u and $j_e \cong 1A/cm^2$ we have

$$\frac{\Delta H_\perp}{H} < 3,10^{-5}$$

Now we have discuss the conditions foe electron beam diameter. As it was pointed above it is planned to accumulate $\sim$ 10^9 particles in the ring. This value is 1-2 orders of magnitude less than the limit due to the space charge (4). Then ion beam would be compressed as a result of cooling. An acceleration leads to further decrease of beam diameter. So it would be reasonable to use thinner electron beam when the target is to operate. That has been foreseen in primary project (4). However in many cases the beam lifetime is determined by the processes of single scattering:

$$\tau_t \cong \frac{A_i \upsilon_s^2 Q_g}{4\pi r_p^2 Z_i^2 Z_A t_z f_s c^4}$$

where Q_g is extremely permissible angle of deviation of ion colliding with target atom. Obviously, this value should be no more than is ordered by ring acceptance. Additionally the scattering ion must stay at electron beam aperture when it pass through the cooling section. So the using of thicker electron beam leads leads to increasing of Q_g and τ_t. But in ourcase it is unlike to make the beam diameter more than 3 cm because of technical limitation. The magnetic field stability conditions are connected with a problem to introduce the ion beam into cooling section and to take it out. It is required $\tilde{H}/H < 10^{-4}$ for the monofrequent instability should be suppresed.

5. CONCLUSION.

Thus, the cooling system with a compensated by charge electron beam should be used to enlarge research possibilities of projecting storage ring.

The cooler shall have the following parameters:

$H \cong 3kOe$	$l_e = 3$ m	$d_e \cong 3cm$
$\Delta H_\perp / H \cong 3 \cdot 10^{-5}$	$E_e \cong 3\text{–}110$ keV	$\Delta V_{e\perp}/V \cong 10^{-3}$
$\tilde{H}/H \cong 10^{-4}$	$j_e \cong 0{,}2 - 1A/cm^2$	$\tilde{E}_e / E_e \cong 10^{-5}$

REFERENCES.

1. G.I. Budker, Atomonya Energya, 22, 346 (1967).
2. D. Möhl et al., Phys.Rep. 58, 75 (1980).
3. G.I. Budker, A.N. Skrinsky, Uspekhi Physicheski Nauk, 124, 561 (1978).
4. Ion Storage Rings with Electron Coolin. Collection of Proposal and Projects in the URSS. Moskow, 1990, IAE.
5. N.S. Dikansky et al. Preprint INP-56, Novosibirsk, 1979.
6. Y. S. Derbenev, A.N. Skrinsky, Plasma Physics (USSR), 4, N 3, 492 (1978).
7. N.S. Dikansky et al. Preprint INP-102, Novosibirsk,1987.
8. W. G. Grahm, J. Phys. Rev., A 30, 722 (1984).
9. A.A. Kolomensky, Physical bases of charged particle acceleration methods, Moskow, 1980, MGV.
10. A.V. Burov et al. Preprint INP-116, Novosibirsk, 1989.
11. N.S. Dikansky et al. XIII Intern. Conf. on High Energy Part. Accel.V.I., p.333, Novosibirsk, 1987.
12. G.I.Budker et al. Preprint INP-71, Novosibirsk, 1977.

BEAM COOLING FOR THE ADRIA PROJECT

Alessandro G. Ruggiero

Brookhaven National Laboratory, Upton (NY)

Introduction

Two techniques of beam cooling are available nowdays: electron and stochastic cooling. Both of them are used in low and medium energy storage rings for heavy ions, where exotic ion fragments are stored and continuosly cooled to increase the phase space density. Electron cooling does not depend on the beam intensity, is very effective at low energy and affects mostly the core of the beam. Stochastic cooling works at any energy; it depends on the number of particles in the beam and squeezes effectively the tails to the center of the distribution. electron cooling is limited by electron beam density and its temperature, whereas stochastic cooling is limited by the electronic bandwidth and power. Electron cooling operates on the beam momentum spread and betatron emittance at the same time while different systems of stochastic cooling are required for momentum and betatron cooling.

A project which requires electron cooling is ADRIA, a complex of accelerators and storage rings for the production of exotic elements, being proposed for construction at the Laboratori Nazionali di Legnaro. The cooling requirements for this project exceeds by far those of other similar projects; the ion specific energy is around few GeV/u and the required electron beam intensity about 10 A over a beam cross-section of one centimeter diameter. The power in the electron beam is remarkably high: 5 MWatt, which makes the energy recovery a problematic issue since a recovery efficiency of at least 99% is required.

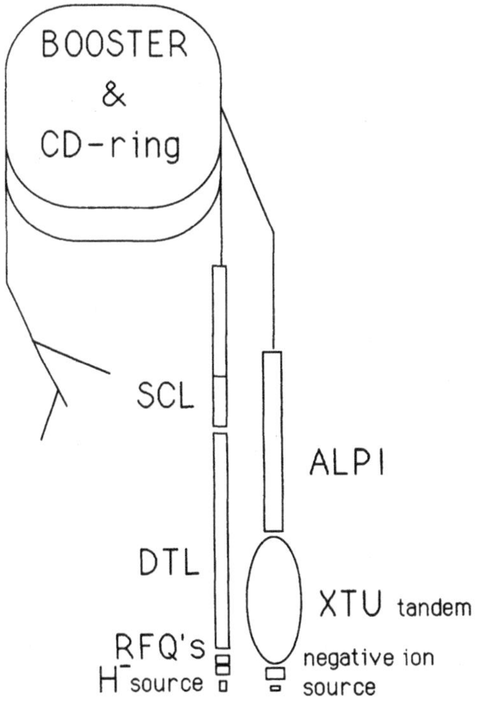

Fig.1 : Block diagram of the proposed ADRIA complex.

The ADRIA Project

The ADRIA project is in a proposal stage and is meant to be build at the Laboratori Nazionali di Legnaro as a future development for heavy ions. Using the Tandem-ALPI complex as the injector, beams of heavy-ions, partially stripped, are accelerated to kinetic energies in excess of 1 GeV/u in a Booster ring, which is a fast synchrotron operating at 10 pulses per second. Beam intensities in excess of 10^{11} ions per second can be accelerated.

Table 1: Beam parameters at Injection in the Booster ring.

		Deuterium	Carbon	Sulfur	Copper	Iodine	Gold	Uranium
Mass Number		2	12	32	63	127	197	238
Atomic number		1	6	16	29	53	79	92
Charge State		1	6	16	27	40	51	54
β		0.2616	0.2224	0.1853	0.1483	0.1140	0.0988	0.0908
Kinetic Energy	(MeV/u)	33.8	23.9	16.4	10.4	6.1	4.6	3.9
Revol. Period	(μs)	3.4	4.0	4.8	6.0	7.8	9.0	9.8
Harmonic number		17	20	24	30	39	45	49
Ions per bunch		$3.4 \cdot 10^9$	$2.5 \cdot 10^9$	$8.3 \cdot 10^8$	$6.1 \cdot 10^9$	$2.0 \cdot 10^8$	$1.5 \cdot 10^8$	$1.4 \cdot 10^8$
Transv. Emittance	(π-mm mrad)	32.5	27.4	22.6	18.0	13.8	11.9	10.9
Bucket Area	(meV-s/u)	73.0	19.1	8.6	5.5	3.5	2.6	2.4

Table 2: Beam parameters at the Extraction from the Booster ring.

		Deuterium	Carbon	Sulfur	Copper	Iodine	Gold	Uranium
β		0.9627	0.9632	0.9632	0.9509	0.9143	0.8802	0.8516
Kinetic Energy	(GeV/u)	2.53	2.53	2.53	2.08	1.37	1.03	0.85
Revol. Period	(μs)	0.924	0.923	0.923	0.935	0.973	1.011	1.04
Bunch Length (rms)	(ns)	1	1	1	1	1	1	1
Transv. Emittance	(π-mm mrad)	9.1	7.6	6.3	5.8	6.1	6.4	6.7

Table 3: Accelerator Parameters.

Max. Magnetic Rigidity	22.25	Tm
Circumference	266.6667	m
Bending Radius	17.599	m
Max. Dipole Field	12.7	kG
Dipole vertical gap	10	cm
Max. Quad gradient	0.9	kG/cm
Quad bore radius	7	cm
Periodicity (with mirror symm.)	4	
Transition Energy (γ_t)	4.59	
Betatron tunes:		
Horizontal plane ν_h	5.8	
Vertical plane ν_v	3.8	
Natural Cromaticity:		
Horizontal plane ξ_h	-5.5	
Vertical plane ξ_v	-4.9	
Repetition Rate:		
Heavy Ions	10	Hz
Protons	50	Hz

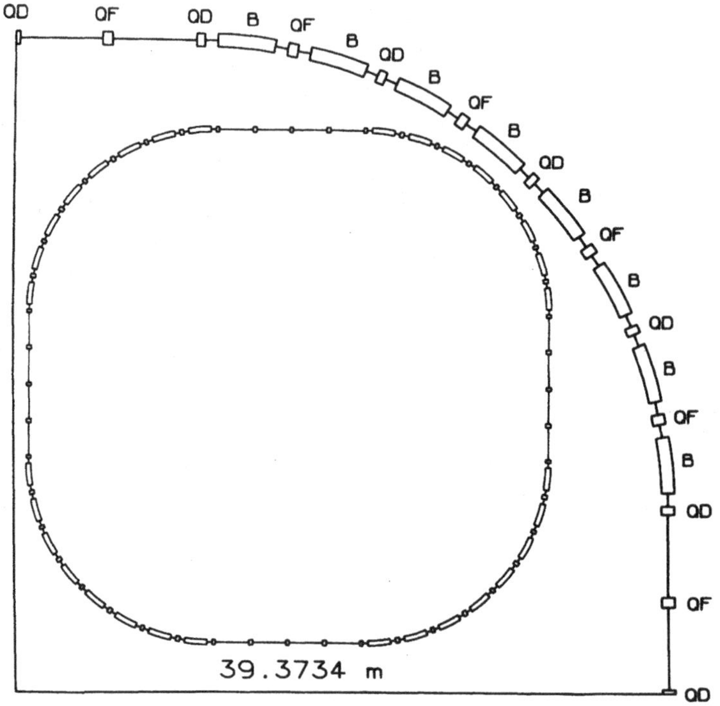

Fig.2 : Lattice of the Booster ring.

Fig.1 gives an overall view of the facility, which includes the Tandem-ALPI injector, the Booster ring, a Transfer Line, where beams of exotic fragments are produced, and the Decelerator/Cooler ring. The two rings have the same size and shape. More details of the Booster ring are shown in Fig. 2 and machine parameters are summarized in Table 3. The CD-ring has a lattice very similar to the Booster ring with the exception of the long straight-sections, shown in Fig.3, where quadrupoles are arranged so as to leave a long drift (10 m)

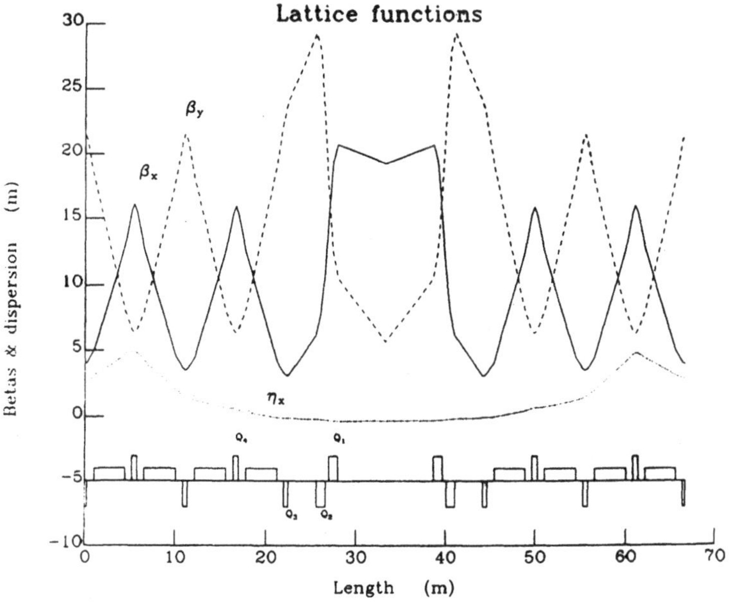

Fig.3 : Lattice functions of the CD-ring (one period).

for the installation of the electron cooling device. The two rings are joined by the Transfer Line which has the length of about half of the Booster ring. Beam parameters at injection are described in Table 1, taking few standard elements as example. Beam parameters at the end of acceleration in the Booster are given in Table 2. At the end of the acceleration in the Booster, the beam is extracted and sent on a target for the production of exotic

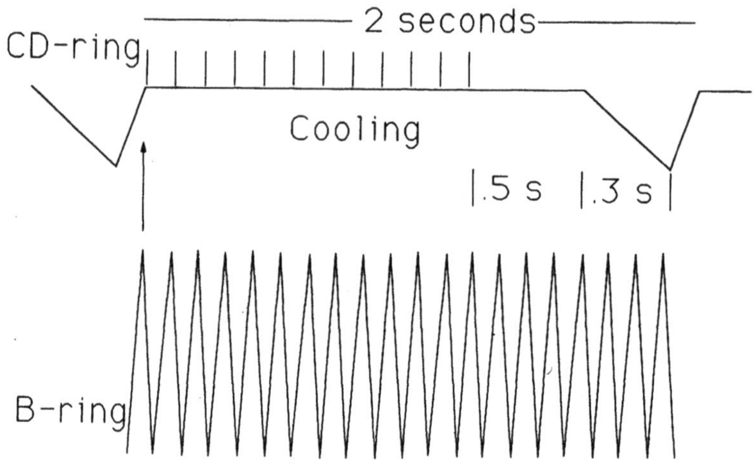

Fig.4 : Heavy Ion cycle in the proposed ADRIA complex.

elements; the large energy of the primary beam is necessary to improve the probability of production of the required fragments. Particles of the wanted species are selected with a complex transport channel including degrader targets and collimators; typically, production rates of 10^{-5} to 10^{-4} can be expected. At production, a full momentum spread of 0.7 % is captured; the production angle is taken to be 7.5 mrad corresponding to a transverse momentum spread which matches the momentum width. The production target is placed in a location

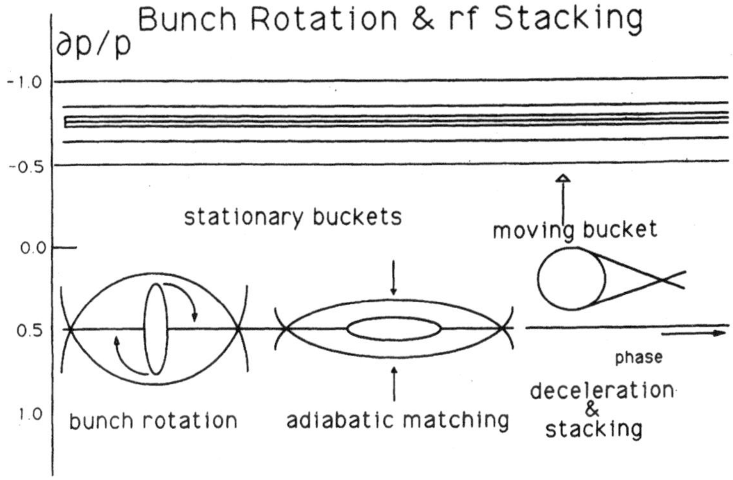

Fig.5 : Bunch rotation and rf stacking scheme.

of the Transfer Line where a waist with $\beta \sim 1$ m is produced ; this corresponds to a betatron emittance of 17 π mm · mrad for the secondary beam.

After production, the beam of fragments are immediately injected in the Cooler/Decelerator ring. A typical mode of operation is shown in Fig. 4 which shows a cycle 2 seconds long. Twelve pulses of ion fragments are transferred to the CD-ring, 100 ms apart from each other. The bunches of each pulse are captured with standing-by stationary

buckets and rotated. The momentum spread of the beam will change therefore from 0.7 %, at the moment of capture, to 0.1 % after the rotation. The beam bunches are then displaced by a second system of rf cavities to a new momentum trajectory where they are stored. On the stacking orbit, electron and stochastic cooling operate on the beam, and this operation continues for another 0.5 s after all 12 pulses have been accumulated. Fig. 5 illustrates the use of the momentum aperture and the location of the beam during the cycle with the relative beam dimensions. At the end of cooling, the beam momentum spread and the betatron emittance are reduced by an order of magnitude. Finally, the beam stack is captured by one more system of rf cavities which decelerates it down to energies of about 10 MeV/u. The deceleration cycle is linear and lasts 0.3 s to which another 0.1 s is added to reset the guiding field to its initial value. At this point the cycle repeats.

Deceleration of the beam to very low energies, corresponding to the Coulomb barrier of the ions, is required for the study of fusion processes which are an important part of the experimental program that can be accomplished with this facility.

Cooling Requirements

The requirements for beam cooling in the CD-ring are as follows: the beam momentum spread has to be continuously reduced to make space for twelve pulses of fragments during the accumulation process. The beam momentum spread must be sufficiently small at the end of the accumulation so that it can be captured and decelerated by a reasonable amount of rf voltage. During deceleration the beam momentum changes by a factor of ten, so that momentum spread and betatron emittance will increase by the same factor; as a consequence the beam dimensions are to be sufficiently small at the end of cooling to avoid an excessive

Table 4: Electron Cooling Parameters.

Specific Kinetic Energy	1.0	GeV/u
β	0.80	
Mass number	200	
Betatron emittance	19.0	π mm mrad
Momentum spread	0.28%	
β^* (h,v)	20, 6	m
Length of e-beam	8	m (3%)
Coulomb log	15	
Beam semi-axis	10	m
e-beam current	12.5	A
e-beam current density	7	A/cm^2
θ, total	1.3	mrad
Cooling time	0.3	s
E-beam energy	0.6	MeV
e-beam power	7.1	MW
Final momentum spread	$\pm$0.01%	

Table 5: Stochastic Cooling Parameters.

Number of Ions	$1{\cdot}10^7$	
Mass number	200	
Charge state	80	
Specific Kinetic Energy	1.0	GeV/u
β	0.80	
Bandwith	1-2	GHz
Method	Notch Filter	
Nr of Pickups	16	
Nr of kickers	32	
Schottky Power (init.)	1.0	kW
Thermal Power	negligible	
Amplifier Gain	160	dB

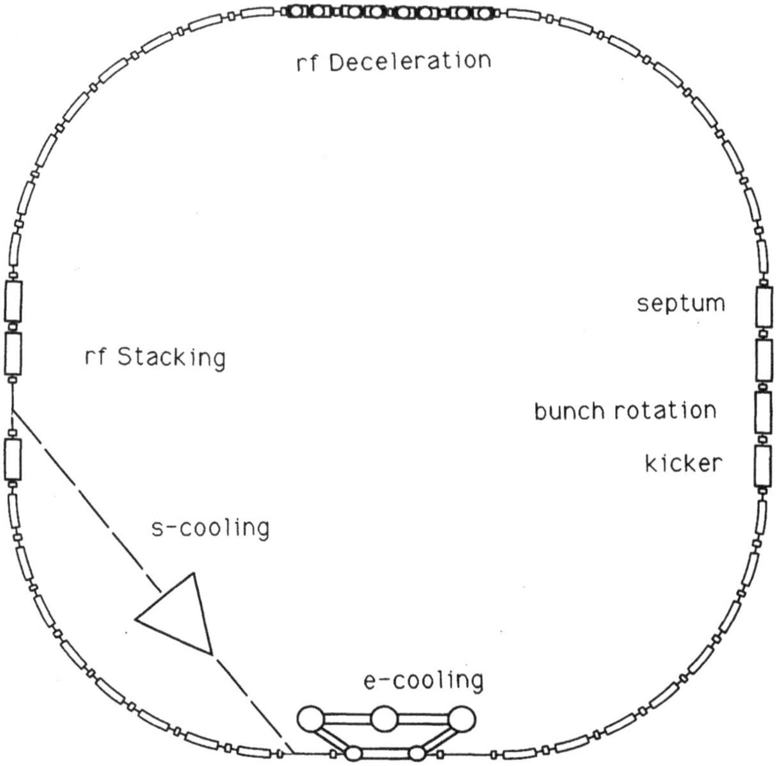

Fig.6 : Arrangement of components in the CD-ring.

increase of the beam dimensions at lower energies. Cooling may also be useful at the end of the cycle, to obtain a smaller momentum spread as it may be required by users.

The most demanding requirements for cooling are imposed by the accumulation process. The momentum spread δ of the stack varies in average according to the following equation

$$\frac{d\delta}{dt} = -\lambda\delta + D \tag{1}$$

where λ is the cooling rate and D is the increase of 0.2 % in momentum spread due to the addition of a new pulse to the stack every 100 ms. By setting the overall cooling time to 150 ms, the average momentum spread in the stack, during the accumulation procedure, is maintained to 0.3 % . To achieve the required cooling rate, both electron cooling and momentum stochastic cooling are required, each providing a cooling time of 0.3 s. With a second cooling period of 0.5 s, following the stacking cycle, at the same rate, the total beam momentum spread is finally reduced to 1×10^{-4}.

Electron cooling alone is adequate for the reduction of the beam emittance. Betatron cooling proceeds at twice the rate of momentum cooling; thus, over a period of 0.5 s, the betatron emittance can be reduced by at least one order of magnitude. Stochastic cooling for the betatron motion is not required; it is in any case more difficult to achieve because of the poor mixing situation due to the relatively small momentum spread of the beam.

Electron Cooling

The design of the electron cooling apparatus is based on the following formula for the cooling rate

$$\lambda = \frac{4\pi Q^2 r_e r_p L\eta J/e}{A\beta^4\gamma^5\theta^3} \tag{2}$$

where r_e and r_p are respectively the electron and proton classical radius; Q is the charge state and A the mass number of the particles being cooled; η is the ratio of the electron beam length to the ring circumference; L is the Coulomb logarithm; β and γ are respectively the velocity and the energy relativistic factors of the beam; J is the electron beam density

$$J = \frac{I}{\pi a^2} \tag{3}$$

with I the electron beam current and πa^2 the cross-section. Finally,

$$\theta^2 = \theta^2_{ion} + \theta^2_{electron} \tag{4}$$

where θ_{ion} and $\theta_{electron}$ are relative velocity spreads respectively in the ion and electron beam; for each of them there is a longitudinal contribution ($\theta_{\parallel}$) and a transverse one ($\theta_{\perp}$), so that

$$\theta^2 = \theta_{\parallel}^2 + \theta_{\perp}^2 \tag{5}$$

Because of the large intensity of the electron beam, the beam is confined to the required dimensions with the use of a solenoidal field along the entire interaction length. If the field is strong enough the trajectories of the electrons are essentially parallel lines directed along the main beam axis and the transverse temperature is effectively reduced, that is $\theta_{\perp} \sim 0$ for the electron beam. With a resonable cathode design we can expect on the other side $\theta_{\parallel} = 0.001$. The longitudinal temperature of the ion beam is given by its momentum spread

$$\theta_{\parallel} = \frac{1}{\gamma}\frac{\Delta p}{p} \tag{6}$$

and the transverse temperature by the beam emittance

$$\theta_{\perp}^2 = \epsilon/\beta^* \tag{7}$$

where β^* is the value of the lattice function at the center of the interaction region.

The parameters which correspond to a cooling time of 0.3 s are given in Table 4. The most important condition to be satisfied is that the electron beam velocity equals in magnitude and direction the velocity of the ion beam. For instance, for cooling of 1 GeV/u ions, the electron energy is about 500 keV which is the largest so far conceived for any application. Moreover, if one considers that the beam current is 10 A, it is seen also that

the beam power is 5 MW continuous; this is a very large value that poses severe constraints on the energy recovery system which requires an efficiency of better than 99 % .

Stochastic Cooling

Momentum stochastic cooling is required to operate in parallel to electron cooling. To release constraints on the two cooling systems, both of them are designed for a cooling time of 0.3 s . The components for stochastic cooling are shown schematically in Fig. 6. The Schottky current signal is detected by pickups made of pair of striplines located in location at zero dispersion; the signal is amplified and filtered through a device called the "notch filter": a cable with the length of half the ring circumference, shorted at the far end. The signal then travels to the next location, half of the ring circumference away, where it is divided evenly to a system of kickers which are also made of pairs of striplines identical to the pickups and are located in zero-dispersion places.

An important parameter which describes the performance of stochastic cooling is the frequency bandwidth of the electronic system which is chosen to optimize "mixing" between pickups and kickers; The mixing between different samples depends on the ring transition energy γ_t and on the beam momentum spread. For the ADRIA CD-ring the optimum bandwidth is 1-2 GHz, since the width of the Schottky band in the middle of this range is about half of the revolution frequency, and there is overlapping between neighboring bands at the upper end.

The main parameters of the stochastic cooling are given in Table 5. The distance between pickups and kickers allows for the adjustment of the electronic delay up to 150 nsec which accommodates any resonable delay of the amplifier and the fine adjustment to synchronize the electronic signal with the beam motion. The total power required of 1 kW is due to the Schottky signal after amplification; thermal power due to thermal noise at the exit of the pickups is low and of no consequences to the cooling performance.

EFFICIENT RECUPERATOR OF INTENSE ELECTRON BEAM ENERGY

(Presented by N. S. Dikansky)

A.N. Sharapa, A.V. Shemyakin
Institute of Nuclear Physics 630090, Novosibirsk URSS

ABSTRACT.

An intense electron beam energy recuperator of high effeciency is proposed. Transformation of a cylindrical beam into a disk-shaped one is shown to provide an opportunity of a cardinal increase in the collector perveance. The use of an intermediate collector reduces the losses caused by paraxial electrones down to $3 \cdot 10^{-4}$ of the total current. Also presented in the paper are main experimental characteristics of a recuperator with a perveance of $400 \mu A / V^{3/2}$, designed for the electron cooling device.

Energy recuperation of an intense electron beam makes it possible to reduce the power on the electron collector and to increase the efficiency of the electron beam device. Energy recuperation is especially efficient in electron cooling devices /1 - 3/.

The feature of these devices consists in that a low transverse velocity cylindrical beam is formed and transformed with the help of a longitudinal magnetic field. As is shown in Ref. /4/, the transformation of a cylindrical beam into a tube-shaped one permits one to drastically increase the efficiency of energy recuperation of an electron beam of high intensity. But sometime the transformation of a cylindrical beam into a disk-shaped one turns out more convenient. This transformation can be performed with a magnetic field of two solenoids with oppositely directed fields (Fig. 1).

One of the main problems arising at a beam transformation in a recuperator, consists in reducing the current losses caused by paraxial electrons. These losses are resulted of violation of the adiabaticity condition near a zero point of the magnetic field /4/.

If one approximates the magnetic field in this region by the first terms of a paraxial expansion, then for the case of equipotential space, one can show that the motion equations become dimensionless and the geometry of trajectories depends on a single parameter:

$$\alpha = \frac{o}{\rho_o^{3/4} L^{1/4}} \qquad (1)$$

where

$$\rho_o = \frac{mc\,V_o}{e\,H_o}\ ,\quad L = \sqrt{R_s^2 + a^2}\ ,$$

Υ_o and H_o are the particle radius and the strenght of the magnetic field in a homogeneous field region, respectively, V_o is the total velocity, e , m are the charge and mass of an electron respectively, c is the velocity of light, R_s is the solenoid radius, a is half the gap between the solenoids.

An important characteristic for recuperation problems is the energy share of transverse motion occuring after transformation $W_\perp / W_o$. This value reaches its maximun along the trajectory at the region of the field maximun in the gap between the solenoids (for $\Upsilon \approx R_S$). According to numerical calculations (see Fig. 2) it is the parameter α which determines $W_\perp / W_D$ with a good accuracy: for a given magnetic field geometry the curves, corresponding to different Larmour radii, practically coincide. It is shown, that the estimate the minimun initial radius Υ_m, which corresponds to a sufficiently weak motion disturbance ($W_\perp / W_o \ll 1$) one can specify:

$$\Upsilon_m \approx 0.6 \cdot \rho_o \cdot (L/\rho_o)^{1/4}$$

An efficient means of reducing Υ_m is the potential decrease in the transformation region. It does not mean that the potential decrease is obligatory over the whole region, it is enough to perform this only in the vicinity of point A (Fig. 1). In this connection, it seems useful to consider such system, in which a beam is transformed into a disk-shaped one by electric field.

Similarly to the solution given in Ref. /5/ for a flat beam, one can obtain an analytical solution for electrostatic transformation of an axisymmetric beam. It has a constant charge density throughout the whole space, its equipotentials are ellipsoid-shaped (Fig.3):

$$U = \beta \cdot (\Upsilon^2 + 4Z^2)$$

where β is a constant, and trajectories $\Upsilon \cdot \sqrt{|Z|} = \text{const.}$

To solve the external problem, it is convenient to use dimensionless values:

$$\Upsilon_1 = \Upsilon / a_o,\ Z_1 = Z / a_o\ ,\ U_1 = U / U_o\ ,\ U_o = 0.24\,\frac{mc^2}{e} \cdot I / I_o$$

$$I_o = mc^3 / e\ .$$

where I is the beam current, a_o is the value with the lenght dimension, determining the boundary trajectory

$$\Upsilon = a_0^{3/2} / Z^{1/2}$$

In the left half-space potentials are calculed analytically:

$$U_1 = \Upsilon_1^2 - 2Z_1^2$$

Due to the complicated geometry of the beam boundary, restoration of the equipotentials in the right-hand half-space is bound up with considerable difficulties. Thus, Harker's method does not allow to find a unique conform transformation for theregion, located at rather a larger distance from the beam boundary. Neverthless, this difficulty is not very important for determining the form of electrodes, since the equipotentials near the beam are determined rather accurately.

Strictly speaking, this solution is incompatible with the beam deceleration, as it implies a radial growth of potential. Neverthless, if the system of electrodes, responsible for the beam transformation is supplemented by a ring collector with a lower potential U_C, thus forming an efficient recuperator (Fig. 3).

An important peculiarity of the solution illustrated in Fig. 3 consists in the fact, that it allows, in principle, the mentioned above transformation without beam losses, resulting from the reflection of paraxial electrons. This peculiarity of the solution can also be used for transformation in a magnetic field. For this purpose one can combine the electrostatic transformation with the magnetic field one. This combination appears to be consistent since one can show that the electron trajectories in the electrostatic transformation system coincide with the magnetic field lines near the axis (Fig.1). But one should keep in mind that an ideal transformation is possible only for strictly specified electrode potentials and a given current. Besides, one has to produce electrodes, exatly imitating the form of equipotentials. In reality, every recuperator, due to the deviation from the exact solution, has some noticeable electron flux which does not reach the ring collector.

In a recuperator (Fig.3) the flux which does not reach the ring collector is determined by particles of 2 types: i) the particles reflected from a code electrode (6) at zero potential and ii) the particled reflected from a magnetic mirror in a radially increasing magnetic field. As it was proved experimentally, the share of these particles can be reduced down to 10^{-3}.

However, electrons reaching large radii, can be captured, if electrode 1 is placed in the beam. For an efficient capture of electrons it is necessary that the width Δ of that part of the beam, which is incident on electrode 1, would be:

$$\Delta \gg \rho$$

where ρ is the Larmour radius, calculed from the electrode 1 potential and the magnetic field near it.The main disadvantage of this recuperator consists in that the electrons reflected from the cone electrode cannot be captured in it. More efficient for the capture of particles, which have not into the ring collector, turned out the recuperator given Fig. 4. This recuperator has no cone electrode, and besides, the form of its electrodes differs from that corresponding to the ideal solution. In such a recuperator the share of electrons which have not reached the ring collector I_{int} is higher than that in the one presented in Fig.3 However, this fact is of no great importance, because all the electrons which have not arrived at the ring collector are captured in an intermediate collector, and secondary electrons (like in a conventional collector, such as "Faraday cup" /6/), are captured by a rather deep potential minimun.

The power W on the intermediate collector, the latter being at potential U_{int}, is equal to:

$$W = U_{int} \cdot I_{int}$$

To decrease this power, one has to maximally reduce U_{int} limited by the space charge. In experiments U_{int} was reduced down to the values, at which perveance $P = I / U^{3/2}_{int}$ reached $20\mu A / V^{3/2}$, I.E. was closed to the limiting perveance of the Faraday cup.

In the recuperator with an intermediate collector the losses ΔI on the anode were considerably smaller and collector perveance was noticeably higher.

MAIN EXPERIMENTAL RESULTS AND BEAM PARAMETERS.

$I = 3.2$ A	$U_{int} = 3$kV
$U_c = 400$ V	$P_{int} = 20\ \mu A / V^{3/2}$
$P_c = {}^{I}/_{U_c}{}^{3/2} = 400\mu A / V^{3/2}$	${}^{I_{int}}/_{I} = 4\%$
$\Delta I/I = 3 \cdot 10^{-4}$	$H_o = 0.09$ T

The perveance P_c is approximately 20 times higher than that for a conventional " Faraday cup" recuperator /5/. Neverthless, these results should be considered preliminary, since the present paper does not provide for the limiting capabilities for recuperators of this type. It is accounted for by the fact, that the main aim of the work was to reduce the losses caused by paracial particles. To obtain the maximun value of one has to bring the dimension of the ring collector (Δ R, L in Fig. 4) in correspondence with the thickness of transformed beam h . The maximun value is attained under the condition /7/:

$$h \sim L \sim \Delta R .$$

In experiments a 30 mm in diameter emitter was used. For this emitter the thickness of the transformed beam h is approximately 10 times smaller than the characteristic dimension of the collector ($h \approx 5$ mm, $\Delta R \approx 50$ mm, $L \approx 50$mm). As a result, the experimental value of perveance is considerably smaller that the maximun one for a ring collector with given dimensions.

MAIN RESULTS:

a) The recuperator with a cylindrical beam transformation into a disk-shaped one ermits one to cardinally increase the collector perveance, as compared to the Faraday cup.

b) With the introduction of the intermediate collector practically all the electrons, that escaped the main collector, can be captured and, thus the anode current losses are decreased down to the values of $3 \cdot 10^{-4}$.

The authors acknowledge the fruitful discussion with I.N. Meshkov and the assistance in calculations of T.N. Andreeva.

REFERENCES.

1. Kudelainen V.I., Meshkov I.N., Salimov R.A. INP Novosibirsk preprint 72 - 70 (1970); Preprint CERN 77 - 08 (1977).
2. T. Ellison et al., IEEE Trans. Nucl. Sci. NS - 30 (1983) , 2636 - 2638.
3. M. Bell et al. Nucl. Instr. Meth., 1981, v. 190, p. 237 - 255.
4. V.I. Kokoulin, I.N. Meshkov, A.N. Sharapa, Sov. J. Tech. Phys., 1980, v. 50, N7, p.1475 - 1479.
5. Kirstein P.T., Kino G.S. , Waters W. Eà. Formation of electron beams, Moskow, Mir, 1970, p. 102 - 104.
6. A.I. Arenshtam, I.N. Meshkov, V.G. Ponomarenko, R.A. Samilov, A.N. Skrinsky, B.M. Smirnov, V.G. Fainshtein, Sov. J. Tech. Phys., 1971, 41, N 2, p. 336 - 338.
7. A.N. Sharapa. Candidate thesis, Novosibirsk, 1982, p. 124.

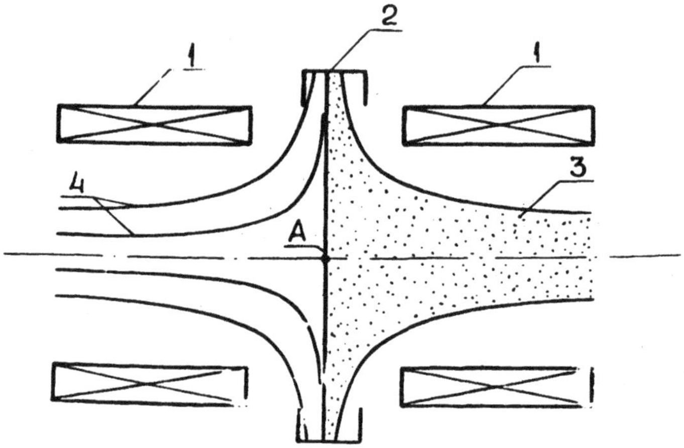

Fig.1 - Schematic transformation of a cylindrical beam into a disk-shaped one. 1 - solenoid, 2 - ring collector, 3 - beam, 4 - magnetic field lines.

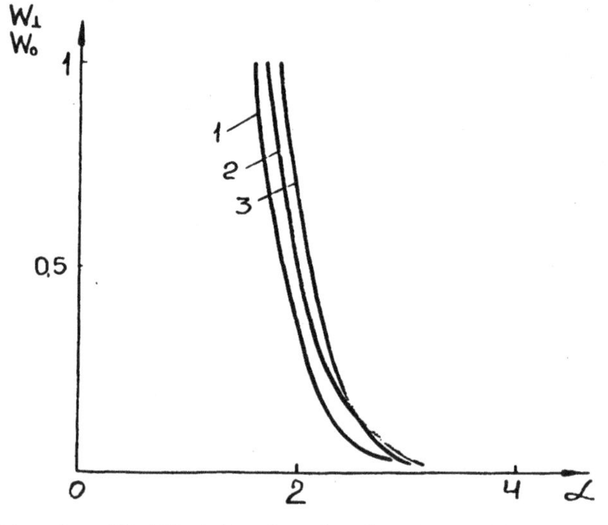

Fig.2 - Dependence $W_\perp / W_0\ (\alpha)$ at the point of maximun value of magnetic field ($r \approx R_c$), R_c = 140 mm, α = 50mm, 1 - ρ_0 = 2.1 mm, 2 - ρ_0 = 1.1 mm, 3 - ρ_0 = 0.7 mm.

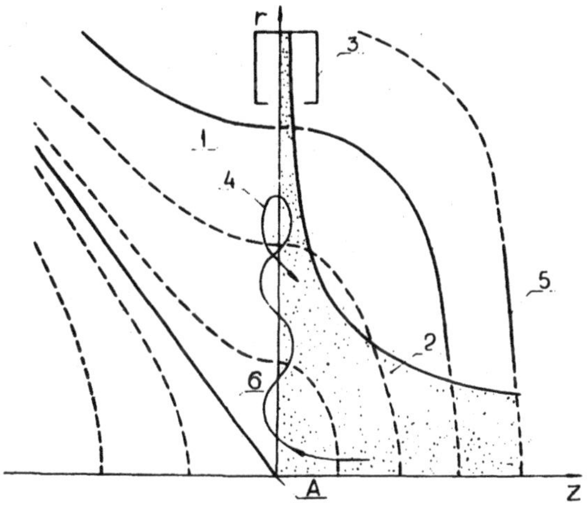

Fig.3 - Recuperator with electrostatic transformation of a cylindrical beam into a disk-shaped one. 1 - electrode, 2 - beam, 3 - ring collector, 4 - trajectory of the electron reflected from a magnetic mirror, 5 - anode, 6 - cone electrode.

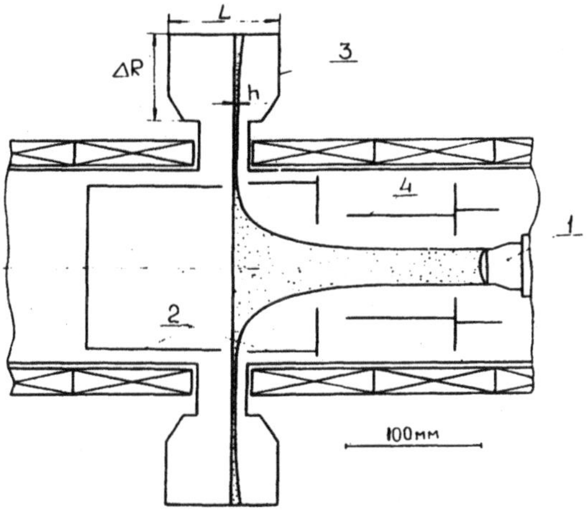

Fig.4 - Recuperator with an intermediate collector. 1 - electron gun, 2 - intermediate collector, 3 - ring collector, 4 - anode.

RELATIVISTIC LASER COOLING

D.Habs, V.Balykin[1], M.Grieser, R. Grimm, E.Jaeschke,
M.Music, W.Petrich, D.Schwalm, A.Wolf
Physikalisches Institut der Universität Heidelberg and
Max-Planck-Institut für Kernphysik,
D-6900 Heidelberg, Federal Republic of Germany

G.Huber
Institut für Physik der Universität Mainz,
D-6500 Mainz, Federal Republic of Germany

R.Neumann
Gesellschaft für Schwerionenforschung,
D-6100 Darmstadt, Federal Republic of Germany

[1] on leave from Academy of Sciences, Troitzk, Moskow Region, USSR

Laser cooling is studied in the regime of large relativistic γ factors, which will become accessible at the high energy accelerators SPS, RHIC and LHC. For partially stripped ions x-ray transitions in the atomic shell are studied. For naked ions the much slower nuclear transitions are discussed. Reasonable cooling times are obtained in both cases.

Introduction

Recently Be^+ and metastable Li^+ ion beams at about 5% velocity of light have successfully been laser cooled at the Heidelberg Test Storage Ring (TSR) to longitudinal temperatures in the rest frame of below 6 K [1]. Very low temperatures in the μK region are expected in the so-called Doppler limit [2]. An acceleration of 10^5 times the gravitational acceleration g is typical for the light interaction in the ions' rest frame.

Presently in heavy-ion storage rings ion velocities $\beta = \frac{v}{c} < 0.1$ and relativistic factors $\gamma \simeq 1$ are explored and the same rather small number of ion species for laser cooling

Ion	Li^+	Be^+	Be^{2+}	Mg^+	Sr^+	Hg^+
Energy [eV]	2.26	3.96	3.33	4.42	2.94	6.39

Table 1: Energies of laser cooling transitions for slow ions

are available, as have been used in experiments at $\beta \simeq 0$ in ion traps [3] . These ions and their transition energies are listed in table 1.
In this paper we address the question, how the region of laser cooling can be extended to more species at velocities very close to the velocity of light ($\beta \simeq 1$ and $\gamma \gg 1$). Due to the Lorentz transformations low-energy laser photons from the laboratory system are converted to much larger photon energies in the ions' rest frame. Thus highly charged ions become accessible to laser cooling. Then perturbing reactions like stripping in the rest gas or in colliding ion beams are considered. The short lifetime of partially-stripped ions allows for a principal investigation of laser cooling at $\gamma \gg 1$ and laser diagnostics, but rules out a useful application for collider operation. Therefore the less efficient laser cooling scheme using a low-energetic nuclear transition is considered, thus avoiding the stripping losses by working with naked ions.

The Lorentz Transformation

The starting point of all considerations is the relativistic transformation [4] of the laser photon energy $\hbar\omega_-$ in the laboratory system to the inner rest frame $\hbar\omega_0$ of ions counterpropagating the laser beam with a velocity $\beta = \frac{v}{c}$ and a relativistic factor $\gamma = \frac{1}{\sqrt{(1-(\frac{v}{c})^2)}}$:

$$\hbar\omega_0 = \hbar\omega_-(1+\beta)\gamma$$

For the conditions $\beta \simeq 1$ and $\gamma \gg 1$, considered in this paper, the transformation formula can be simplified to:

$$\hbar\omega_0 = 2\gamma\hbar\omega_- \qquad (\gamma \gg 1)$$

For large γ factors large transition energies become accessible.

[1]for laser cooling during 10% of circumference
[2]for ions with charge $Z_i = 80$

	TSR	RHIC	SPS	LHC
circumference	0.055 km	3.8 km	6.9 km	26.7 km
revolution time	$\sim 4\mu sec$	$\sim 13\mu sec$	$\sim 23\mu sec$	$89\mu sec$
energy	0.001 GeV/N	100 GeV/N	220 GeV/N	3152 GeV/N
γ	1.001	108.	237.	3388.
number of stored particles	10^9	$5*10^{10}$	$8*10^{11}$	10^{14}
ion beam density	$5*10^5/cm^3$	$10^{11}/cm^3$	$10^{11}/cm^3$	$10^{12}/cm^3$
$\hbar\omega_-$	4eV	4eV	4 eV	4 eV
$\hbar\omega_0$	4eV	0.86keV	1.9 keV	27.1 keV
$\hbar\omega_+$	4eV	0.19 MeV	0.9 MeV	184. MeV
lifetime of transition(lab)	10000 psec	25 psec	11 psec	0.7 psec
saturation intensity (lab)	0.01 W/cm^2	320 kW/cm^2	3.4 MW/cm^2	12.2 GW/cm^2
cooling force(lab)	0.7 eV/m	3 MeV/m	37 MeV/m	110 GeV/m
number of cooling cycles	$5*10^5$	$5*10^4$	$2*10^4$	$2*10^3$
cooling time[1]	100 msec	25 μsec	4 μsec	30 nsec
minimum energy spread(lab)	$\sim 10^{-8}$	$\sim 10^{-6}$	$\sim 10^{-6}$	$\sim 10^{-4}$
synchrotron energy loss[2]		0.11 eV/turn	2.5 eV/turn	136 keV/turn

Table 2: Storage rings for relativistic ions and achievable photon properties in comparison to TSR

The reemitted photon $\hbar\omega_+$ emerges at a small angle θ_{lab} relative to the ion beam and its energy is given to a good approximation by:

$$\hbar\omega_+ = \frac{4\gamma^2\hbar\omega_-}{1+\theta_{lab}^2\gamma^2}$$

The symmetric emission in the rest frame is compressed into a narrow region about $\theta_{lab} = 0^0$ and is given by the Lorentz transformation [4]:

$$tan\theta_{lab} = \frac{sin\theta_0}{\gamma(cos\theta_0 - \beta)}$$

Therefore an ion looses in the laboratory approximately the energy $\hbar\omega_+$ in one absorption emission cycle, which approximately is given by:

$$\hbar\omega_+ = 4\gamma^2\hbar\omega_- \qquad (\gamma \gg 1; \theta_{lab} \simeq 0)$$

Only photons being emitted in the rest frame in the forward hemisphere are relevant for the cooling process. For slow ion beams the symmetric spontaneous emission leads to a spatial diffusive heating of the ions like in a random walk [2]. For relativistic beams the relevant emission in forward direction strongly reduces this diffusive heating.

We apply these formulae for the storage rings of relativistic heavy ions, which are compiled in table 2. The data of the low energy Heidelberg Test Storage Ring TSR are given for comparison. Presently the super proton sychrotron SPS at CERN is available for the acceleration of light heavy ions, like sulfur, to 220 GeV/N. Its injector system will be improved to accelerate ions as heavy as lead [5]. The construction of the heavy ion collider RHIC [6] at Brookhaven has been started and design studies are performed for the large hardron collider LHC [7] at CERN. In the first section of table 2 typical properties of these storage rings are given.

Present cw-laser systems supply maximum photon energies $\hbar\omega_-$ of about 4 eV with sufficient intensities. The calculated rest-frame energies $\hbar\omega_0$ are then in the keV region. For ions with partly filled shells up to quantum number n, a simple hydrogenic scaling of the E1-transitions gives:

$$\hbar\omega_0 = 13.6eV * Z^2\left(\frac{1}{n^2} - \frac{1}{(1+n)^2}\right)$$

Therefore highly charged ions become accessible to laser cooling at sufficiently large ion velocities.

The reemitted photons reach very high energies $\hbar\omega_=$ in the MeV region, which can be easily detected. This opens up new possibilities for beam diagnostics or for high-energy light sources.[8,9]

In the next chapter the quantities of table 2 which are relevant for laser cooling will be discussed.

Laser Cooling

All basic relations for laser cooling of slow beams [2] can be taken over after transforming the components of the specific tensor quantities from the laboratory system to the rest frame system of the ions. In the rest frame the transitions are in the keV-region. For these large energies many other related quantities, like the linewidth or the saturation power, change in magnitude. Since in laser physics all these quantities are well known for low-energy optical transitions, we perform for convenience and understanding of the scaling laws a second transformation in the rest frame of the ions, where we express all quantities corresponding to keV photon energies, which

quantity	rest frame(0)	lab. frame(-)	lab. frame(+)	opt. system(opt)
frequency	ω_0	$2\gamma * \omega_-$	$\frac{1}{2\gamma} * \omega_+$	$2\gamma * \omega_{opt}$
wave vector	k_0	$2\gamma * k_-$	$\frac{1}{2\gamma} * k_+$	$2\gamma * k_{opt}$
absorption cross sec.	σ_0	$\frac{1}{4\gamma^2} * \sigma_-$	$4\gamma^2 * \sigma_+$	$\frac{1}{4\gamma^2} * \sigma_{opt}$
cooling section length	l_0	$\frac{1}{\gamma} * l_-$	$\frac{1}{\gamma} * l_+$	l_{opt}
dipole moment	d_0			$\sqrt{(2\gamma)} * d_{opt}$
width	Γ_0	$\gamma * \Gamma_-$	$\gamma * \Gamma_+$	$(2\gamma)^2 * \Gamma_{opt}$
lifetime	τ_0	$\frac{1}{\gamma} * \tau_-$	$\frac{1}{\gamma} * \tau_+$	$\frac{1}{(2\gamma)^2} * \tau_{opt}$
electric field	E_0	$2\gamma * E_-$	$\frac{1}{2\gamma} * E_+$	$\sqrt{(2\gamma)} * E_{opt}$
light energy flow	I_0	$4\gamma^2 * I_-$	$\frac{1}{4\gamma^2} * I_+$	$2\gamma * I_{opt}$
saturation intensity	$I_{sat,0}$	$4\gamma^2 * I_{sat,-}$	$\frac{1}{4\gamma^2} * I_{sat,+}$	$(2\gamma)^5 * I_{sat,opt}$
saturation parameter	G_0	G_-	G_+	$(\frac{1}{32\gamma^4}) * G_{opt}$
maximum cool. force	$F_{max,0}$	$2\gamma^2 * F_{max,-}$	$2 * F_{max,+}$	$8\gamma^3 * F_{max,opt}$
photon absorp. rate	$\frac{dN_0}{dt_0}$	$\gamma * \frac{dN_-}{dt_-}$	$\gamma * \frac{dN_+}{dt_+}$	$\frac{1}{(2\gamma)^2} * \frac{dN_{opt}}{dt_{opt}}$

Table 3: Relations between quantities in the rest system (0), laboratory systems (-,+) and the optical system. All quantities of one line are equal.

have an index '0' ,by the equivalent quantities for optical transitions, which have an index 'opt'. We select the optical transition energy $\hbar\omega_{opt}$ identical to the laser energy in the laboratory system $\hbar\omega_-$ and thus have a well defined transformation in the rest frame. In order to clearly separate between the Lorentz transformations and this internal transformation, we will use ω_{opt} and ω_- independently though we required the identity:

$$\omega_{opt} \equiv \omega_-$$

All relations between the relevant quantities in the rest frame and the optical system are deduced in the appendix. For easy reference they are also compiled in table 3. First we define the quantities relevant for laser cooling in the rest frame. The radiation pressure force for a plane electromagnetic wave is given by [2]:

$$\vec{F}_0 = \hbar\vec{k}_0\Gamma_0 \frac{G_0}{1 + G_0 + \frac{[\Omega_0 - k_0 v_{z0}]^2}{\Gamma_0^2}}$$

Here $\Omega_0 = \omega_0 - \omega_{00}$ is the detuning of the laser frequency ω_0 from the transition frequency ω_{00}, v_{z0} is the small ion velocity in the rest frame of the beam along the beam axis, $\vec{k}_0$ is the k-vector of the photon in the inner rest frame. The force is maximum, when the laser frequency is Doppler shifted into resonance. The spontaneous decay probability W_0 of the upper ionic state is :

$$W_0 = 2\Gamma_0 = \frac{1}{\tau_0} = \frac{4}{3}\frac{d_0^2\omega_0^3}{\hbar c^3}$$

For an estimate of the dipole moment of this ionic transition we assume that only one electron is taking part in the laser-induced transition and that in all cases the oscillator strength f_0 is approximately close to 1. The oscillator strength is given by:

$$f_0 = g_{if} * \frac{2 * m_e}{3e^2\hbar} * d_0^2 * \omega_0$$

Here g_{if} is a spin dependent weighting factor. This scaling of the dipole moment d_0 with $\omega_0^{-1/2}$ in atomic physics differs from the usual scaling in nuclear physics, where a constant dipole moment is used.
The saturation parameter G_0 depends on the electric field strength E_0, the dipole moment d_0 and the width Γ_0:

$$G_0 \equiv \frac{1}{2}\left(\frac{d_0 E_0}{\hbar\Gamma_0}\right)^2 = \frac{I_0}{I_{sat,0}}$$

with the saturation intensity:

$$I_{sat,0} \equiv \frac{\hbar\omega_0\Gamma_0}{\sigma_0}$$

the light intensity:

$$I_0 \equiv \frac{E_0^2}{8\pi} * c$$

and the photon absorption cross section:

$$\sigma_0 = \frac{\lambda_0^2}{4\pi}$$

For very large saturation parameters we obtain a saturated maximum cooling force:

$$\vec{F}_{max,0} = \hbar\vec{k}_0\Gamma_0$$

The rate of absorption of photons in the rest frame is given by:

$$\frac{dN_0}{dt_0} \equiv \Gamma_0 G_0 = \Gamma_0 * \frac{I_0}{I_{sat,0}}$$

Due to time dilatation the absorption rate of laser photons in the laboratory is given by:

$$\frac{dN_-}{dt_-} = \frac{1}{\gamma} * \frac{dN_0}{dt_0}$$

For most of the laser cooling quantities the Lorentz transformation for $\beta \simeq 1$ to the laboratory system could be given in table 3 with transformation properties known from textbooks for tensor components [4]. For e.g. the saturation intensity the detailed definition by ω_0, Γ_0, and σ_0 is irrelevant, but it is important that it transforms like the $T^{0,0}$ component of the field stress tensor of a plane wave.
Next we investigate the scaling with γ of several quantities by comparing the laboratory system values and values for the optical system. The lifetime scales with $\frac{1}{\gamma}$:

$$\tau_- = \frac{1}{4\gamma} * \tau_{opt}$$

The saturation power increases by γ^3:

$$I_{sat,-} = 8\gamma^3 * I_{sat,opt}$$

as well as the maximum cooling force:

$$F_{max,+} = 4\gamma^3 * F_{max,opt}$$

The absorption rate is reduced by γ^2:

$$\frac{dN_-}{dt_-} = \frac{1}{8\gamma^2} * \frac{dN_{opt}}{dt_0}$$

With these relations the values of table 2 have been calculated. For large γ values one finds very short lifetimes and very large saturation intensities with very large cooling forces.
Since typical saturation intensities for optical dipole transitions are of the order of $10\frac{mW}{cm^2}$ the γ^3 dependance makes it difficult to reach saturation for large γ- values. Due to small ion-beam diameters intensities of $1\frac{kW}{cm^2}$ can be realized. For very large γ values, like those of LHC, retarded transitions with orders of magnitude smaller width Γ_0 are advantageous, because they allow for cooling to lower temperatures, while at the same time the reduced saturation intensities are still hard to reach experimentally.
The optimum achievable energy spread $\frac{\Delta E_+}{E}$ in the laboratory can be estimated from the Doppler limit [2] for the minimum temperature T_{min0} in the rest frame:

$$k_B * T_{min0} = \frac{\hbar\Gamma_0}{2}$$

This relation is obtained from the fluctuation dissipation theorem considering the balance between diffusive heating and laser cooling. For fast E1 transitions the ratio of the width $\hbar\Gamma_0$ and the recoil energy $\frac{(\hbar k_0)^2}{2M_0}$ is approximately:

$$\frac{\hbar\Gamma_0}{\frac{(\hbar k_0)^2}{2M_0}} \simeq A$$

where A is the mass number of the stored ion. Therefore the achievable absolute energy spread is:

$$\Delta E_+ \simeq \frac{\hbar\Gamma_0}{\frac{(\hbar k_0)^2}{2M_0}} * \hbar\omega_+ \simeq A * \hbar\omega_+$$

and :

$$\frac{\Delta E_+}{E} \simeq \frac{4 * \gamma^2 \hbar\omega_-}{\gamma * GeV} \simeq 2 * \gamma * 10^{-8}$$

Compared to present day energy spreads of high energy mashines of typically 10^{-3} several orders of magnitude lower energy spreads appear possible. Transitions with a reduced width allow for even smaller energy spreads.

Similar to results observed at TSR, the very high particle densities $\sim \frac{10^{12}}{cm^3}$ in relativistic beams may lead to a heat exchange between the longitudinal and transverse degrees of freedom via intrabeam scattering, thus allowing also for an indirect transverse laser cooling.

To estimate typical cooling times for an energy spread of 10^{-3} and ions with a mass number of 100, we calculate the number of necessary cooling cycles by dividing the absolute energy spread by the energy transferred in one absorption emission cycle. They are tabulated in table 2 and decrease linearly with increasing γ. The cooling time is obtained by multiplying this number of cycles with twice the lifetime of the transition in the laboratory. Since the cooling section is assumed to be only 10% of the ring, a further factor of 10 has to be taken into account. The cooling times for large γ values become very short, but they are deduced for unreasonable large laser powers at saturation. Cooling times of 10 seconds can be realized for practicable laser powers in all cases.

Until now we discussed only a decelerating laser force. It leads to a compression of the velocity distribution, if the laser frequency is swept to lower frequencies. To reach a stationary situation with a fixed laser frequency requires a counteracting accelerating force, which does not heat up the narrow distribution achieved by laser cooling. It seems possible to obtain this even with a conventional rf-system, if a proper locking by beam pick ups is done to the actual bunch structure.

The exploration of these high quality beams, the limiting heating processes and the laser cooling process itself justify machine developing time.

Perturbing Processes

At the TSR the lifetimes and stripping cross sections have been measured for non naked ions. Within less than a factor of 2, the theoretical cross sections [10] reproduce the measurements:

$$\sigma = 4\pi a^2 (\frac{v_0}{v_i})^2 \frac{Z_a^2 + Z_a}{Z_i^2}$$

where a is the Bohr radius, v_i the ion velocity, $v_0 = \alpha * c$, Z_i the ion charge and Z_a the nuclear charge of the rest-gas ion. Since here velocities $v_i \simeq c$ are considered, while we had $v_i \simeq \frac{c}{10}$ at the TSR, two orders worse vacuum in the region of 10^{-8} mbar will allow for lifetimes in the region of about a second. Better residual gas pressures are planned for SPS and LHC, while 10^{-8} mbar prevails in the AGS injector of RHIC. Ionisation cross sections of the order of kbarn can be calculated [11] for colliding heavy ions at relativistic energies, which are many orders of magnitude larger than the small cross sections of interest. Therefore partially stripped ions cannot be used in collider experiments. New laser cooling schemes with much slower and less efficient nuclear transitions are considered in the next section which, however, can be used for naked ions.

The energy loss E_{syn} by synchrotron radiation per turn for an ion with charge Z_i is given by:

$$E_{syn} = \frac{4\pi Z_i^2 e^2 \gamma^4}{3\rho}$$

The values calculated for $Z_i = 80$ in table 2 are still very small compared to the loss introduced by laser cooling.

Nuclear laser cooling

For a colliding beam operation naked ions have to be considered, because non-naked ions would be stripped very soon by the interaction with the ions of the other beam [11]. For the laser cooling of naked ions one has to rely on nuclear transitions. In table 4 we compiled stable nuclei with low-energetic first excited levels. The known lifetimes for neutral atoms have been corrected for conversion when calculating the lifetimes of naked ions. These lifetimes are increased by an additional factor of γ when transforming from the rest system to the laboratory system. In the following we want to consider a typical naked ion with a nuclear transition of 10 keV and a lifetime of $1\mu sec$ and a nuclear mass number of 100. All results will be compiled in table 5. If one compares the lifetime in the laboratory of 3 msec with the revolution frequency of 89 μsec at LHC a deexcitation occur on the average every 34 turns.

nucleus	transition energy [keV]	multipolarity	lifetime neutral [ns]	lifetime fully stripped [μs]
235 U	0.075		1.6*10^{12}	
201 Hg	1.58			
187 Os	9.75	M1	2.38	0.55
181 Ta	6.21	E1	6.8 *10^{3}	479
169 Tm	8.41	M1	4.04	1.09
83 Kr	9.396	M1	147	2.5

Table 4: Low energetic first excited levels of stable nuclei

This situation is very different from the atomic laser cooling, where several cooling transitions occur in one cooling section. For an initial relative energy spread of 10^{-3} the absolute energy is about 3*10^5 MeV. Dividing this number by the energy of reemitted photons $\hbar\omega_+$, one obtains the required number of cooling cycles $5 * 10^3$. Multiplying this number with the time per cycle, one obtains a cooling time of 30 seconds. If we compare for the LHC the given atomic lifetimes of 0.7 psec with the nuclear lifetime of 3 msec we obtain approximately an additional hindrance of about 10^9. Reducing all laser cooling values, given in table 2, by this factor leads to small saturation intensities ($10W/cm^2$) and cooling forces (100 eV/m). However, the absorption line width of the nucleus is:

$$\Delta\nu_{nucl,0} = \frac{1}{2\pi\tau_{nucl,0}} \simeq 10^5 Hz$$

It is very narrow compared to a typical linewidth of a laser in the rest frame of the ion :

$$\Delta\nu_{laser,0} = \Delta\nu_{laser,lab} * 2\gamma \simeq 2\gamma * 10^6 Hz$$

On the other hand the frequency shift $\Delta\nu_{recoil}$ by one absorption in the rest frame amounts to:

$$\Delta\nu_{recoil} = \frac{(\hbar k_0)^2}{2M} * \frac{1}{\hbar} \simeq \gamma * 10^7 Hz$$

In the estimate of the cooling time it was assumed that the laser line width in the rest frame equals the recoil frequency shift. Therefore the saturation intensity has to be increased by the ratio $\frac{\Delta\nu_{recoil}}{\Delta\nu_{nucl,0}} \simeq 100 * \gamma$, leading to a value of about 1 MW/cm^2. Increasing the cooling time to a few hours leads to still practicable laser powers.

typical nuclear transition energy	10 keV
Typical nuclear lifetime (at rest)	1 μsec
Lifetime for $\gamma = 3388$(lab)	3.3 msec
energy $\hbar\omega_+$	6.74 MeV
number of cooling cycles	$5*10^4$
cooling time	30 sec
saturation intensity	1 MW/cm^2

Table 5: Laser cooling properties for a typical nuclear transition at LHC

Depending on other heating mechanisms, very narrow energy distributions could be achieved. If these narrower distributions are to be maintained much shorter cooling times are required.
As a by-product laser spectroscopy of nuclear levels can be started.

Conclusion

We have shown that all ionic and nuclear systems with low-energetic short lived intrinsic excitation can be laser cooled, enlarging the number of species for laser cooling significantly. For somewhat larger γ values than at LHC, the nuclear laser cooling scheme would become much more favourable because many more nuclei would become available and their lifetimes are much shorter.
It would be interesting to find also laser cooling schemes for elementary particles like electrons or protons. Perhaps the application of external fields like in a FEL may supply possible excitations which may be used also for these particles.
If these possibilities of phase-space cooling can be verified experimentally, these methods due to the increased luminosity will become very importance.

Appendix
The transformation between the rest system and the optical system.

For the transformation to optical transitions in the rest frame we apply the conservation of the oscillator strength:

$$\omega_{opt} \equiv \omega_{-} = \frac{1}{2\gamma}\omega_0$$

$$f_{opt} = f_0$$

$$d_{opt} = d_0\sqrt{(\frac{\omega_{opt}}{\omega_0})} = d_0\sqrt{(\frac{1}{2\gamma})}$$

$$\Gamma_{opt} \equiv \frac{2}{3}\frac{d_{opt}^2\omega_{opt}^3}{\hbar c^3} = \frac{1}{(2\gamma)^2}\Gamma_0$$

The decay probability W_{opt} is:

$$W_{opt} = 2\Gamma_{opt} = \frac{1}{\tau_{opt}} = \frac{4}{3}\frac{d_{opt}^2\omega_{opt}^3}{\hbar c^3} = \frac{1}{4\gamma^2} * \frac{4}{3}\frac{d_0^2\omega_0^3}{\hbar c^3} = \frac{1}{(2\gamma)^2} * W_0$$

Since we assume that the number of photons stays the same while we reduce the photon energy from $\hbar\omega_0$ to $\hbar\omega_{opt}$ in the rest frame, we obtain:

$$I_{opt} = \frac{I_0}{\gamma}$$

$$E_{opt} = \frac{E_0}{\sqrt{(\gamma)}}$$

$$\sigma_{opt} \equiv \frac{\lambda_{opt}^2}{4\pi} = 4\gamma^2\frac{\lambda_0^2}{4\pi} = 4\gamma^2\sigma_0$$

$$I_{sat,opt} \equiv \frac{\hbar\omega_{opt}\Gamma_{opt}}{\sigma_{opt}} = \frac{1}{4\gamma} * \frac{\frac{1}{2\gamma}\hbar\omega_0 * \frac{\Gamma_0}{\gamma}}{4\gamma^2\sigma_0} = \frac{1}{(2\gamma)^5} * I_{sat,0}$$

For the saturation parameters we obtain:

$$G_{opt} \equiv \frac{I_{opt}}{I_{sat,opt}} = 32\gamma^4 G_0$$

The maximum cooling force is given by:

$$F_{max,opt} \equiv \hbar k_{opt} * (\frac{2}{3}\frac{d_{opt}^2\omega_{opt}^3}{\hbar c^3}) = \frac{F_{max,0}}{8\gamma^3}$$

$$\frac{dN_{opt}}{dt_0} \equiv \Gamma_{opt} * \frac{I_{opt}}{I_{sat,opt}} = 8\gamma^2 * \frac{dN_0}{dt_0}$$

REFERENCES:

[1] S.Schöder et al.
Phys. Rev. Lett. 64 (1990) 2901
W.Petrich et al. this conference
[2] V.S.Letokov and V.G.Minogin, Phys.Rep.73 (1981) 1
V.I.Balykin, V.S.Letokhov and V.G.Minogin,Sov.Phys.Usp.28 (1985)803
[3] D.Wineland, R.E.Drullinger, F.L.Walls
Phys. Rev. Lett. 40 (1978) 1639 and W.Neuhauser, M.Hohenstatt, P.Toschek, H.G.Dehmelt
Phys.Rev.Lett. 41 (1978) 233
[4] J.D. Jackson, Classical Electrodynamics, John Wiley & Sons, New York, 1975
[5] Concept for a lead-ion accelerating facility at CERN. R.Billinge, E. Boltezar, D. Boussard, E. Brouzet, R. Cappi, B. de Raad, N. Doble, P. Grafström, H. Haseroth, C.E. Hill, K.H. Kissler, J. Knott, T. Linnecar, F. Nitsch, A. Poncet, U. Raich, N. Rasmissen, H. Schönauer, T.R. Sherwood, N. Siegel, U. Tallgren, P. Tetu, D. Warner and M.Weiss, CERN 90-01,28.2.1990, Proton Synchrotron Division.
[6] H.Hahn, The Relativistic Heavy Ion Collider at Brookhaven, European Particle Accelerator Conference, Rome, June 7 -11, 1988, World Scientific, Singapore, p 109
[7] G.Brianti, D.Boussard, L. Burnod, G. Drouet, J.B. Jeanneret, J. Gaeeyte, Ph. Lebrun, D. Leroy, M. Morpurgo, R. Perin, W. Scandale and E. Weisse, "The Large Hardron Collider in the LEP tunnel", Particle World, 1 (1990) 104
[8] N.G.Basov, A.N. Oraevkii, B.N. Chichkov, Sov.Phys.JEPT 62 (1986) 37
[9] A.M.Sandorfi et al.IEEE Transactions on Nuclear Science NS-30(1983)3083 ((laser-electron gamma-source legs))
[10] D. Krämer, G. Bisoffi, M. Blum, A. Friedrich, Ch. Geyer, B. Holzer, H.W. Heyng, D. Habs, E. Jaeschke, M. Jung, W. Ott, R.E. Pollock, R. Repnow, F. Schmitt and M. Steck, NIM A287 (1990) 268
[11] G. Deco,K. Momberger and N.Grün, J.Phys. B: 23 (1990) 2091
[12] H.C.Bryant et al. "Atomic physics with relativistic beams", Atomic Physics 7 , Plenum Press New York, 1980, p. 29

DESIGN AND EXPERIMENTAL VALIDATION OF PICK-UP FOR TRANSVERSE TEMPERATURE MEASUREMENT IN ELECTROON COOLING EXPERIMENT

Giuseppe Di Massa*, Maria Rosaria Masullo

Istituto Nazionale di Fisisca Nucleare, Sez. di Napoli
Mostra d'Oltremare Pad. 20, 80125 Napoli, Italy
*Dipartimento Elettrico, Universitá della Calabria
87036 Rende (CS) Italy

ABSTRACT

A pick-up with high sensitivity and wide bandwidth, working in the microwave region is suggested in order to measure the transverse electron temperature for the E-cooling device in preparation at National Laboratory of INFN in Legnaro (Italy). The study, design and laboratory results are reported.

1. INTRODUCTION

Electron cooling is a fast process to compress the phase-space density of stored charged-particle beams without loosing particles. This method was proposed by G. Budker in 1966[1] and experimentally confirmed in 1974-75 at Novosibirsk[2]. E-cooling experiments were also performed at CERN and at Fermilab. One of its most exotic application in accelerator Physics is the possibility of increase the beam luminosity in colliding beam machines.

An e-cooling device to cool (anti)proton beams at intermediate energy is in preparation at National Laboratory of INFN in Legnaro in collaboration with different INFN institutes[3,4]. The energy of the e^- beam is varied between 0.1

and 0.7 MeV with electron current of 3.5 A and transverse beam temperature of 0.5-1 eV.

The implementation of an adequate diagnostic system is required to have high quality beams (necessary in this kind of experiments) and to achieve satisfactory cooling performances. The transverse electron temperature is one of the most important information to be picked up. It can be determined from the electromagnetic field, at microwave frequency, radiated by e^- spiralling in the magnetic field of a solenoid [5]. The e-m field excites a synchronous pick-up (PU) where the phase velocity of the wave concides with the beam velocity. The PU proposed in this paper is an inhomogeneous waveguide with metallic corrugated walls. The presence of metallic corrugations changes the surface impedance of two opposite walls of the guide so that slow wave ($\beta_w < 1$) can propagate in the structure[6].

The PU is terminated, on both sides, with a transition to an output ridged waveguide (a uniform rectangular guide with a centered rectangular ridge on both its wide side). In this region, where the coupling between the beam and the guide is weak because there is no more synchronism betweem beam and induced wave, the main problen is to minimize the reflections for reducing the power losses to the ouput guide.

2. THE TRANSVERSE PICK-UP SENSITIVITY

The electron beam, moving along spirals with the Larmor radius, r_o, in the magnetic field of the solenoid where the PU is located, generates an electromagnetic field of frequency ω_c. The voltage induced by the beam and detected by the PU isr_o[4]:

$$V_{PU} = S r_o I_F \tag{1}$$

where S is the pick-up sensitivity and I_F is an equivalent current depending on the PU length and on the number of electrons. The equation (1) shows that the voltage (or equivalently the Power detected) is proportional to the PU sensivity and to the electron transverse temperature through r_o.

The transverse PU sensitivity S_Δ, for the transverse direction x (fig. 1) is defined as [7,8].

$$S_\Delta = \frac{1}{I_b}\frac{\partial V}{\partial x}\bigg|_{x=x_o} \qquad (2)$$

where

$$\underline{I}_b = I_b e^{jk_o z/\beta_p}\delta(x-x_o)\delta(y-y_o)\hat{z} \qquad (3)$$

is a current line along $\hat{z}$ describing the beam, k_o is the propagation constant in free space, β_p the relative beam velocity, x_o and y_o the transverse coordinates of the beam and V the voltage induced by the beam on the output termination R_0.

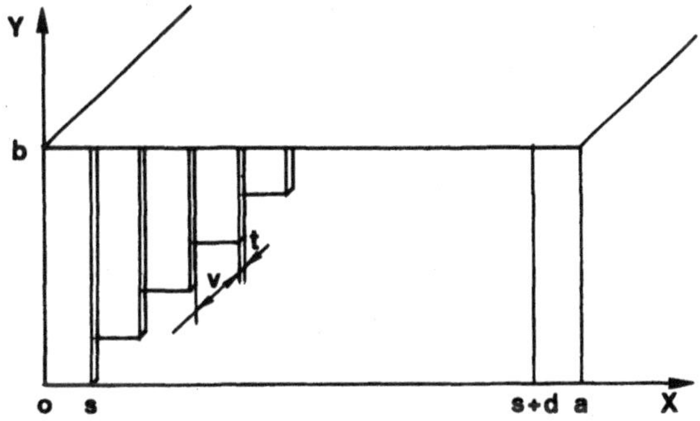

Fig. 1. Corrugated waveguide

Eqn. (2) leads to

$$S_\Delta = \frac{1}{2}\sqrt{\frac{R_o}{2P_T}}\left[\frac{\partial E_z(x,y_o)}{\partial x}\right]_{x=x_o} L\left|\text{sinc}\left\{k_o\frac{L}{2}\left(\frac{1}{\beta_p}-\frac{1}{\beta_w}\right)\right\}\right| \qquad (4)$$

where sinc(x)=sin(x)/x, R_0 is the PU load, P_T the power carried by the wave, E_z the z-component of the field when an odd mode is propagating, L the PU length and β_w the relative phase velocity of the electromagnetic field.

For the corrugated waveguide of fig. (1) the fundamental mode is Longitudinal Section Electric (LSE) [6]. The relation between wavenumbers, for a slow wave ($\beta_w < 1$), is given by:

$$q^2 = \frac{k_0^2}{\beta_w^2}(\beta_w^2 - 1) - k_y^2 \tag{5}$$

where q and k_y are the wavenumbers along x and y respectively. The matching of the fields on the vacuum corrugations interface leads to the following trascendental equation for the antisymmetric solution of the potential equation

$$\beta \tan \beta s = q \tanh q \frac{d}{2} \tag{6}$$

where β is the propagation constant for the TE_{10} mode into the corrugation and s the depth of corrugation (fig. 1). Equation (6) combined with (5) provide the dispersion relation of the corrugated waveguide in the limiting case of many corrugation per wavelength.

Selecting the even solution for the potential equation (odd for e_z) the fields and consequently the power P_T flowing in the waveguide can be calculated. The following expression for S_Δ results [8]:

$$S_\Delta = \left\{ \frac{R_o}{\omega \mu_o \gamma_n \beta^2 \frac{b}{2} \frac{(qd + \sinh qd/2)}{2q}} \right\}^{1/2} \bullet$$

$$j\omega\mu_o \cosh q \left(x - \frac{a}{2}\right) \cos\left(\frac{m\pi}{b} y\right) L \, \mathrm{sinc} \left| k_o \frac{L}{2} \left(\frac{1}{\beta_p} - \frac{1}{\beta_w}\right) \right| \tag{7}$$

$\gamma_n = \frac{k_0}{\beta_w}$ being the wave propagation constant. In (7) only the first odd mode is cosidered. The normal modes propagating in this inhomogeneous waveguides are hybrid modes (combination of TE and TM modes). The general solution of the potential equation is a superposition of an even and odd solution with respect to the transverse dimension. The even solution as well as the high mode for the odd solution are not synchronous with the beam, that is the phase velocity β_w (5) of those waves is different from the beam one β_p. As result they are not excited from the beam.

3. DESIGN OF TRANSVERSE PICK-UP

The required beam energy variation implies a frequency range between 3.56 and 7.05 GHz. When we consider the dispersion diagram of a corrugated waveguide we have for every beam velocity (propagation constant) a central working frequency and a band, computed through the (7), too little for our purpose.

The design consists in adjusting the dimensions of the PU in order to optimize the relation between the frequency and the wave velocity β_w (dispersion diagram) to match it, closely as it is possible, to the required relation between the relative beam velocity and the frequency.

The dispersion relation for the antisymmetric solution when we take into account the finite thickness, t, of the corrugation and its pass v is

$$\frac{v+t}{v} q \tanh q\frac{d}{2} = \beta \tan \beta s \tag{8}$$

In fig. (2) are reported several curves, for different PU dimensions, compared with the required dispersion relation $\beta_p = \beta_p(f)$.

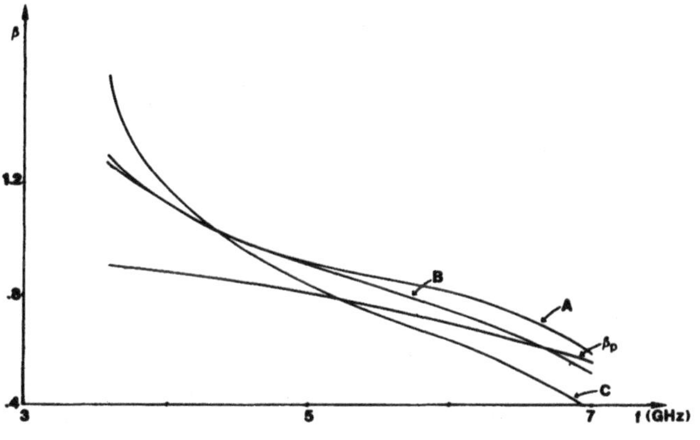

Fig. 2. Dispersion relation relevant to different PU dimensions compared with the particles dispersion relation β_p. Velocity as a function of frequency.Curve A: $s = 8.5; \frac{d}{2} = 28; v = 2; t = 1; l = 130$. Curve B: $s = 8.5; \frac{d}{2} = 28; v = 2; t = .5; l = 130$. Curve C: $s = 9.6; \frac{d}{2} = 25; v = 2; t = .5; l = 150$. Dimensions in mm.

The argument of the sinc function in (7) is assumed as parameter. The condition

$$k_o \frac{L}{2} \left(\frac{1}{\beta_p} - \frac{1}{\beta_w} \right) \leq \frac{\pi}{2} \tag{9}$$

is verified with the dimensions relative the diagram A of fig. (2)

$$s = 8.5mm; \quad \frac{d}{2} = 28mm; \quad v = 2mm; \quad t = 1mm; \quad L = 130mm$$

In fig. (3) the sensitivity (7) over the entire band is reported (curve A) when β_p varies between .91 and .55 (frequency between 3.5 and 7 GHz). The curve B is relevant to a fixed value of particles velocity β_p =.7 (f = 6 GHz).

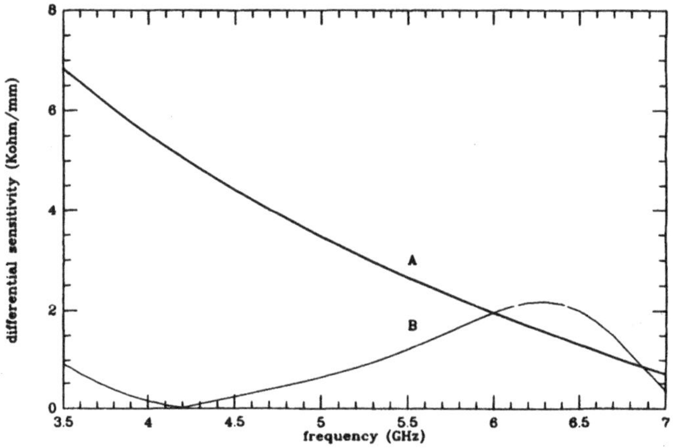

Fig. 3. Differential sensitivity as a function of frequency. Curve A: sensitivity when β_p varies between .55 and .91. Curve B: sensitivity for a fixed value of β_p = .7. The dimensions are relevant to curve A of fig. 2.

Two types of transition from the corrugated waveguide to the outside are taken into account:

a) a corner reflector antenna[9)] consisting of a wide band triangular dipole with flare angle of 40^0 which is bent 90^0 along its axis so that the dipole is parallel to both sides of the relector. The distance between the driven $\frac{\lambda}{2}$ dipole s and the corner is selected in order to have a radiation resistence of about 50Ω over the entire band;

b) a transition formed by a trough waveguide that couples on both sides of the corrugated waveguide to two ridged waveguides.

In the following only the results relevant to the second transition are presented. The first transition does not cover the whole frequency band. With the computed dimensions and the second transition the PU of fig. (4) results.

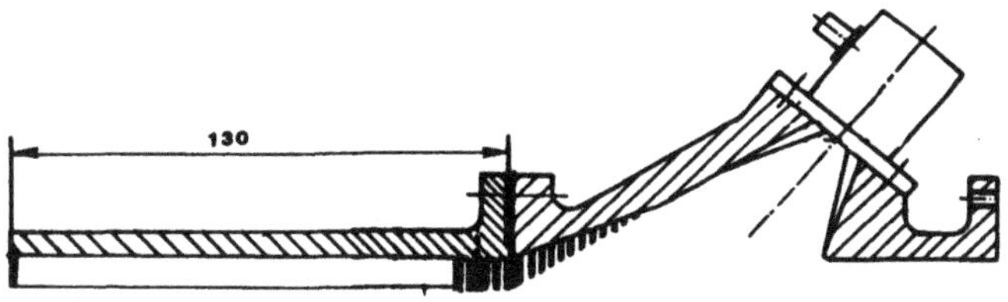

Fig. 4. Corrugate PU with transition to ridged waveguide.

4. MEASUREMENT PROCEDURE

In order to characterize the designed pick-up and to verify its experimental validation we have measured its sensitivity with a bench method. In the chosen technique the beam is represented by a strip-line stretched in the middle of half pick-up (fig. 5) on which a microwave signal is sent and that will excite a transverse PU mode. The phase velocity, β_w, of the surface wave propagating on the strip is function of the chosen dielectric and of the system geometry. This means that in order to represent the beam we have to control the β_w value so that it results in the required e^- velocity band.
The geometrical and dielectric characteristics of the strip line used, made of kapton (ϵ_r =3.7), are reported in fig. (5). The resulting phase velocity (0.6) is well inside

inside e^- velocity band ($\beta_e = 0.55 - 0.91$).

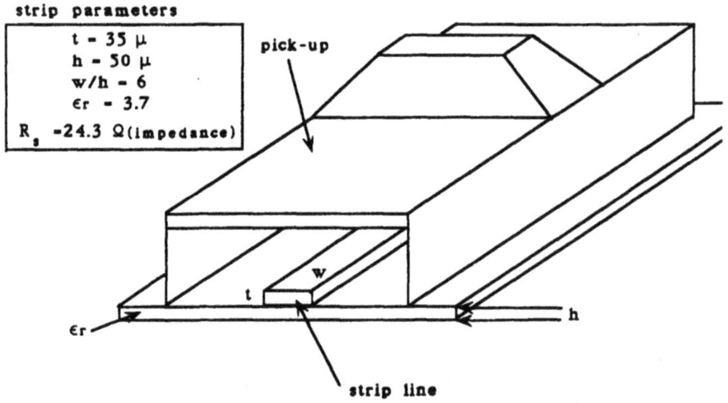

Fig. 5. Half PU and microstrip transmission line.

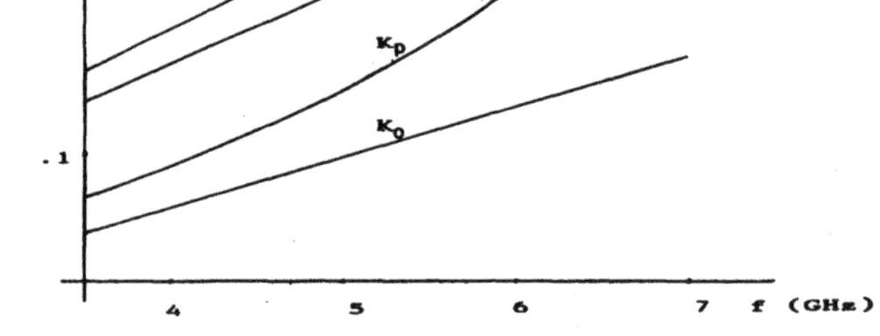

Fig. 6. Experimental dispersion diagram for the microstrip. Propagation constant as a function of the frequency.

The fig. (6) shows the dispersion curves for a wave propagating on the strip and for the particles in the frequency region of interest. The two diagrams intersect at a frequency of roughly 6.8 GHz. Following the theory previously discussed, we expect to find a maximun of the PU sensivity at this frequency value for which the phase velocity of the slow wave propagating in the PU is roughly the same as that of the beam (or equivalently of the wave on the strip).

The required sensitivity is obtained from the measured transmission coefficient of the system microstrip- half pick-up.

The fig. (7) shows the simplified scheme of the method in the frequency domain, meanwhile in fig. (8) a detailed set-up of all the instruments is reported.

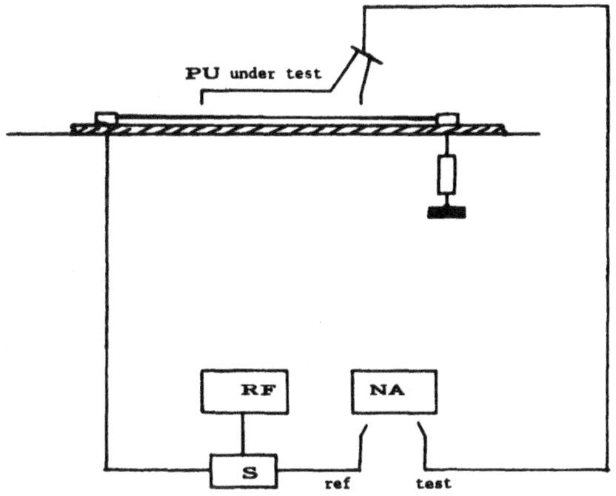

Fig. 7. Test set for transverse sensitivity measurements. S: splitter , RF: generator HP8350B, NA: network analizer HP8410C.

The measurements have been performed through a Network Analyzer (N.A.) Hp-8410C connected to the pick-up through a Reflection-Transmission Test Unit, Hp 8743B. The Sweep Generator Hp 8350B provides the input signal, sent on the strip-line. The response detected from the pick-up, through an antenna localized in the transition, is sent to N.A. and displayed there (Cartesian and/or

Polar display). Then it goes splitted in module and phase, through two channels of an ADC module to the computer for the storage and analysis.
The same computer (Hp 9000, series 300) is used to control the Sweep Generator.

A program performs the whole acquisition from the choice of the sweep to the storage(optional) and the data treatment.

The complete analysis is performed simulating a Time Domain Reflectometry (indirect TDR) by means of a Discrete Fast Fourier Transform. The f-domain measured response is taken at harmonic intervals of the principal frequency, derived from the initial sweep sent in the test object. After a time-domain transform (inverse FFT) the data represent the reflected waveform which would result from the excitation of the test object by a pulse. Apart from the main spike (pulse), any discontinuity in the measured system will result in a short pulse returned at the beginning and delayed in time by the double path lenght involved.

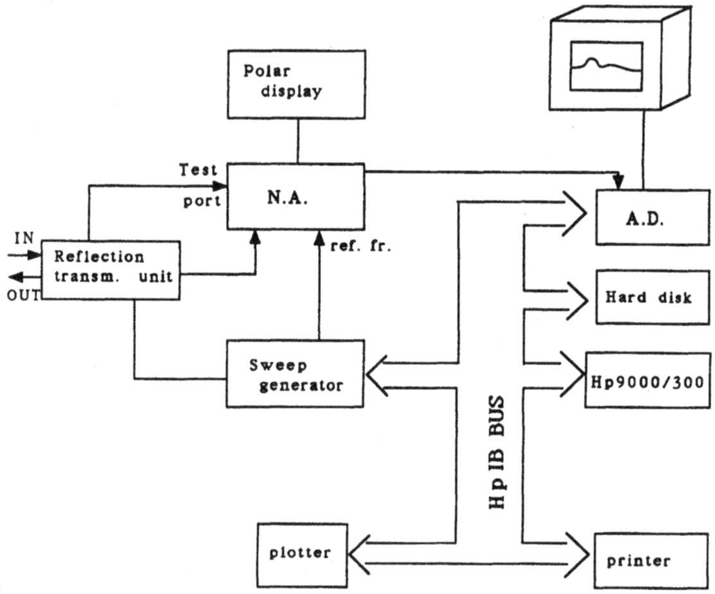

Fig. 8. Detailed set-up for PU measurement.

This technique leads to a good description of the system under test using the figure of merits of both domains. The choice of the frequency domain for the acquisition easily allows to provide good instruments in the microwave region and the possibility of varying the f-band width can results in a better S/N value leading to a higher spectral power density than in time measurements; on the other hand the choice of the time domain for the analysis allows to identify any reflection along the test system, otherwise hidden in a frequency response, and, clearing the signal, to improve the results.

The data processing is completed by adding two other tools.

In a computer Fourier Transform (FT) the data cannot extend from zero frequency to the infinity. The truncation done on the band, or equivalently after some terms of the Fourier series, produces in the reconstructed time domain response new side-lobes on each side of the main lobe and this for each discontinuity. To eliminate this effect or to reduce it a pulse shaping (called Windowing) is applied to the frequency data.

In our program the window function can be choosed between various mathematical functions, like $\cos x, \cos^2 x$, etc. This large choice comes from the fact that the Windowing, effecting also the main lobe, can decrease the resolution. So a better compromise between the side-lobe presence and the resolution has to be looked for.

The other tool included in our program is the technique called "Gating", that's a kind of filtering, performed on the time domain data, that allows to eliminate the undesiderable effect of multiple reflections on the final response. When the "gate" is applied the signal outside the choosed range is eliminated leaving intact the reflections of interest. It has to be mentioned that this method can produce big errors or meaningless results in the f-domain, after the FFT, if the cutting is not carefully applied.

Before measuring the transmission coefficient of the pick-up, we have measured the same parameter for each part of the whole system (strip-line, filter, transition,etc.) in order to better recognize the contributions of each single part.

The fig. (9a,b) shows the result obtained for the whole system in the band $3 - 7GHz$. A filter, put after the pick-up, cut everything outside this band.

In the shown transmission coefficient a maximum at roughly 6.4 GHz is evident (6.8 expected). The total response is made of contributions coming from all the parts of the system (cables, strip-line, etc) as can be easily seen in the time transform of the frequency data (fig. 9c).

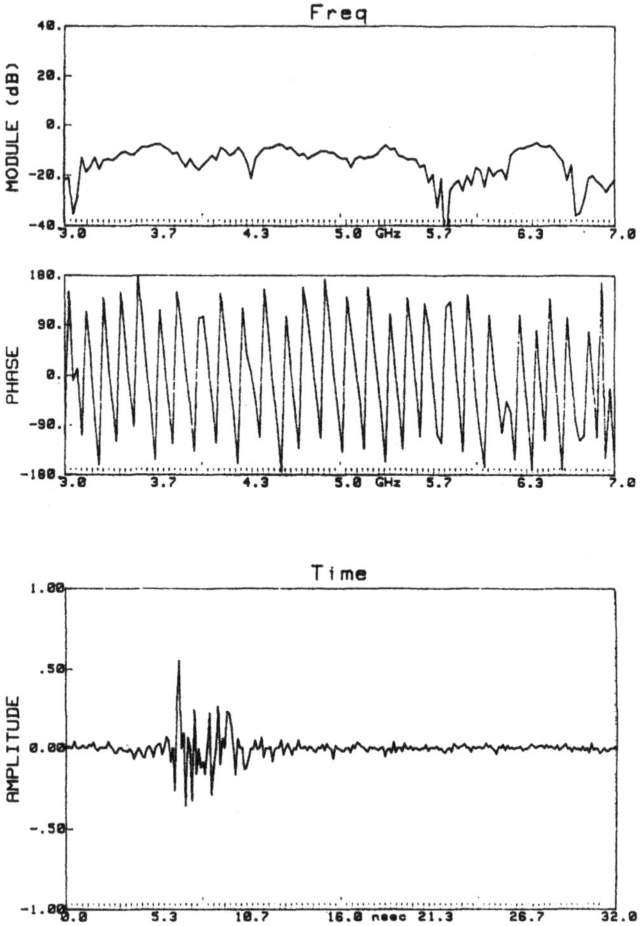

Fig. 9. Transmission coefficient. F-domain: a) module; b) fase. c) T-domain transform.

The main peak at $6.5nsec$ is due to the discontinuity seen by the pulse in the transition from the strip-line to the P.U. antenna. Every thing that proceeds is due to the path of the signal along the strip (where was put our reference origin), meanwhile all the other discontinuities after the main peak (after 10 GHz) are due to the path from the exit P.U. connector to the N.A. The main peak width probably comes from the non perfect signal transmission in the transition region from the strip to the antenna and to the external.

In order to eliminate the contributions not belonging to the P.U., we applied a double gate on the time response ($0 - 6nsec; 10.7 - 32nsec$) obtaining in the frequency domain the result shown in fig. (10). It's evident the "cleaning" of the response due to the Gating.

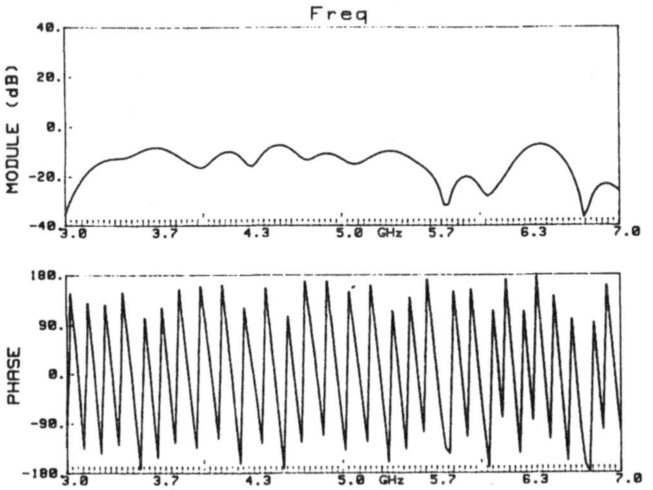

Fig. 10. F-domain transform of T-domain data (fig. 9c) after a double gate.

Fig. (9a,b) shows a different signal level at the band extremities. In order to eliminate the consequences of this effect we have applied different kind of windows to the data. In fig. (11 a,b) the use of a $\cos^2 x$ function is shown. The effect of this window can be seen from the time domain transform (fig. 11c). The main peak is now narrowed and some secondary peaks near by it are disappeared.

Appling again the same gate operation as before the IFFT curve of fig. (12) is obtained.

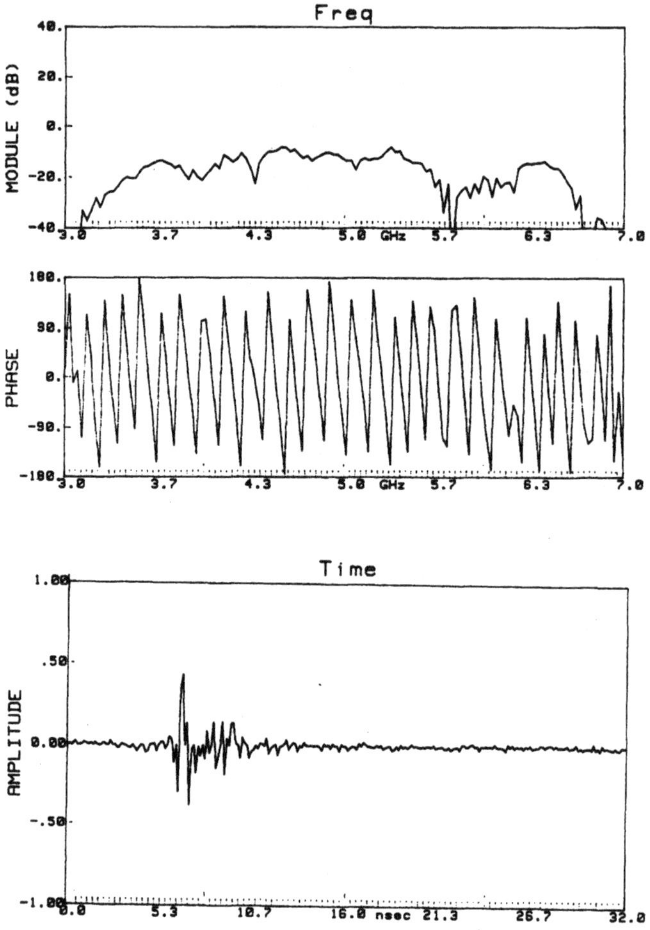

Fig. 11. Transmission coefficient. F-domain: $\cos x$ window application (a,b). c) T-domain transform

All the measurements done, also varying the acquisition band, present a main peak at around 6.4 GHz that's different from the expected value of 6.8 GHz. The difference between the two values can be mainly attributed to two facts. One is the difficulty of a precise alignment of the pick-up on the strip, that affects the coupling between the two systems; the other is the evalutation of the right dispersion curve for the wave propagating on the strip that depends on the choosed value for the kapton ϵ_r taking into account the propagation of part of the field out of the strip itself. Keeping in mind these two effects we can say that the results are in reasonable agreement with the theoretical values.

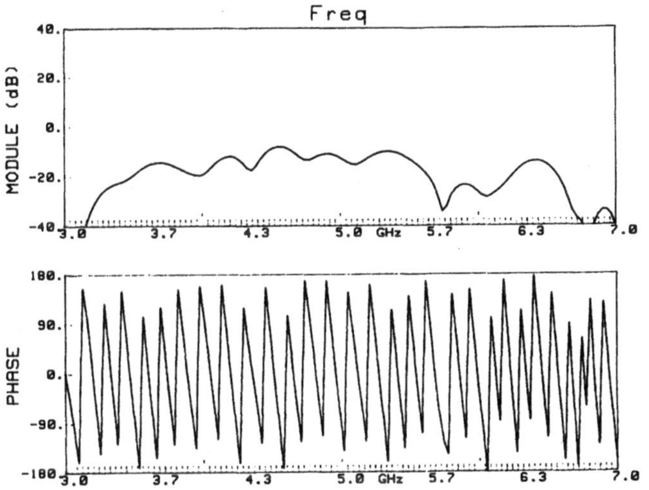

Fig. 12. F-domain transform of T-domain data (fig. 11c) after windowing

5. CONCLUSIONS

The development of a syncrhonous wave PU using a corrugated waveguide to measure the transverse temperature in electron cooling experiment is presented. A design procedure to find the PU dimensions is indicated and the application of such a design to the considered problem is also reported.

A new bench method to measure the PU is reported in the paper. The transmission coefficient between the strip line, representing the beam, and half PU is detected and worked out trought an Indirect Time Domain Reflectometry.

The test results show the possibility to use such a kind of PU to measure the electron transversal temperature. The wide frequency spectrum of the PU allow to work with the same PU at any beam energy.

ACKNOWLEDGMENT

The authors wish to express their gratitude to the many people of *Laboratori Nazionali di Frascati dell'INFN* for their cooperation in design and costruction of the PU prototipe.

REFERENCES

1. Budker,G.I., At. Energ., 22, 346, (1967);
2. Budker,G.I. et al., Part. Acc., 7, 197,(1976);
3. Bizzarri U.,Conte M.,Tecchio L.,Vignati A.,Nuovo Cimento A, 73, 425, (1983);
4. Di Massa G. et al, Nuovo Cimento A, 102, 823, (1989);
5. Rubbia C., CERN-EP-Int 77-4, (1977)
6. Dybdal R.B.,Peters L, Peake W.H., IEEE Trans. Micr. Tech., *MTT − 19*, 2, (1971);
7. Boussard D., Di Massa G., CERN SPSS/86-4, (1986)
8. Di Massa G., Nuovo Cimento A, 100, 249, (1988);
9. Jasik H.," Antenna engineering handbook", New York, Mc Graw-Hill (1961).

LONGITUDINAL TEMPERATURE MEASUREMENT IN ELECTRON COOLING EXPERIMENT

D.Babusci, M.Castellano, A.Ghigo

INFN, Laboratori Nazionali Di Frascati, P.O. Box 13

00044 FRASCATI (RM) ITALY

tel. (396) 94031

N.Cavallo and F.Cevenini

INFN-Sezione di Napoli and

Dip. di Scienze Fisiche dell'Università di Napoli, pad.19-20, Mostra d'Oltremare,

I-80125 NAPOLI - ITALY

tel.(3981)7253400

ABSTRACT

The system for the longitudinal temperature measurement which will be used in an electron cooling experiment is presented. Counting rates and experimental techniques are determined by means of a Montecarlo simulation.

1. - INTRODUCTION

A project of a high-energy electron cooling device is actually in progress as an INFN collaboration. Several prototypes have been successfully tested and the final version has been constructed in the Frascati Laboratories. The first tests will be performed in the Legnaro Laboratory.

A detailed description of the whole apparatus has been already published [1], while the main characteristics of the device are summarized in Table 1.

The longitudinal energy spread (temperature) and the transverse beam density can be measured from the spectral width of a laser light backward scattered from the electron beam [2].

Table 1. - Characteristics of the electron cooling device.

Electron beam energy	$(0.1 \div 0.7)$ MeV
Electron current	3.5 A
Electron beam diameter	3 cm
Electron momentum spread	10^{-3}
Electron transverse temperature	1 eV
Collector voltage	$(0.5 \div 6)$ kV
Magnetic field	3 kG
Drift region length	1.5 m

A feasibility study of this type of measurement on our gun has been already published [3], showing that, given an adequate laser power, a longitudinal momentum spread of the order of 10^{-4} can be measured.

In this paper we give a complete description of the whole experimental set-up. A Montecarlo code has been used to determine the geometrical and spectral properties of the scattered radiation. Using the Montecarlo results and the real characteristics of commercially available lasers, the more suitable measurement techniques and instrumentation have been defined, allowing us to give a precise evaluation of the energy range that can be covered and of the ultimate resolution achievable.

2. - PHOTON SCATTERING FROM FREE ELECTRONS

The basic formalism of the photon scattering from free electrons and its use as a diagnostic tool in plasma. and electron beam physics has been extensively reported in [3], [4]. We will limit ourselves to a brief summary of the main features of the process which are fundamental to the technical choices of the experimental set-up.

The process, schematically shown in Fig. 1, is a backward Compton scattering, and the differential cross section, in the laboratory frame, is given by

$$d\sigma(\theta,\chi,\varphi) = \frac{r_o^2}{\gamma^4}\,\frac{1+\beta\cos\theta}{(1-\beta\cos\theta)^4}\left\{(1-\beta\cos\chi)^2\,\gamma^2 + \sin^2\varphi\,\sin^2\chi\right\}\,d\Omega \qquad (1)$$

with $d\Omega = \sin\chi \, d\chi \, d\varphi$, where β is the electron relativistic velocity factor, γ its energy in rest mass units, r_0 the classical electron radius, θ and χ the incident and scattered photon angles, respectively.

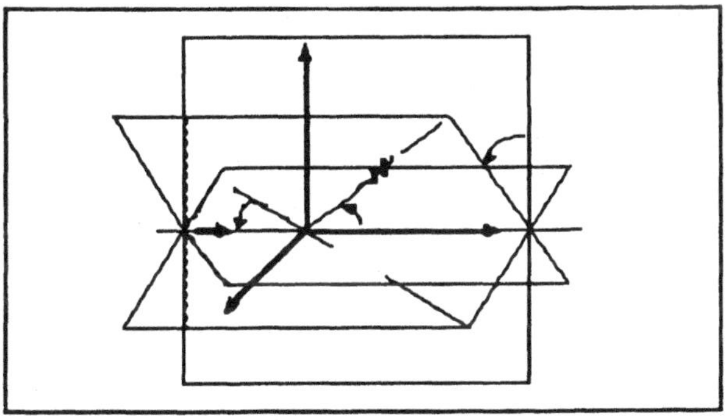

Fig. 1 - Scheme of Compton back scattering.

The wavelength λ_1 of the scattered photon is given by

$$\lambda_1 = \lambda_0 \frac{1 - \beta \cos \chi}{1 + \beta \cos \theta} \tag{2}$$

in which λ_0 is the wavelength of the incident photon.

The differential scattering cross section is strongly peaked in the forward direction (χ=0) so that we must measure the radiation in a small solid angle around this direction.

In order to assure the interaction between photons and electrons at the same radial coordinate along the whole straight section, the laser light must be parallel to the particle beam (θ=0).

The broad electron energy range generates a scattered radiation spectral range which is too wide to be easily measured if a single frequency laser is used as light source. The use of two different laser beams (Nd at 1.06 μm and CO_2 at 10.6 μm) confines the radiation in the visible spectrum, much eagerly measurable, for a relatively large energy range, as is shown in Fig. 2.

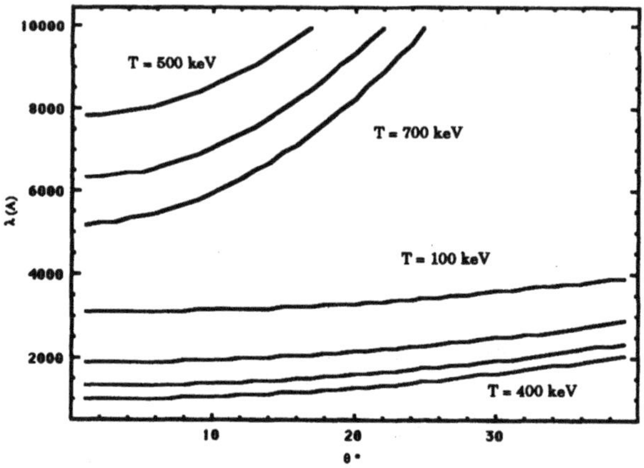

Fig. 2 - Backscattered radiation Wavelength as function of the scattering angle and the electron energy for Nd-YAG and 10.6 μm CO_2 line.

The low energy side, between 100 and 150 keV, which is of extreme interest at least in the first stage of the gun operation as a full current test of the energy recovery system, will be covered by a Nd-YAG laser.

The upper part of the energy range, between 500 and 700 keV, will be spanned by the 10.6 μm line of a CO_2 laser.

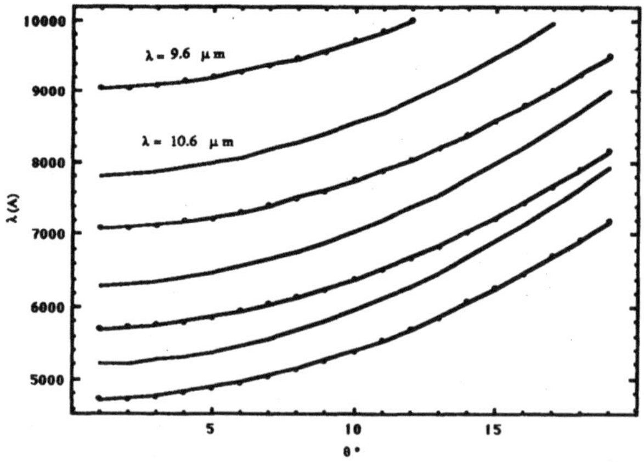

Fig. 3 - Backscattering radiation wavelength for the 10.6 and 9,6 μm CO_2 lines.

We can partially close the gap between these two ranges operating the same CO_2 laser on the 9.6 μm line, in this way going down to 400 keV, as can be seen in Fig. 3.

The scattered radiation wavelength is a function of the electron velocity, and this is the effect we want to exploit, but it is also a function of the scattering angle χ.

The measured line width is thus due to both the electron energy spread and the finite solid angle accepted by the measuring instrumentation.

A counting rate evaluation was necessary to define the measurement techniques and instrumentation, and to evaluate the signal to noise ratio and thus the ultimate resolution achievable.

We decided to develop a Montecarlo computer code simulating the whole process and taking into account the real geometry to give an answer to all the aforementioned questions.

3. - GENERAL LAYOUT

A schematic view of the accelerator and of the measuring instrumentation is shown in fig. 4.

The measurements will be performed as follows: the laser radiation, focused in the centre of the straight section, interacts with the incoming electron beam. The back-scattered radiation is splitted off from the laser and electron direction by a plane dichroic mirror, tilted at 45°, and is focused into the monochromator.

A parallel displacement of the laser beam, respect to the trajectory main axis, allows the transverse scanning of the electron beam profile.

In order to perform these measurements, the main vacuum pipe is provided with two extensions aligned to the electron beam in the 1.5 m interaction section. The first of them, placed upstream with respect to the electron motion, is closed by an optical window. The downstream extension consists of a chamber containing the dichroic mirror which is used to extract the back-scattered radiation.

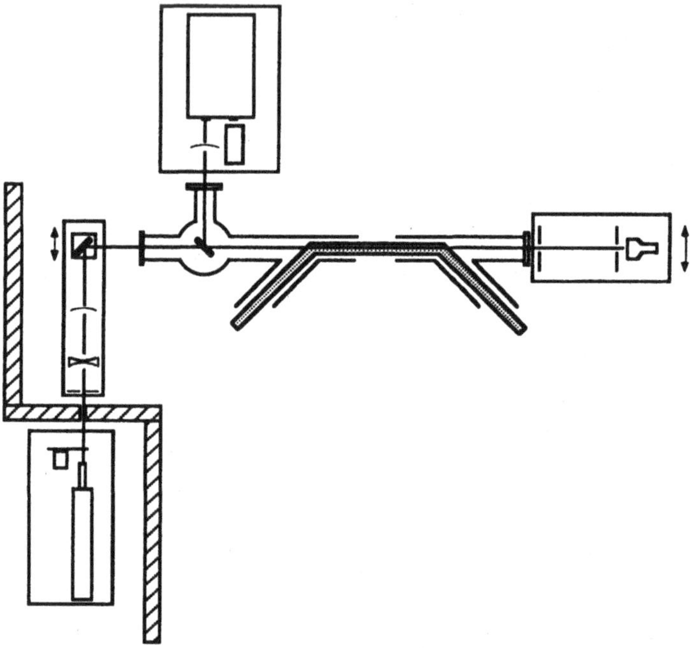

Fig. 4 - Experimental layout.

This chamber has two optical windows. The input infrared radiation passes through the first window, suitable centred, and interacts with the electron beam. The second window is perpendicular to the beam axis and allows the scattered radiation, which is reflected by the dichroic mirror, to be extracted. A precision high repeatibility mechanical movement allows to move the dichroic mirror away from the primary IR laser beam path.

We decided to develop a computer simulation to evaluate the input laser power, the monochromator resolution, the dichroic mirror size as well as the photomultiplier and acquisition system performances. This simulation takes into account the constraints given by the aforementioned experimental apparatus and measurement characteristics.

One of the main constraint in the measurement set-up was the dimension of the dichroic mirror, both due to practical reasons and the diverging cost, the largest affordable size being a 75 mm diameter, reduced to an effective 50 mm in the horizontal dimension due to the angular tilting. The Montecarlo should give us the answer if this dimension is large enough to collect a sufficient number of photons, and, on the other side, if a smaller iris is needed to reduce the effect of large angle, smaller energy, photons.

4. - NUMERICAL SIMULATIONS

The electrons were supposed uniformly distributed in a cylinder with 3 cm diameter and 150 cm length. The laser beam has a gaussian distribution on the transverse plane. The optical system is planned to achieve a laser beam waist of 5 mm diameter located at the centre of the straight section.

A detailed description of the Montecarlo results will be published elsewhere [5], here we will give only some example in order to show the experimental consequences.

All the simulations have been performed assuming a nominal mean laser power of 20 W, the results must be scaled with the real laser performance.

For the electron energy of 100 keV, using the Nd-YAG laser, the large scattering cone makes the measured linewidth insensitive to the iris aperture up to 20 cm diameter, at least for the anticipated momentum spread $\Delta p/p = 10^{-3}$. If a smaller electron energy spread can be reached, the measuring set-up must be changed, as discussed in the last section.

In fig. 5 it is shown the Montecarlo result for the collected photons through a 5 cm iris from a 20 W Nd-YAG laser scattered by a 100 keV beam with an electron density as given in Table 1.

The continuous curve is a gaussian fit which gives a sigma value of 3.4 Å and a peak value of 3.4 photons/second · Å · W.

For an electron energy of 700 keV, with the CO_2 laser, the results are quite different, due to the more peaked scattered photon cone.

For the momentum spread of $\Delta p/p = 10^{-3}$, an iris diameter of 5 cm is just the limit for obtaining a correct measurement of the electron longitudinal temperature, as is shown in fig. 6, from which we derive a linewidth of 9.6 Å and a peak rate of 46.37 photons/seconds · Å· W.

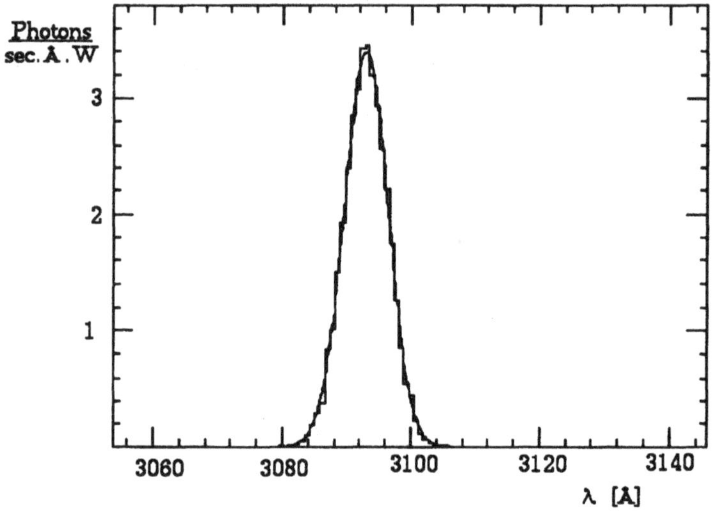

Fig. 5 - Linewidth of the scattered radiation of a Nd-YAG laser by a 100 keV electron beam trough a 5 cm iris.

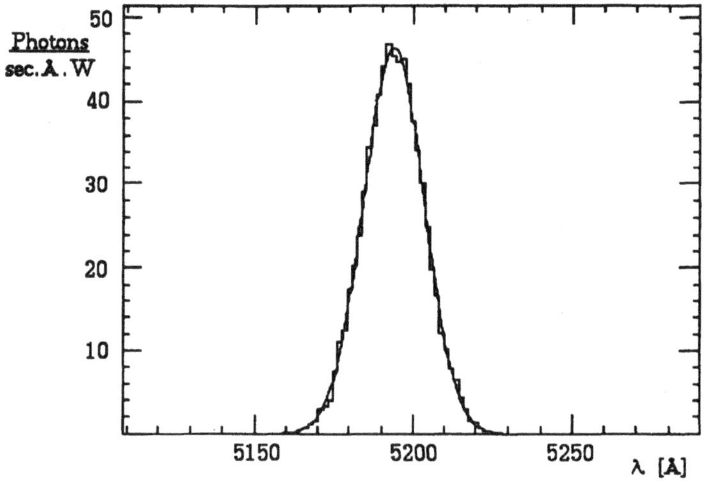

Fig. 6 - Linewidth of the scattered radiation of a CO_2 laser by a 700 keV electron beam trough a 5 cm iris.

For a narrower energy spread we must decrease the iris diameter. With an electron beam of 10^{-4} momentum spread, a 2 cm iris is required.

5. - LONGITUDINAL TEMPERATURE MEASUREMENTS

5.1 - Experimental set-up.

The proposed measurement of the longitudinal temperature will be carried out as in the general scheme of fig. 4, where the main components are shown.

The control of the laser beam direction is performed by means of a beam steering and a movable 45° tilted mirror, which is also used to scan the transverse profile of the electron beam.

The back-scattered radiation is analysed by a system composed of a focusing lens, a monochromator and a photomultiplier detector.

The alignment procedure of the optical elements and of the monochromator on the scattered radiation is made very difficult by its low intensity, and will be carried out using an He-Ne laser and movable irises, profiting of the possibility of removing and repositioning the dichroic mirror with high precision. A detailed description of the procedure will be published elsewhere [5].

5.2 - Low energy (100 keV) electron beam measurement.

A pulsed Nd:YAG Quantel YG 585/50 laser will be used for this measurement. The energy per pulse is about 280 mJ with 50 Hz repetition frequency. Other characteristic are shown in table 2.

A beam splitter is placed just in front of the Nd:YAG laser on the same optical table and a very fast photodiode is used for the acquisition system timing.

The telescope lenses as well as the input and output windows of the vacuum pipe are BK7 made and A.R./A.R. coated at the laser wavelength of 1.06 μm.

Table 2. - Characteristics of the laser sources.

pulsed Nd:YAG (QUANTEL YG585-50)			CO_2 c.w. (EDINBURGH PL5)	
Wavelength	1.06	μm	9.6-10.6	μm
Energy per pulse	280	mJ		
Pulse duration	10-12	nsec		
Repetition frequency	50	Hz		
Average power	14	W	50	W
beam diameter	7	mm	7.5	mm
Divergence	0.7	mrad	<2	mrad
Amplitude stability	2.5%	max	1%	max
Frequency stability			2	MHz
Quasi gaussian beam profile			Gaussian beam profile	

The optical glasses of the window have been chosen of 75 mm diameter; in this way the flange mounting reduces the optical aperture to about 50 mm. The same argument has been taken into account for the herasil dichroic mirror. This one is 45° tilted so, in order to have the physical horizontal dimension equal to the window effective aperture, a 75 mm diameter has been chosen.

The dichroic mirror is coated in order to have the maximum transmissivity at 1.06 μm as well as the maximum reflectivity between 295 and 325 nm. This is because the back-scattered radiation is Doppler-shifted at about 310 nm.

The herasil output window has a 50 mm effective diameter and it is A.R./A.R. coated in order to have the maximum trasmissivity between 300 and 700 nm.

The monochromator is a Jobin Yvon HRS-2 with a minimum bandwidth of 0.2 Å. A Thorn-EMI 9893-QB/350 photomultiplier with high gain ($\sim 8\times 10^7$ typical) and very low dark current (150 cps maximum in between -30° C and 60° C of temperature) is used as radiation detector.

To determine the counting rate, and so the more suitable measuring technique, we start from the Montecarlo result of fig. 5. If we want to scan the linewidth with a monochromator resolution of 1 Å, taking into account the laser power of 14 W, the overall optical losses evaluated of the order of

31 %, and the photomultiplier cathode quantum efficiency of 20 %, we obtain a peak counting rate of 6.6 counts/second, that, for the laser frequency of 50 pulses/second, corresponds to 0.13 counts/pulse.

This counting rate allows us to use a simple photon counting technique. In fact the probability of a double photon events is as low as 0.8 %, while the photomultiplier background counting can be kept at the level of 2.5 10^{-3} counts/pulse by the use of a 50 ns gate on the counting unit. The background due to the cathode blackbody radiation and to the electron bremsstrahlung on the residual gas is estimated to be even lower.

The acquisition system is shown in fig. 7.

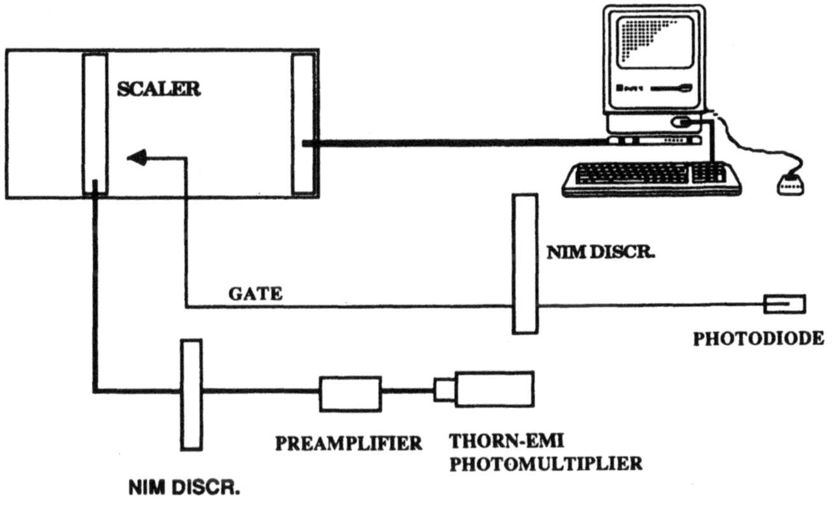

Fig. 7 - Acquisition system with the Nd-YAG laser.

5.3 - High energy (500-700 keV) electron beam measurement.

For the high energy measurement, a c.w. CO_2 (Edinburgh Instruments PL5) will be utilized. It delivers over 50 W c.w. output power on the single line.

The extracavity PL5-SF spatial filter cuts higher laser modes and a opto-galvanic stabilizer provides a long term frequency stability of the order of 2 MHz.

In order to match the whole optical system to the CO_2 wavelength, the optical elements will be changed respect to the low energy measurements. In particular both the windows at the end of the vacuum chamber will be made of ZnSe with 10.6 μm A.R./A.R. coating.

The dichroic mirror is made of Ge with maximum transmissivity at the same wavelength and maximum reflectivity centred in the visible region because the back-scattered radiation covers the spectral range between 780 and 520 nm, corresponding to 500 and 700 keV kinetic energies.

For the 9.6 μm line the back-scattered wavelength will be a bit lower (of the order of 700 and 470 nm respectively for 500 and 700 keV electron energies).

The peak flux given by the Montecarlo calculation, as shown in fig. 6, suggests us to use a different measuring technique.

The use of a high power cw laser can be best exploited by a synchronous demodulation of the laser signal, mechanically chopped, by a lock-in amplifier. The acquisition system is shown in fig. 8.

The measurement technique consists of a modulation of the input beam, which is accomplished by a chopper, together with a synchronous demodulation of both the incident radiation signal, detected by the photodiode, and the back-scattered radiation signal, detected by the photomultiplier. For this purpose two Lock-in amplifier (EG&G 5206) will be used.

The relatively low gain of the RCA C31034 photomultiplier (4.8×10^5 typical), which is preferred for its flat response in the range of 300-900 nm wavelength, is compensated by the high gain of the KEITHLEY 428 current amplifier (larger than 10^6).

Starting from the 50 W laser power, and taking into account the 50% transmission efficiency of the chopper, the total optical losses of 34%, and the photomultiplier quantum efficiency of 20%, we arrive to a final counting rate of 153 pulses/second with a monochromator resolution of 1 Å.

The linewidth of the backscattered radiation allows us to use a measuring resolution of 2 Å, obtaining a maximum counting rate of 306 pulses/second.

A few seconds lock-in integrating time is thus enough to reach the desired measuring accuracy. The high selectivity of the demodulation technique makes the background completely negligible.

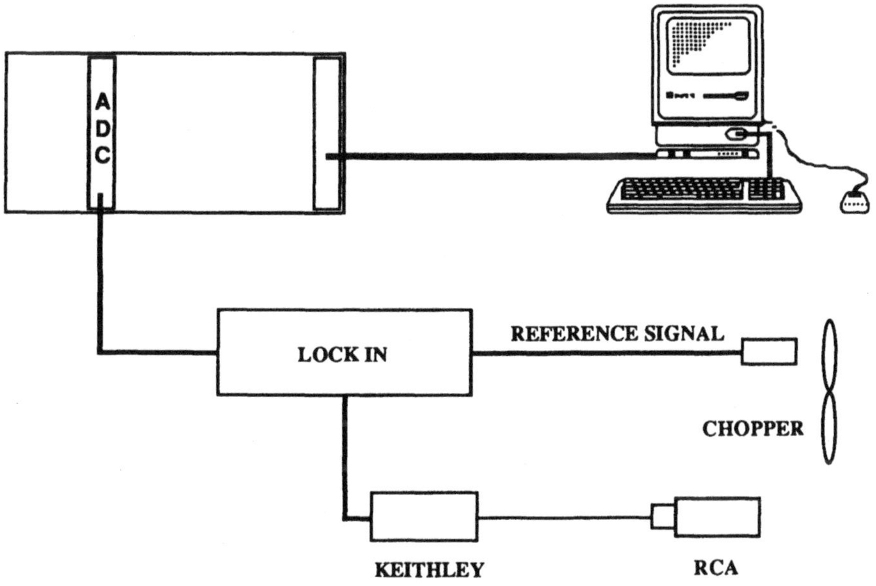

Fig. 7 - Acquisition system with the CO_2 laser.

6. - CONCLUSIONS

The results of the numeric simulations and the analysis of the measuring instrumentation capability show that an electron beam momentum spread of 10^{-3} can be effectively measured.

For the higher electron energies, with the use of the CO_2 laser, the proposed measuring technique has enough sensitivity to measure a momentum spread one order lower, but with an electron energy of 100 keV the detection of a momentum spread in the 10^{-4} range requires a different instrumentation.

This is true both for the laser source, whose linewidth is comparable to the backscattered radiation one, and for the monochromator, whose resolution becomes too large for an effective analysis of the radiation itself.

The use of an intracavity frequency filter for the laser, and of a Fabry-Perot as measuring instrument, can allow also the analysis of this extremely favourable situation.

Reference.

[1] U.Bizzarri, M.Conte, R.Scrimaglio, L.Tecchio and A.Vignati:
High-Energy Electron Cooling at LEAR.
Nuovo Cimento A, **73**, 425 (1983);

R.Calabrese, F.Petrucci, M.Savriè and L.Tecchio:
Design of the Beam Line and Energy Recovery System for High-Energy Electron Cooling Device.
Nuovo Cimento A, **100**, 763 (1988);

[2] B.Kells:
Laser diagnostic for electron cooling beam.
Fermilab Rep. TM-771 1503.000, (1978);

C.Habfast, H.Poth, B.Seligmann, A.Wolf, J.Berger, P.Blatt, P.Hauck, W.Meyer and R.Neumann:
Measurement of laser light Thomson-scattered from a cooling electron beam.
CERN-EP/87-71,1987;

[3] L.Tecchio, C.Salvetti, L.Busso, F.Tosello, R.Calabrese, F.Petrucci, M.Savriè, M.Conte and G.Simone:
Diagnostics with a Laser Beam in Electron Cooling Experiment.
Nuovo Cimento B, **85**, 217 (1985);

[4] U.Schumacher:
Electron ring diagnostic with light scattering.
Max-Plank Institute fur Plasmaphysik, IPP 0/36 (1977);

[5] D. Babusci, M. Castellano, A. Ghigo, N. Cavallo, F. Cevenini
Laser Diagnostics of Low Energy Electron Beams.
To be published.

THE LONGITUDINAL SCHOTTKY NOISE OF INTENSE BEAM

D. V. Pestrikov
Institute of Nuclear Physics, Novosibirsk 630090, USSR.

ABSTRACT

Specific properties of the longitudinal current fluctuations in a coasting ion beam, moving in a storage ring with the beam cooling, are reviewed in the paper. As the cooling changes both the debunching mechanism and the dielectric constant of the beam, it modifies the longitudinal Schottky noise spectra of both low-intensity and high-intensity beams. At the statistical equilibrium the longitudinal beam spectra are described by the universal formula, where contributes the beam dieletric constant only. This enables one to create the fitting code for the diagnostic of the beam parameters as well as its interaction with surrounding electrodes in storage rings with the beam cooling.

1. INTRODUCTION

The measurement of Schottky noise spectra provides the powerful diagnostic tool for (especially coasting) beams in storage rings[1]. Since it is based on the analysis of the noise signals, which the beam induces in pick-up electrodes, this is non-destructive method and results of the measurements generally reflectes the mutual processes in the beam. Nevertheless the interpretation of such measurements is simple only for low-intensity uncooled beams, when the noise spectrum coincides with the revolution frequency distribution function in the beam and the power with the beam number of particles N.

For high-intensity or cooled beams the noise spectra can be distorted by either collective effectes or by the cooling and accompanying diffusion. Thus the interpretation of experimental data becomes more complicated and generally needs some fitting code though can yield much more information concerning beam parameters and its interaction with surroundings. Reffering to ion beams in storage rings with electron cooling this was first pointed out in refs. [2,3].

In this paper we shall review some phenomena, which can be observed in the longitudinal Schottky spectra in storage ring with electron cooling as well as some technical details, which can be useful for the models of fitting.

2. THE LONGITUDINAL SCHOTTKY NOISE OF THE NORMAL BEAM

Let us discuss first the basic properties of the current noise in the uncooled, coasting beam. Below we shall call such a beam as a normal one. The linear microscopic density in the coasting beam as the form

$$\rho(\vartheta, t) = \sum_{a=1}^{N} \delta[\vartheta - \vartheta_a(t)] \tag{2.1}$$

where ϑ is the azimuth and $\vartheta_a(t)$ is its instantaneous value for a-th particle, N is number of particles in the beam, $\delta(\vartheta)$ is the periodic δ-function. Using:

$$\delta(\vartheta) = \sum_{n=-\infty}^{\infty} \frac{\exp(in\vartheta)}{2\pi} \tag{2.2}$$

one can get for Fourier harmonics of ρ:

$$\rho(\vartheta, t) = \sum_{n=-\infty}^{\infty} \frac{\rho_n(t)\exp(in\vartheta)}{2\pi} \tag{2.3}$$

$$\rho_n(t) = \sum_{a=1}^{N} \exp[-in\vartheta_a(t)] \tag{2.4}$$

Due to the thermal motion of particles both ρ and $\rho_n(t)$ are random functions of the time with average values:

$$\langle\rho\rangle = N, \quad \langle\rho_n(t)\rangle = 0, \qquad n \neq 0 \tag{2.5}$$

The r.m.s. value of $|\rho_n|^2$, which can be measured from pick-up electrodes is connected with two-time correlation function:

$$K_n(t_1, t_2) = \langle\rho_n(t_1)\rho_n^*(t_2)\rangle \tag{2.6}$$

Generally, the calculation of $K_n(t_1, t_2)$ can be very complicated problem. Therefore, some simplifying assumptions become necessary to find out this function in particular cases. For instances, for the beam of low intensity one can neglect the interaction of particles and, hence, write:

$$\vartheta_a(t) = \omega(\Delta p_a)\,t + \varphi_a = \omega_s t + \omega_0' \Delta p_a t + \varphi_a,$$

$$\omega_0' = \frac{\omega_s \eta}{p_s}, \quad \eta = \gamma^{-2} - \alpha, \quad \Delta p_a = p_a - p_s \tag{2.7}$$

where α is the momentum compaction factor of the ring. The substitution of these formulae into eq. (2.6) after the averaging over initial phases φ_a yields:

$$K_n(\tau) = \left\langle \sum_{a=1}^{N} \exp(-in\omega_a \tau) \right\rangle, \quad \tau = t_1 - t_2, \tag{2.8}$$

Introducting the momentum distribution function:

$$f_0(\Delta p) = \left\langle \frac{1}{N} \sum_{a=1}^{N} \delta(\Delta p - \Delta p_a) \right\rangle, \tag{2.9}$$

one can rewrite eq (2.8) in the form:

$$K_n(\tau) = N \int_{-\infty}^{\infty} d\Delta p\, f_0(\Delta p) \exp[-in\omega(\Delta p)\,\tau] \tag{2.10}$$

The spectrum analyzer measures the Fourier amplitudes of $K_n(\tau)$:

$$K_n(\omega) = \int_{-\infty}^{\infty} d\tau\, K_n(\tau) \exp(i\omega\tau) \tag{2.11}$$

For the normal beam of the low intensity this coincides with the beam frequency distribution function:

$$K_n(\Delta\omega) = 2\pi N \int_{-\infty}^{\infty} d\Delta p\, f_0(\Delta p)\, \delta(\Delta\omega_n - n\omega_0' \Delta p), \tag{2.12}$$

$$\Delta\omega_n = \omega - n\omega_s$$

Generally, $K_n(\Delta\omega_n)$ is centered around $n\omega_s$ and has the width proportional to the revolution frequency spread in the beam $n\Delta\omega$. The power of the noise spectrum (2.12) is:

$$W_n = K_n(0) = \int_{-\infty} \frac{d\Delta\omega_n}{2\pi} K_n(\Delta\omega_n) = N \tag{2.13}$$

As can be seen, for the low-intensity beam the correlator K_n depends only on the difference of times $\tau = t_1 - t_2$, which means that K_n describes the stationary noise. Thus, during the measurements K_n can be stored.

For beams with the interaction between particles the Schottky noise becomes stationary in the statistical equilibrium, but interaction generally distorts the shape of the noise spectra from that described by eq. (2.12). For typical measurements, when harmonic numbers n are not too high, one may expect that such interaction can be associated with fields, which the beam induces in surrounding electrodes. Then for the Fourier amplitudes:

$$\rho_n(\omega) = \int_0^{\infty} dt \rho_n(t)\, e^{i\omega t} \quad , \quad \mathrm{Im}\,\omega > 0$$

the calculations based on the linearised Vlasov's equations yield (see refs. [2,4] for details):

$$\rho_n(\omega) = \frac{\rho_n^{(0)}(\omega)}{\varepsilon_n(\omega)} \tag{2.14}$$

Here $\rho_n^{(0)}(\omega)$ is the Fourier harmonic of $\rho_n(t)$ calculated without the interaction between particles,

$$\varepsilon_n(\Delta\omega_n) = 1 + \frac{\Omega_n^2}{n\omega_0'} \int_{-\infty} d\Delta p \frac{\partial f_0 / \partial \Delta p}{\Delta\omega_n - n\omega_0' \Delta p} \quad , \quad \mathrm{Im}\,\omega > 0 \tag{2.15}$$

is the dielectric constant of the beam, which by means of the dispersion equation:

$$\varepsilon_n(\Delta\omega_n) = 0 \tag{2.16}$$

yields eigen-frequencies of longitudinal coherent oscillations in the beam, and

$$\Omega_n^2 = n^2 \frac{N (ze)^2 \omega_0 \omega_0'}{\Pi} (iZ_n / n) \tag{2.17}$$

is the squared longitudinal coherent tune shift, ze is the charge of ions, Π is the perimeter of the orbit. These equations imply that the interaction of the beam with surrounding electrodes is desribed by the longitudinal broad band impedance:

$$E_n(\omega) = -\frac{Nze\omega_0}{\Pi} Z_n \rho_n(\omega) \quad , \tag{2.18}$$

where $E_n(\omega)$ are harmonics of the longitudinal electric field induced by the beam. Using eq. (2.14) we can rewrite eq. (2.18) in the form

$$E_n(\omega) = \frac{E_n^{\,0}\,(\Delta\omega_n)}{\varepsilon_n\,(\Delta\omega_n)}\,,\quad E_n^{(0)} = -\frac{Nze\omega_0 Z_n}{\Pi}\,\rho_n^{(0)}\,(\Delta\omega_n) \qquad (2.19)$$

which demonstrates that collective reaction of the beam reduces external fields in $\varepsilon_n\ (\Delta\omega_n)$ times.

From eq. (2.14) one can find out for $K_n\,(\omega)$ (see in ref. [4]).

$$K_n(\Delta\omega_n) = \frac{K^{(0)}\,(\Delta\omega_n)}{|\,\varepsilon_n(\Delta\omega_n)\,|^2}\,, \qquad (2.20)$$

$$K_n^{(0)}\,(\Delta\omega_n) = 2\pi N \int_{-\infty}^{\infty} d\Delta p\; f_0(\Delta p)\;\; \delta\;\, (\Delta\omega_n - n\omega_0'\,\Delta p)$$

Equation (2.20) describes so-called Schottky noise suppression in high-intensity beams and was calculated in many papers (see, for instance, in ref. [2,4,5,6]). It is caused by the propagation of coherent oscillations along the beam and the effect is as stronger as longer is the life time of such oscillations. The usage of eq. (2.20) to fit Schottky spectra measured for normal beams enables one to get the number of particles in the beam, its frequency spread $\Delta\omega$, both parts of $Z_n = Z_n + iZ_n$ and, in principle, the momentum distribution function.

Some general properties of the noise spectra for intense beams can be seen directly from eq. (2.20). It may have sharp peaks in the close vicinity of roots of eq. (2.16) provided decrements of coherent oscillations are small enough. In fact, this can be realized in two cases:

1. if coherent oscillations are stable and the beam is cold enough;

2. if coherent oscillations can be unstable and the beam is tuned close enough to threshold of the instability.

Since the impedance Z_n generally has both real and imaginary parts, even for symmetrical distribution functions, the spectrum $K_n\,(\Delta\omega_n)$ can be asymmetrical in $\Delta\omega_n$ and such asymmetry increases, if the beam intensity tends to any thresholds of coherent instabilities.

For the low and medium energy ion storage ring one may expect the dominant contribution into the impedance Z_n from Coulomb forces:

$$(-iZ_n\,/\,n) = \frac{1}{v\gamma^2}\begin{cases} \ln(b/a)+1/2, & n \ll n_o = R_o\,/\,a \\ (n_o/n^2, & n \gg n_o\,, \end{cases} \qquad (2.21)$$

where b and a are respectively vacuum chamber and beam radii, $R_0 = \Pi\,/\,2\pi$. With this impedance and for Gaussian momentum distribution in the beam

$$f_0(\Delta p) = \frac{1}{\sqrt{2\pi}\ \sigma} \exp(-\Delta p^2/2\sigma^2) \tag{2.22}$$

eq. (2.20) can be written in the form $(\varepsilon = \varepsilon' + i\varepsilon'')$

$$K_n(\Delta\omega_n) = \frac{2N}{\zeta} \frac{\varepsilon_n''(\Delta\omega_n)}{\Delta\omega_n |\varepsilon_n(\Delta\omega_n)|^2}, \tag{2.23}$$

$$\zeta = \Omega_n^2 / n^2 \Delta\omega^2 \quad , \quad \Delta\omega = \omega_0' \sigma$$

For cold beams ($\zeta >> 1$) one has in eq. (2.23) $|\varepsilon''| \ll |\varepsilon'|$ for all $\Delta\omega_n$ except those, which satisfy the dispersion equation

$$\varepsilon'_n(\Delta\omega_n) \cong 1 - \Omega_n^2/\Delta\omega_n^2 = 0$$

Then, using the substitution

$$\lim_{\Delta \to 0} \frac{\Delta}{x^2 + \Delta^2} = \delta(x)$$

one can replace eq. (2.23) by

$$K_n(\Delta\omega_n) = (N_c/2)[\delta(\Delta\omega_n - \Omega_n)] \tag{2.24}$$

$$N_c = \eta\, (\sigma/p_s)^2 \frac{2\pi p v}{z^2 e^2 \omega_s} (n/-iZ_n)$$

describing the double-peak spectrum. The power of this spectrum obviously is N_c and thus is proportional to the longitudinal temperature of the beam[2]. Let us also note that the power of the spectrum (2.23) can be calculated exactly using the dispersion relation for the beam response function:

$$\chi(\Delta\omega_n) = 1 - 1/\varepsilon_n(\Delta\omega_n) = \chi' + i\chi''$$

$$\chi'(\Delta\omega_n) = PV \int_{-\infty}^{\infty} \frac{d\omega'}{\pi} \frac{\chi''(\omega')}{\omega' - \Delta\omega_n} \tag{2.25}$$

where PV $\int$ denotes the calculation of the principal value of the integral. The result is[4]

$$W_n = N/(1 + \zeta) = N/(1 + N/N_c) \tag{2.26}$$

Let us discuss now some properties of Schottky noise for beams near the thresholds of coherentinstabilities[4]. In this region $\varepsilon_n(\Delta\omega_n)$ can be expanded in powers of $(\Delta N / N)$, $\Delta N = N_{th} - N \ll N$ (N_{th} is the number of particles at the threshold). For the broad band impedance in the first approximations one has

$$\varepsilon_n(\Delta\omega_n) = (\Delta N/N) + i\varepsilon_n''(\Delta\omega_n) \tag{2.27}$$

The substitution of this expression into eq (2.20) yields

$$K_n(\Delta\omega_n) = \frac{2\pi^2 N}{1 - N/N_{th}}\,\delta\,[\varepsilon_n''(\Delta\omega_n)]\int_{-\infty}^{\infty} d\Delta p\, f_0(\Delta p)\,\delta\,(\Delta\omega_n - n\omega_0\Delta p) \tag{2.28}$$

As the position of peaks now is determined by the roots of virtually unstable solutions of eq. (2.16)

$$\varepsilon_n''(\Delta\omega_n) = 0 \quad , \tag{2.29}$$

in contrast with eq. (2.24), the shape of the spectra (2.28) can be very asymmetric in respect to the point $\Delta\omega_n = 0$, if Re $(Z_n) \neq 0$. In particular, it can consist even of one narrow peak. The power of the spectrum (2.28) is

$$W_n = \frac{\pi N}{1 - N/N_{th}}\sum_\beta |\partial\varepsilon''/\partial\omega_\beta|^{-1}\int_{-\infty}^{\infty} d\Delta p\, f_0(\Delta p)\,\delta\,\Delta\omega_n^\beta - n\omega_0'\Delta p) \tag{2.30}$$

where β marks solutions of eq. (2.29). For instance, for the negative mass instability $(d\omega_0/\delta\pi < 0)$ in a Gaussian beam one has

$$N_{th} = N_c,\ \Delta\omega_n = 0\ ,\ \ |\partial\varepsilon''/\partial\omega|_{th} = \frac{\sqrt{\pi/2}}{n\Delta\omega}$$

and, hence,

$$W_n = N/(1 - N/N_c) \tag{2.31}$$

the noise power increases, when N tends to the threshold value. We have to note that the same increase in the noise power can take place below transition energy of the ring, if the impedance has large inductive component.

Since for high harmonic numbers Z_n decreases, asymptotically one has $\Omega^2{}_n \to$ const, $\zeta_n \to 0$, and therefore there will be no signal suppression, $(W_n \to N)$ for harmonic numbers above

$$n = n_D = \sqrt{\Omega_n^2/\Delta\omega^2} \tag{2.32}$$

which corresponds to Debye shielding radius $r_D = R_0/n_D$.

3. THE LONGITUDINAL SCHOTTKY NOISE OF COOLED BEAM.

If the beam is cooled by any mechanism, one more parameter - the cooling decrement λ can affect the shape of its noise spectra. As from λ and $n\Delta\omega$ one can construct the dimensionless parameter

$$q = (n\Delta\omega / \lambda)^2 \tag{3.1}$$

the different behavior for spectra might be expected, depending on whether q is small or large. The physical reasons are the following.

In the coasting beam any distortion of phase density can dilute only due to the relative motion of particles along the beam caused by their energy deviations. If the cooling is weak $q \gg 1$, particles leave and enter the sample of azimuthal size $\sim 1/n$ faster then cooling starts to work. Hence, in this region of parameters one may expect small distortion of the noise spectra by cooling.

If the cooling is strong($q \gg 1$) after the cooling time $1/\lambda$ particles change their azimuthal position by $\Delta\vartheta_a = \Delta\omega_a/\lambda$ and then stop. In this region the dilution becomes possible only due to associated with cooling random kicks (or collisions), which provide the necessary deviations in the particle energies. Once these kicks are random, one has

$$<\Delta\vartheta_a> = <\Delta\omega_a>/\lambda = 0 \quad , \quad <\Delta\vartheta^2_a> = <\Delta\omega^2_a>/\lambda$$

where the solid line means the averaging over collisions. Taking into account that the average frequency of kicks is λ, one can get that the distortion in phases will dilute due to weak diffusion of particles

$$< \frac{d}{dt} < \Delta\vartheta^2 \gg = \frac{\Delta\omega^2}{\lambda} \tag{3.2}$$

Generally, such dilution machanism is specific for that in dense media (see in ref.[8]), when collision frequency is high. It obviously can be realized, if collisions occur more frequently then the thermal motion produces new fluctuations. For the Schottky band near the frequency $n\omega_s$ this requires $\lambda \gg n\Delta\omega$.

The influence of the described phenomenon on the equilibrium Schottky spectra can be calculated using the Fokker-Planck equation[7]:

$$\frac{\partial f}{\partial t} + \omega_0(p)\frac{\partial f}{\partial\vartheta} + zeE(\vartheta,t)\frac{\partial f_0}{\partial\Delta p} = \frac{\partial}{\partial\Delta p}\left[\lambda\Delta p\, f + \frac{d}{2}\frac{\partial f}{\partial\Delta p}\right] \tag{3.3}$$

To calculated values referring to the beam equilibrium state, we can adopt that both λ and the diffusion coefficient d are constants, while $f_0(\Delta p)$ is Gaussian distribution (2.22) with $\sigma^2 = d/2\lambda$. Before making the particular calculations we have to remind that actually eq. (3.3) is not usual kinetic equation because entering there function $f(\Delta p, \vartheta, t)$ is not the conventional distribution function, but is only its fluctuating part averaged over collisions. Thus, its average value over the beam fluctuations is $< f(\Delta p, \vartheta, t) > = 0$, while the second momenta are related to the beam noise characteristics[4]. In fact, $f(\Delta p, \vartheta, t)$ must be calculated from the microscopic equation

$$\frac{\partial f}{\partial t} + \omega_0 (p) \frac{\partial f}{\partial \vartheta} + zeE(\vartheta, t) \frac{\partial f_0}{\partial \Delta p} = -\frac{\partial}{\partial \Delta p} \{ (-\lambda \Delta p + d(t)) f \}$$

where d(t) is the random function describing collisions. Therefore, the use of eq. (3.3) is based on the assumption that the correlation time of external noise is much shorter then that of beam fluctuations.

The simple calculations with eq. (3.3) yield for harmonics of phase density $\rho_n(\omega)$ the same equation as eq. (2.14) (see in refs [4,7] for details)

$$\rho_n(\omega) = \rho_n^{(0)} (\Delta\omega_n) / \varepsilon_n (\Delta\omega) ,$$

but with

$$\varepsilon_n (\Delta\omega_n) = 1 + \zeta e^q \sum_{1=0}^{\infty} \frac{(-q)^1}{1!} \frac{q+1}{\frac{-i\Delta\omega_n}{\lambda} + q + 1} , \tag{3.4}$$

which is quite different from that of eq. (2.15). Therefore, we can use for spectra eq. (2.20) after the calculation of $K^{(0)}{}_n (\Delta\omega_n)$ for cooled, low-intensity beam. The last can be done directly using eqs. (2.4), (2.6):

$$K_n^{(0)} = < \sum_{a=1}^{N} < \exp(-in [\vartheta_a(t+\tau) - \vartheta_a(t)] \gg \tag{3.5}$$

and equations of the particle motion:

$$\omega_a = -\lambda\omega_a + D(t) , \quad \omega_a = \omega_0' \Delta p_a ,$$

$$\vartheta_a = \omega_a , \tag{3.6}$$

where collisions are described by the stationary white noise D(t) :

$$< D(t) D(t) > = < D^2 > \delta(t - t') , \quad < D^2 > = 2\lambda \Delta\omega^2$$

$$< D(t_1) \ldots D(t_{2k+1}) > = 0 , \tag{3.7}$$

$$< D(t_1) \ldots D(t_{2k}) > = \sum_P < [D(t_1) D(t_2)] > \ldots <[D(t_{2k-1}) D(t_{2k})]>$$

and P means the summation over all combinations of couples. The simple calculations yield[4]

$$K_n^{(0)}(|\tau|) = \exp\{-q\psi(\lambda|\tau|)\} < \sum_{a=1}^{N} \exp\left\{\frac{in\omega_a}{\lambda}(1 - e^{-\lambda|\tau|})\right\} >$$

$$\psi(x) = x - 2(1 - e^{-x}) + (1 - e^{-2x})/2,$$

or using eq. (2.9)

$$K_n^{(0)}(|\tau|) = \exp\{-q\psi(\lambda|\tau|)\} \int_{-\infty}^{\infty} d\Delta p\, f_0 \exp\left\{\frac{in\omega'_0 p}{\lambda}(1 - e^{-\lambda|\tau|})\right\} \tag{3.8}$$

In the region $\lambda\tau \gg 1$ and, hence, $\psi(\lambda\tau) \cong \lambda\tau$; this function decays with the time constant

$$\tau_n = \frac{\lambda}{(n\Delta\omega)^2} >> (n\Delta\omega)^{-1}$$

in agreement with eq. (3.2).

For the Gaussian distribution function eq. (3.8) can be rewritten in the form

$$K_n^{(0)}(\Delta\omega_n) = \frac{2N}{\lambda} e^q \sum_{1=0}^{\infty} \frac{(-q)^1}{1!} \frac{q+1}{(\Delta\omega_n/\lambda)^2 + (q+1)^2} . \tag{3.9}$$

The power of this spectrum is the same ($W_n = N$) as that for the normal, low-intensity beam, but widths of Schottky bands are much smaller provided the cooling is strong.

$$K_n^{(0)}(\Delta\omega_n) = 2N \frac{\lambda q}{\Delta\omega_n^2 + (q\lambda)^2} , \quad q << 1. \tag{3.10}$$

The Schottky noise spectrum for the cooled beam can be obtained by the substitution of eqs.(3.4) and (3.9) into eq. (2.20). As the comparison of these equations gives

$$K_n^{(0)}(\Delta\omega_n) = \frac{2N}{\Delta\omega_n} \mathrm{Im}\left\{\frac{\varepsilon_n - 1}{\zeta}\right\} , \tag{3.11}$$

one can write $K_n(\Delta\omega_n)$ using only one function - $\varepsilon_n(\Delta\omega_n)$:

$$K_n(\Delta\omega_n) = \frac{2N}{\Delta\omega_n |\varepsilon_n(\Delta\omega_n)|^2} \mathrm{Im}\left\{\frac{\varepsilon_n 1}{\zeta}\right\} \tag{3.12}$$

It is easy to verify that the normal beams with Gaussian momentum distribution eq. (2.20) coincides with eq. (3.12), which thus gives the universal representation for Schottky spectra of both normal and cooled beams. This fact also can be traced by direct calculation of the asymptotic for $\varepsilon_n(\Delta\omega_n)$ when $q \to \infty$ (refs.[4,8]):

$$\varepsilon_n = 1 + \zeta + iy\zeta\, e^q \sum_{l=0}^{\infty} \frac{(-q)^l}{l!} \frac{1}{-iy + q+l}, \quad y = \Delta\omega_n / \lambda \ ,$$

$$= 1 + \zeta + iy\zeta \int_0^{\infty} ds \exp\{isy - qs + q + (1-e^{-s})\}$$

$$= 1 + \zeta + iy\zeta \int_0^{\infty} ds \exp\{isy - qs + q\,(1 - (1 - s + s^2/2 + ...))\}$$

$$\approx 1 + \zeta + i\Delta\omega_n\zeta \int_0^{\infty} ds \exp\left\{ i\Delta\omega_n t - \frac{(n\Delta\omega)^2 t^2}{2} \right\}, \ q \gg 1 \qquad (3.13)$$

In some particular cases eq. (3.12) can be simplified. If, for instance, $\mathrm{Im}\zeta = 0$ (i.e. $\mathrm{Re}Z_n = 0$), eq.(3.12) gives for the spectrum eq. (2.23) as it was calculated in ref.[2]. Some more useful representations for ε_n can be found in ref.[4]. The behavior of Schottky spectra described by eq. (3.12) for some particular parameters in shown in fig. 1,3. It indicates well pronounced asymmetry in spectra, if $\zeta'' \neq 0$. Recently such effect was detected in measurements at TSR[9].

4. THE LONGITUDINAL COHERENT OSCILLATION OF COASTING BEAM WITH STRONG COOLING.

As for $q \gg 1$ the dielectric constant for cooled beam coincides with that for normal one, we shall concentrate here on the brief discussion of properties of longitudinal coherent oscillations for strongly cooled ($q \ll 1$) coasting beam. Before making calculations, let us mention also that with the given $\Delta\omega$ and λ the comparison of q with 1 separates coherent oscillations (and, hence, Schottky bands) by the harmonic number. From this point of view, below we shall discuss the properties of longitudinal coherent oscillations with relatively small n

$$n \ll \lambda / \Delta\omega . \qquad (4.1)$$

For q « 1 only two firs items in r.h.s of eq.(3.4) are important. Therefore the dispersion equation takes the form

$$\varepsilon_n(\Delta\omega_n) = 1 - \frac{\zeta q}{(y+iq)(y+iq+i)} = 0 \ . \tag{4.2}$$

It has the roots

$$y = -iq - \frac{i}{2} \pm \left\{ \frac{\Omega_n^2}{\lambda^2} - \frac{1}{4} \right\}^{1/2} ,$$

or

$$\Delta\omega_n = \pm\ \Omega - i\delta_\pm , \tag{4.3}$$

$$\Omega = \left\{ \frac{X + U - \lambda^2/4}{2} \right\}^{1/2} , \tag{4.4}$$

$$\delta_\pm = \frac{\lambda}{2} + q\lambda \pm \left\{ \frac{X - U + \lambda^2/4}{2} \right\}^{1/2} , \tag{4.5}$$

$$X^2 = (U - \lambda^2/4)^2 + V^2 , \ \Omega_n^2 = U + iV$$

Using eq.(4.5) one can find out that oscillations will be stable ($\delta > 0$), if

$$U + (n\Delta\omega)^2 > V^2/\lambda^2. \tag{4.6}$$

In particular, this means that the beam cooling does not stabilize the negative mass instability[7]. Nevertheless, it increases the size of the stability diagram in the plane (U, V) along the axis V especially in the region U > 0. It is interesting to note that for fixed value of q (not necessarily small) the stability diagram asymptotically ($U \to \infty$) follows the parabolic law (4.6). This can be seen directly from eq.(3.4) by the calculation of ε_n for the asymptotic region ($\mathrm{Im}\Delta\omega_n = 0$, $\Delta\omega_n \to \infty$)

$$\varepsilon_n(\Delta\omega_n) \approx 1 - \frac{\zeta q}{y^2}(1 + i/y) = 0, \qquad y \to \infty \ ,$$

or

$$(U/\lambda^2) = \frac{y^4}{1+y^2} \ , \quad (V/\lambda^2) = -\frac{y^3}{1+y^2} \ ;$$

$$(V/\lambda^2) = \pm U^{1/2}\lambda \ , \quad U >> \lambda^2 \ .$$

In the region q» 1 there can be the situation, when before reaching the parabolic asymptotic the stability diagram shrinks in V like it does that for the normal Gaussian beam. This time the stability diagram will have the waist as it is shown on fig.4. Respectevely, during the storing it can occur initial deterioration and subsequent enhancement of the beam collective stability, when the beam current increases. Analogous variations in the stability of coherent oscillations have been observed in experiments at TSR[9]. The same behavior of the stability diagram was recenttly described in ref.[6] using quite differnt collision integral.

The threshold current of the beam can be found from the stability condition (4.6)

$$N_{th}/N_c = \zeta/2 + \sqrt{\zeta(1+\zeta/4)} \quad \zeta = Z_n''/qZ_n' \qquad (4.7)$$

In the region $d\omega_0/dp > 0$ and $\zeta > 0$ it can significantly exceed the threshold of negative mass instability N_C

$$N_{th} = N_c\zeta \;\alpha\; \lambda^2 (Z_n''/Z_n') \qquad (4.8)$$

if $\xi >> 1$.

From eq.(4.5) one can see that with strong cooling ($q \ll 1$) Schottky bands of intense beam are broad enough $\Delta\omega \approx \lambda/2$, if the beam is not close to threshold of a coherent instability. Though, if the beam intensity is not so high ($U \ll \lambda^2$), or cooling is very strong, the spectrum will have only one well pronounced peak, which is centered around $\Delta\omega_n = 0$ and has the width (see in ref.[2] and also in fig.1)

$$\delta_- = \frac{(n\Delta\omega)^2 + U}{\lambda}. \qquad (4.9)$$

The splitting of the spectrum 2Ω at high beam intensities $U \gg \lambda^2$ is proportional to $N^{1/2}$, while at $U \approx \lambda^2$ this scaling changes for (see eq.(4.4))

$$\Omega \sim \sqrt{U - \lambda^2/4}$$

Close behavior of the splitting recently has been observed at TSR[9]. The behavior of the spectra close to thresholds of coherent instabilities has been discussed before.

5. DISCUSSION.

The presented results shows that both the spectra and the power of the Schottky noise of the coasting beam are very sensitive to collective behavior of the beam. This can be used for fitting of Schottky noise measurements and recalculation of beam parameters,

parameters of cooling device and beam surroundings. The use of the universal representation for spectra by eqs. (3.12) (3.4) can simplify the fitting.

The sensitivity of Schottky spectra to thresholds of coherent instabilities can be used to study that, making measurements below the thresholds. In the storage rings with beam cooling the distance to the threshold can be controlled by manipulation with the cooling rates.

I am indebted to N. Dikansky, B. Sukhina from INP (Novosibirsk) and to E. Jaeschke, M. Grieser, M. Steck from INP-MPI (Heidelberg) for the stimulating discussion.

REFERENCES.

1. Borer J., Braham P., et al. Proc. of the 9-th Intern. Conf. on High Energy Accel., Stanford 2-7 May 1974, - Stanford: SLAC, 1974, p.53.
2. Parkhomchuk V.V., Pestrikov D.V. Sov. Phys. Tech. Phys., 1980. v.50. p.1411.
3. Dementiev E.I., Dikansky N.S. et al. Sov. Phys. Tech Phys., 1980 v.50, p.1717.
4. Dikansky N.S., Pestrikov D.V. Physics of Intense Beams in Storage Rings. Nakua (Novosibirsk), 1989, 334ps.
5. Chattopadhyay S. CERN 84-11, 1984.
6. Cocher S., Hofmann I. GSI-89-61, 1989.
7. Dikansky N.S., Pestrikov D.V. Proc of the 6-th ALl Union Part. Accel. Conf., Dubna 11-13 Oct.,1978, Dubna, 1979, v.1, p.111.
8. Klimontovich U.L. Statistical Physics. M. Nakua, 1982, 608 ps.
9. Jaeschke E., Bisoffi G., et al. MPI H-1989-V 33.

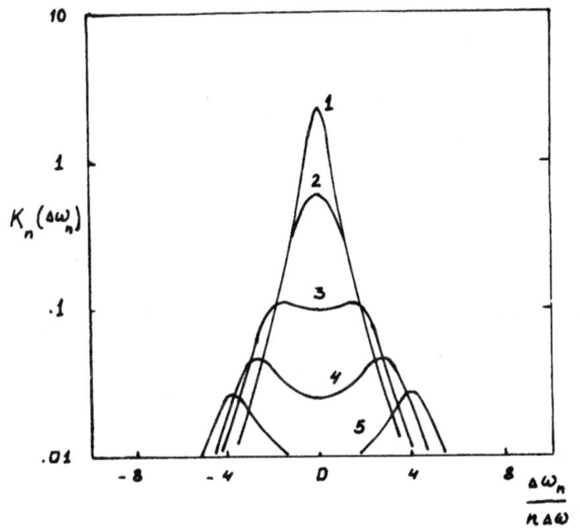

Fig.1 - The Schottky spectra of the cooled beam calculated using eq. (3.13) with $q = .25$, ($\zeta = \zeta' + i\ \zeta''$) $\zeta'' = 0$ and from the top to the bottom $\zeta' = 0;\ 1;\ 4;\ 9;\ 16$.

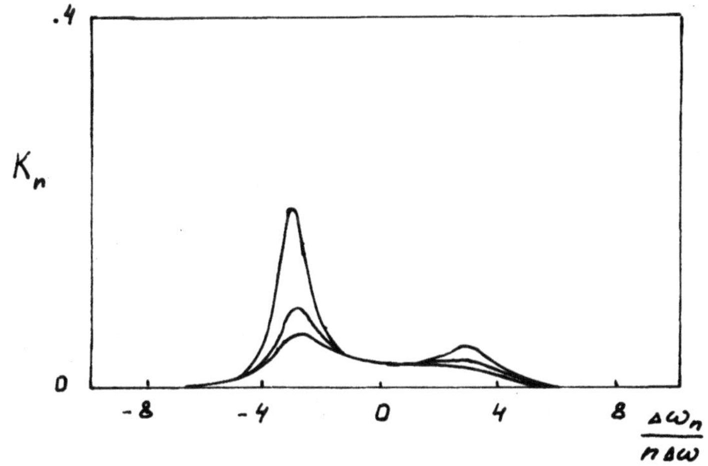

Fig.2 - The dependence of the shape of the Schottky spectra for the cooled beam on ζ'' ; $q = .25$, $\zeta' = 9$ from the top to the bottom $\zeta'' = -4;\ -2;\ -.5$.

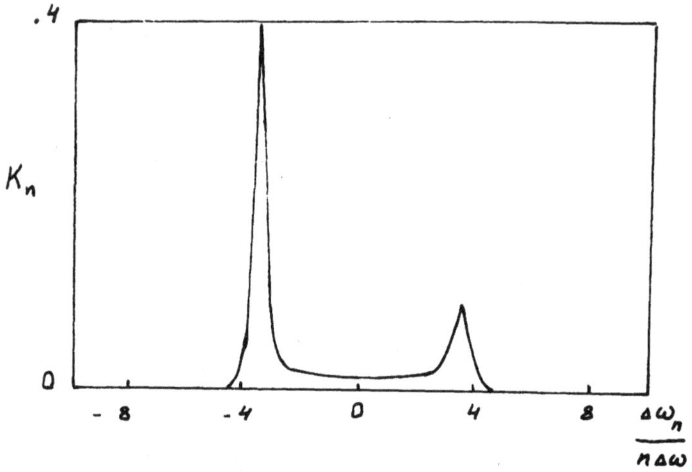

Fig.3 - The same as for fig.2 but with q = 25, ζ' = 9, ζ" = -.5.

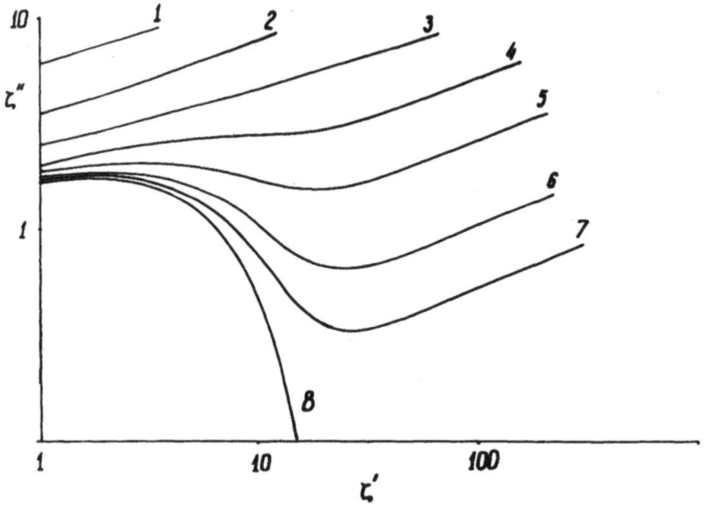

Fig.4 - The upper right quarter of the stability diagram for the cooled beam; from the top to the bottom q=1/16, 1/4, 1, 4, 16, 100, 400; 8: $q \to \infty$.

A HIGH EFFICIENCY ELECTRON MULTISTAGE COLLECTOR

V. Variale, I. Boscolo*, V. Stagno, M. Bianconi**
Istituto Nazionale di Fisica Nucleare, sezione di Bari e Dipartimento di Fisica dell' Università di Bari, Via Amendola 173, 70100 Bari
* Dipartimento di Fisica dell'Universita' di Milano e Istituto Nazionale di Fisica Nucleare di Milano via Celoria 20, 20100 Milano,
** PROEL Tecnologie di Firenze

ABSTRACT
A new type of multistage collector design is presented. This collector promises a very high efficiency electron energy recovery as shown by analytical calculations and computer numerical simulation. The main advantages of this collector with respect to the Santa Barbara multistage collector will be outlined. The collector mechanical design has reported.

1. INTRODUCTION

An Electrostatic Accelerator is capable to provide a high power continuous wave (cw), or at least long pulsed, multiampere electron beam with a few MeV of energy only if it is foreseen a very good recovery of the spent electron beam (eb). This kind of eb is very useful for several applications. For example Electron Cooling (EC) and Free Electron Laser (FEL) devices need this kind of eb. In this paper we will refer only to the FEL device where the eb to be recovered has a large energy spread.

The scheme of an FEL device with electrostatic accelerator is shown in fig.1. The voltage behaviour at the High Voltage (HV) terminal is described by :

$$dV/dt = (I_c + I_r - I_g)/C \qquad (1)$$

where C = HV terminal capacitor, V is the terminal voltage and I_c, I_r, I_g are respectively the charging current , the recovery current and the gun current. To obtain V = Const. we should have : $I_g = I_c + I_r$. The present - day HV technology can provide an I_c of some mA, then if we want an I_g of some Ampère with cw or long pulsed operation we must have $I_g \approx I_r$ that means a very high recovery efficiency. In fact assuming : $I_g = 10$ A, W = 2 MeV (Peb = 20 MW), $I_c = 10$ mA , the recovery efficiency must be $\varepsilon = I_r/I_g = 99.9\%$.The higher is the recovery efficiency the lower can be the charging current I_c . This high recovery efficiency imposes to recover very carefully the I_c electron charge and their kinetic energy.

In the FEL device the eb interacts with a wiggler and then get an energy spread [1]:

$$\Delta\gamma/\gamma \approx 1/2N \qquad (2)$$

where, N = number of wiggler periods. Assuming a $\Delta\gamma/\gamma$ about 1% (N = 50) as trade off to get high output radiation power and very good electron beam recovery, with an energy W = 2 MeV we have a $\Delta W \approx 20$ keV that further complicate the high efficiency recovery. In fact assuming that we have a beam optics (decelerating column, transition region to match the eb with the collector , etc.) capable to convey all the electron to the collector we can say that to get high recovery efficiency the collector must recover all the electrons at their minimum kinetic energy . To do this we need a multistage collector because of the electron energy spread.

2. THE MULTISTAGE COLLECTOR

The principle behind the multistage collector is to spatially disperse the electrons having different energy and collect each separated stream of electrons (with approximately the same energy) inside a kind of Faraday cup (connected to the appropriate voltage). The conceptual scheme of the multistage collector is shown in fig.2. It is a spectrometer: the electron energy spread is transformed into spatial dispersion . The desired dispersive properties for the electrons can be obtained with either an an electrostatic or/and a magnetic field.

The recovery system consists essentially in the decelerating column plus the multistage collector. When the eb reaches the multistage (depressing) collector the 99% of the electron energy has been recovered by the decelerating column, the remanent 1% should be recovered by the multistage collector. We remark that, for an eb of 2 MW, 1% means 20 kW that is still a relevant quantity that imposes a high efficiency multistage collector when we want the cw or long pulsed operation.

The first multistage collector was built at the University of California in Santa Barbara (UCSB) to recover an eb with an energy spread due to FEL interaction [2].

In that collector the spatial dispersion was obtained by applying a purely axial electric field with a transverse component simply by tilting the axis of the collector with respect to the beam propagation direction. The scheme of this collector is shown in fig.3. You can notice that the longitudinal electric field (with respect to the beam axis) continues to decelerate the electrons while the transverse component bent them on the electrodes.

In the fig.4 are shown some particle trajectories obtained by computer simulations. From this simulation the electron energy losses, on the collector electrodes, are assessed around 15%, while the measured losses at UCSB experiment were 25% [3].

From this simulation we can further observe that the electron energy recovery is not very efficient as matter of fact the electrons, entering the collector, are soon bent from the transverse electric field before to lose all their kinetic energy so this energy is wasted impinging the collector electrodes.

To improve the energy recovery efficiency we have study a different collector design. In this collector the dispersive properties are obtained by adding a transverse magnetic field to the decelerating electric field. The new electron trajectories are shown in fig.5. In this case the electrons follow straighter paths till they loose, almost completely, their kinetic energy . In this way the spatial dispersion is improved so we could add other electrodes to increase the energy recovery efficiency.

The physical reason of this behaviour can be easily understood. The force acting on the electrons is the Lorentz force:

$$\mathbf{F} = e\,(\mathbf{E} + \mathbf{v} \times \mathbf{B}) \qquad (3)$$

then we can choose the transverse magnetic field direction so that the magnetic component of the force opposes the transverse electric one and in turn the electron deflexion . This means that electrons will deflect only when their velocity becomes small enough to reduce the magnetic component of the Lorentz force.

In order to get a more quantitative analysis of the problem, it is useful to write down the single particle motion equations neglecting, in a first step the space charge contribution

$$\begin{aligned} m\,\ddot{x} &= -\,e\,E_x + e\,B_0\,\dot{z} \\ m\,\ddot{y} &= 0 \\ m\,\ddot{z} &= -\,e\,E_z - e\,B_0\,\dot{x} \end{aligned} \qquad (4)$$

Here the propagation axis has been chosen as z-axis, the electric field E_o has only component in the (x, z) plane and the magnetic field B_o is parallel to y-axis; let's assume both fields uniform in space and the electric one tilted with respect to the beam propagation axis of a 'tilting angle' θ =atan (Ex/Ez).

The analytical integration of eqs. system (3) gives

$$\begin{aligned} x(t) &= \frac{C}{\omega_c}\left[\sin\phi - \sin\left(\omega_c t + \phi\right)\right] - \frac{E_z}{B_0}\,t \\ y(t) &= y_0\,t \\ z(t) &= \frac{C}{\omega_c}\left[\cos\phi - \cos\left(\omega_c t + \phi\right)\right] + \frac{E_x}{B_0} \end{aligned} \qquad (5)$$

where C is a motion constant given by

$$C^2 = (\dot{x}_0 + E_z/B_0)^2 + (\dot{z}_0 - E_x/B_0)^2 \qquad (6)$$

ω_c is the cyclotron frequency and the phase f is

$$\phi = \text{atan}\left[\frac{\dot{z}_0 - E_x / B_0}{\dot{x}_0 + E_z / B_0}\right] \tag{7}$$

In the fig.6 analytical electron trajectories for different tilting angles θ are reported.

From the figure is evident that it is possible to control the particle trajectories trough θ, then it is possible to force the electrons along straighter trajectories in order to get the maximum particle deceleration before collecting them. It is worth observing that the same results can be obtained fixing the tilting angle and tuning the transverse magnetic field intensity.

In the fig.7 we report some analytical particle trajectories having different initial energy and fixed value of θ and B = 5G, the very good spatial dispersion obtained in this way leads to a very high efficiency recovery. In fact the higher energetic electrons "fill" a more sgnificant magnetic force component , and then are forced to go straight for a longer time with respect to the lower energetic electrons.

In conclusion by suitable choosing the magnetic field intensity B and the tilting angle θ the spatial dispersion and, consequently, the collecting energy for the electrons can be optimized in such a way that the energy recovery efficiency can be better than 90%.

Observing that, from theoretical point of view, it is equivalent to tilt the axial beam with respect to the electric field or the reverse that is tilting the electrodes, we have designed a new multistage collector with the electrodes tilted at a desired angle (θ) [4]. In fig.8 are reported the simulations results of some electron trajectories in the new multistage collector. In that results we compared the case with transverse magnetic field (b). with respect to the case without magnetic field (a). Defining the energy recovery efficiency η as :

$$\eta = (\Sigma_j E_i(j) - \Sigma_j E_f(j))/ \Sigma_j E_i(j) \tag{8}$$

where $E_i(j)$ and $E_f(j)$ are respectively the energy of the electron j at the entrance of the collector and on the collecting electrode, we have evaluated an energy recovery efficiency about 85% for B = 0 (fig.8 a) and 95% for B = 4 G (fig.8 b). This improvement is due to the higher number of collecting electrodes. In fact, when a transverse magnetic field is present, the spatial dispersion for the electrons with different energies increases and so we can increase also the number of the collecting electrodes. Passing from the ideal situation to the real situation with space charge force and stray magnetic field we get results with less efficiency but, anyway, enough high. In fig.9 we have presented the same cases of fig.8 with more electron trajectories with space charge effect included. It is worth observing that in the case B = 0 (fig.9 a) it is visible that one electron is not collected by the electrodes but went out from the collector, while, by using a transverse magnetic field B (fig.9 b) we could tune it to prevent the exit of electrons, at the end we can say that with the magnetic field we have a kind of extra knob to control the recovery efficiency.

In the fig.10 is shown the mechanical design of our new multistage collector where the idea of using tilted electrodes instead of to tilt the all collector has been implemented.

The cooling problem was one of the main concern because we wanted the cw operation but in a first stage we will test the collector in the pulsed regime.

3. CONCLUSION

We can conclude recalling briefly the advantages of our multistage collector:

a) The collector with tilted electrodes maintains the mechanical axis of the decelerating column all along the system, in this way the number of collecting stages can be as large as required from the electron beam energy spread.

b) Fixed the tilted angle θ, one can tune the magnetic field intensity B to match the incoming electron beam qualities. The charge recovery can be very near to 100% owing to the magnetic field action.

REFERECES

[1] See e.g.: Novel Sources of Coherent Radiation, vol.5, eds. S.F. Jacobs, M. Sargent III and M.O. Scully (Addison Wesley, Reading MA, 1978).

[2] L.R. Elias and G.J. Ramian in: Free Electron Generators of Coherent Radiation, vol.9, eds. S.F. Jacobs, G. T. Moore, H.S. Pilloff, M. Sargent III, M.O. Scully and L.R. Spitzer (Addison Wesley, Reading MA, 1982).

[3] L. Elias private comunication.

[4] I. Boscolo et al.,Nucl. Instrum. and meth. in Phys. Res. A275 (1989) 184.

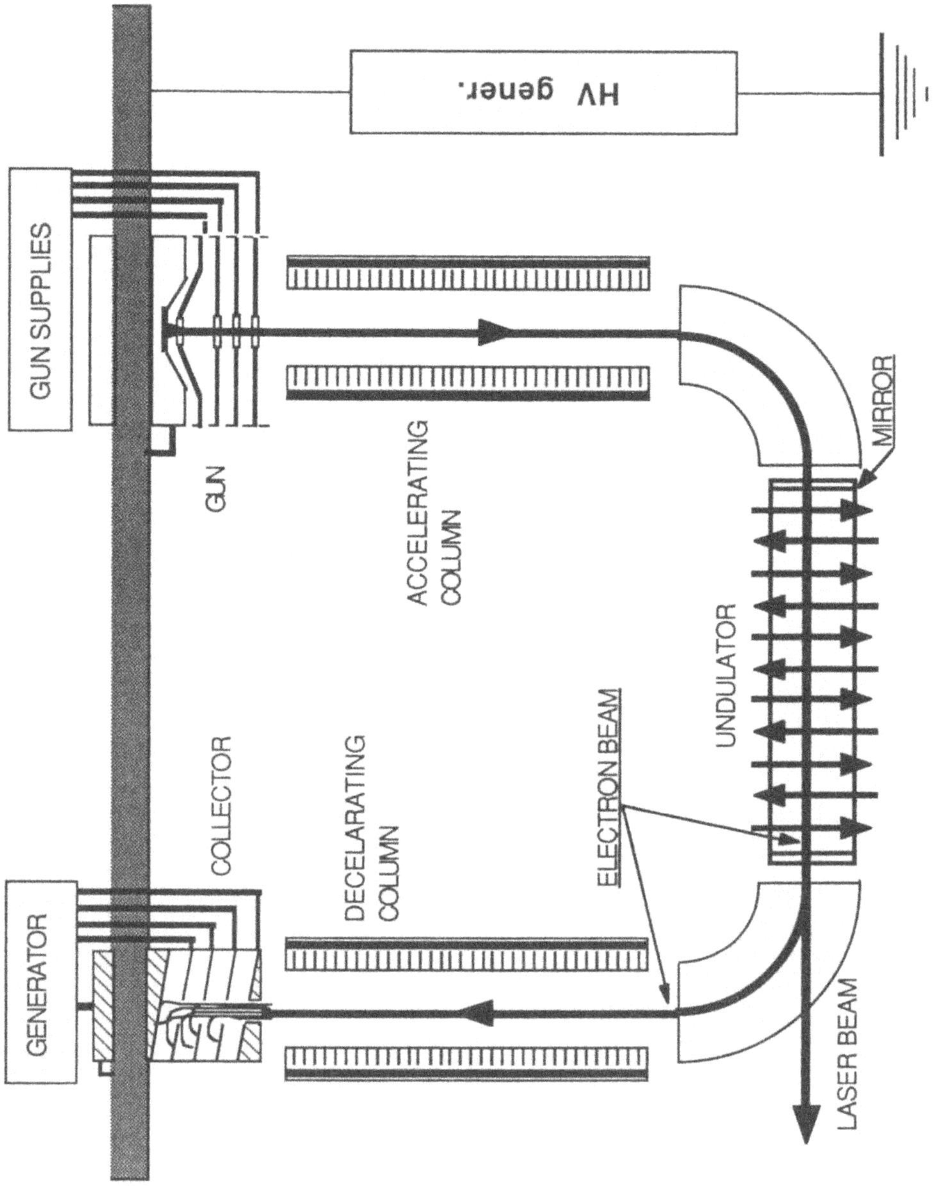

Fig. 1 Scheme of the FEL device with an electrostatic accelerator plus recovery system.

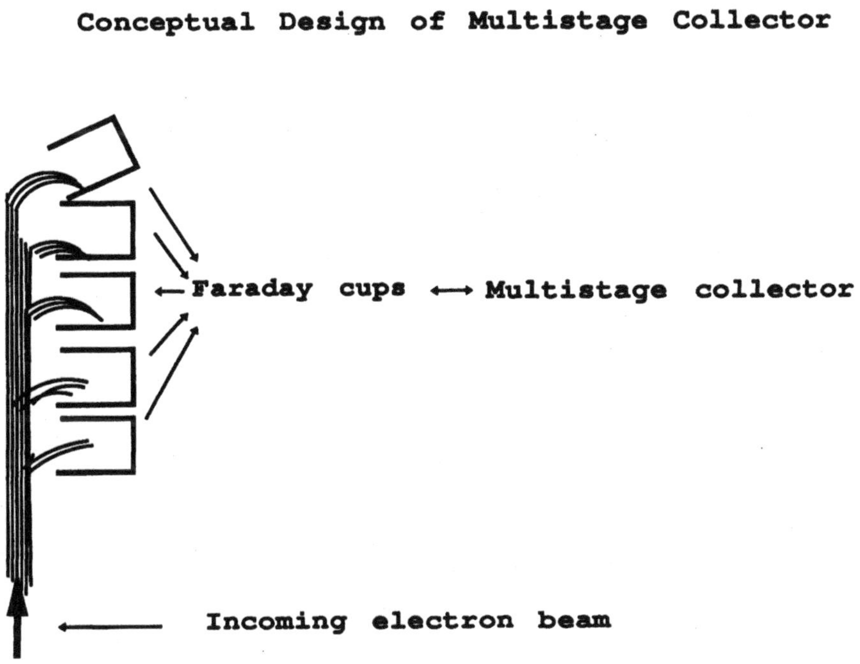

Fig.2 Conceptual Scheme of multistage collector.

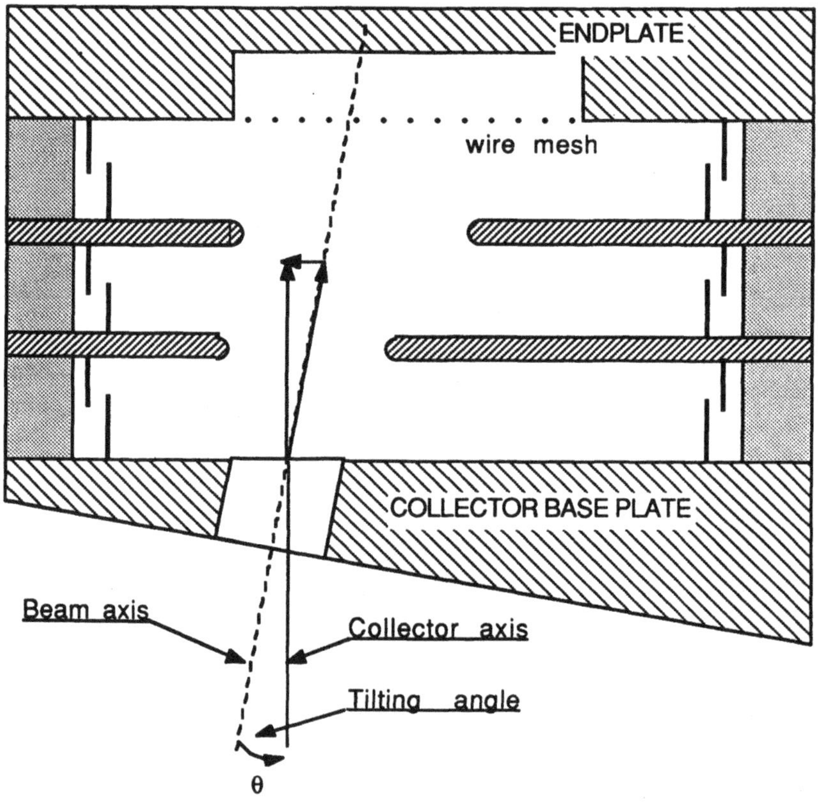

Fig.3 The UCSB multistage collector design . E_t and E_l are respectively the transverse and the longitudinal component of the electric field due to tilted electrodes.

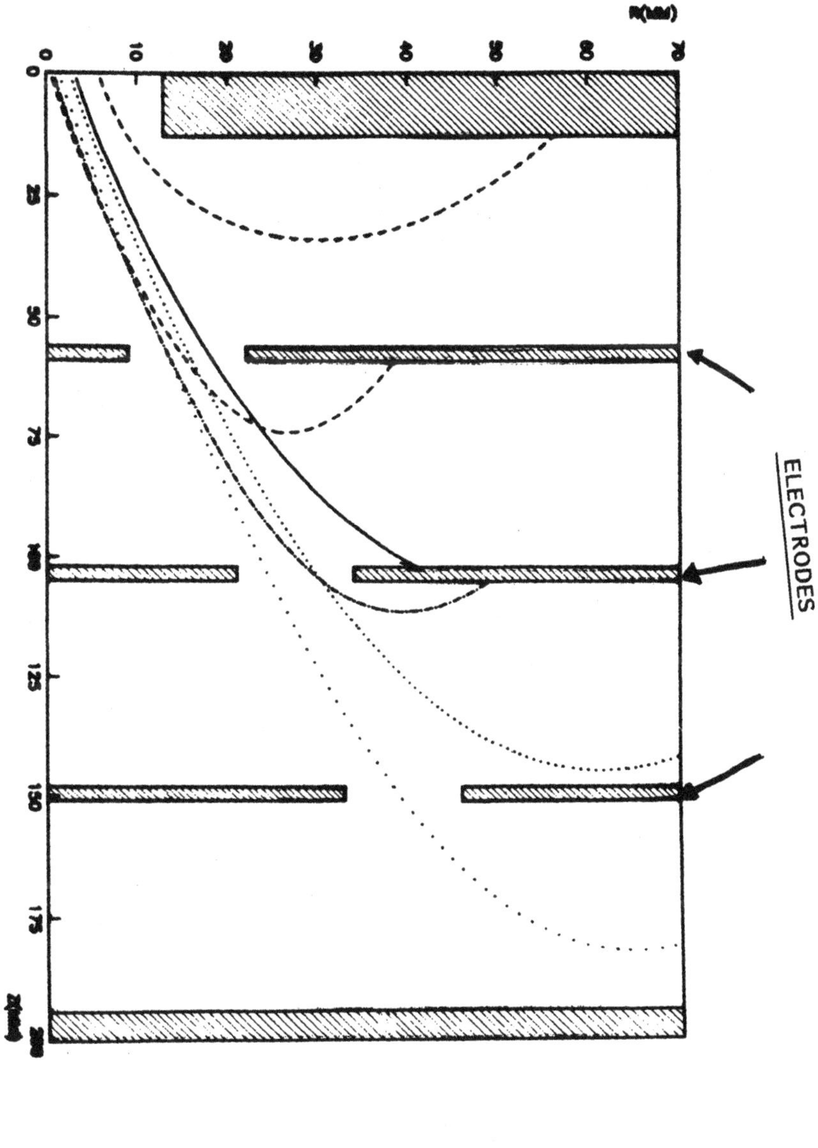

Fig.4 Computer simulations for electron trajectories in the UCSB multistage collector.

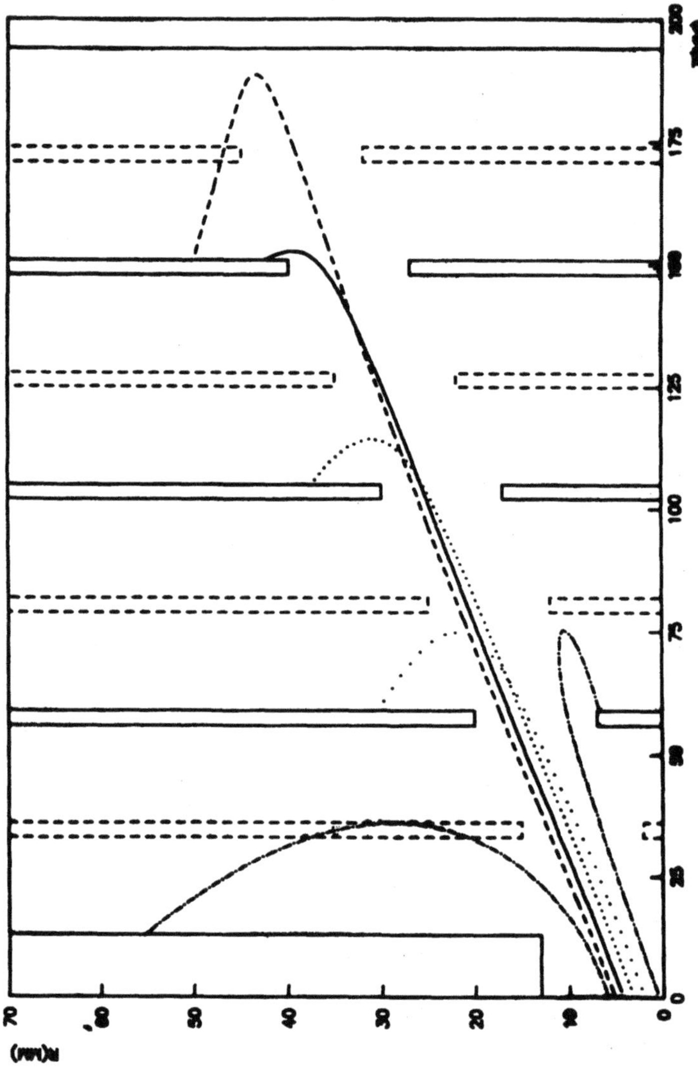

Fig.5 Electron trajectories in UCSB multistage collector with a transverse magnetic field.

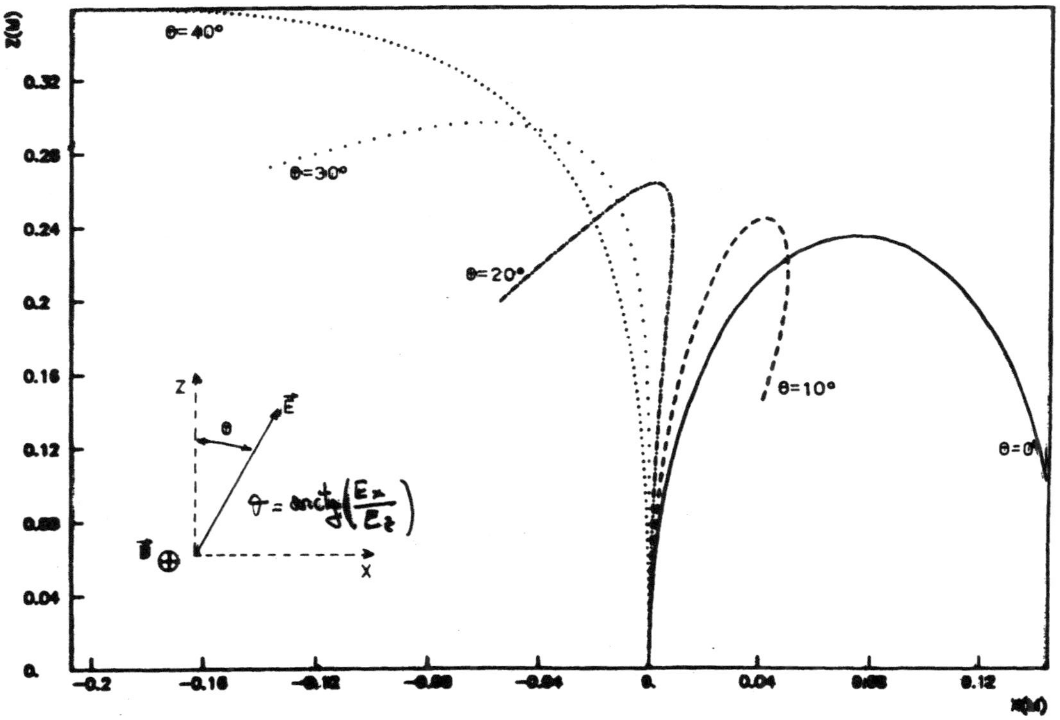

E = 0.8 kV/an; B = 5 G; particle energy: ΔW = 20 keV

Fig.6 Analytical electron trajectories for different tilting angle θ.

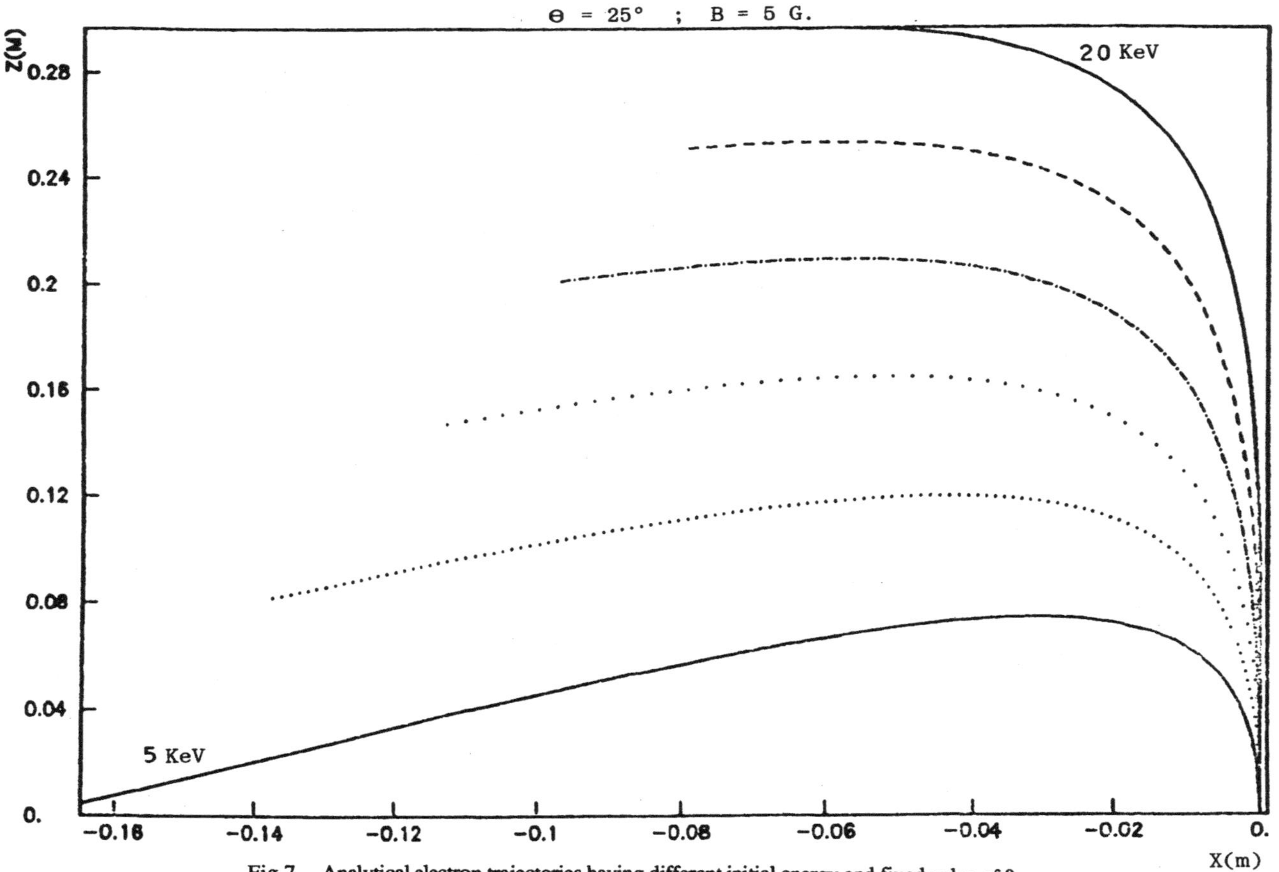

Fig.7 Analytical electron trajectories having different initial energy and fixed value of θ and B.

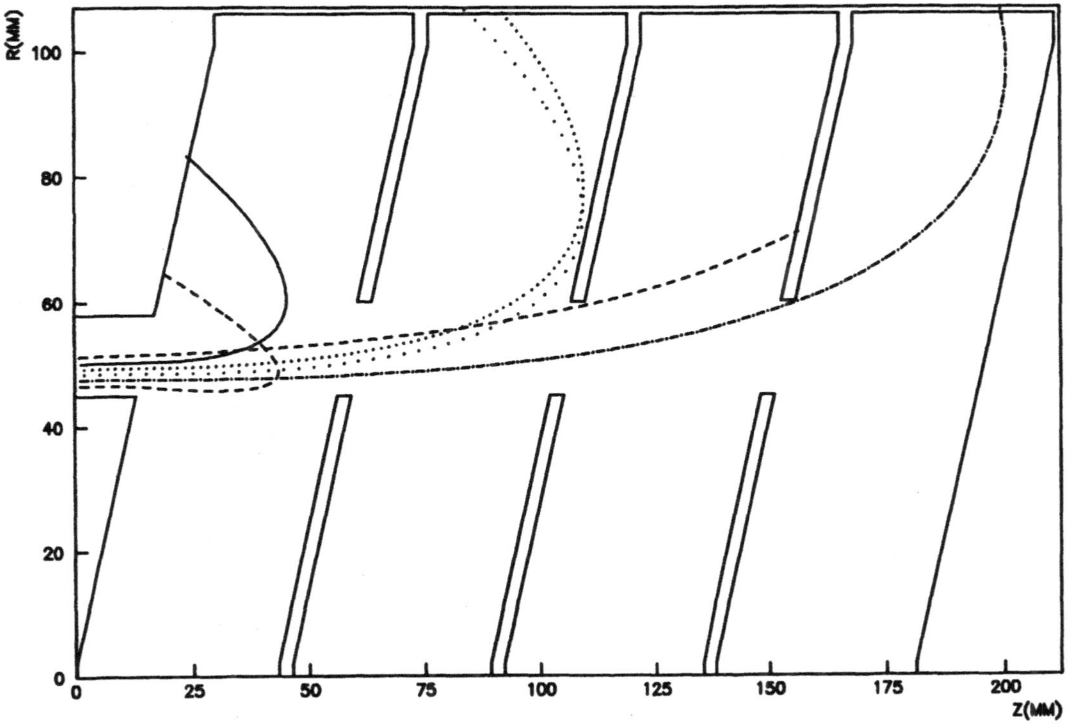

Fig. 8 a) Computer simulations of electron trajectories in the new multistage collector : (a) without transverse magnetic field,

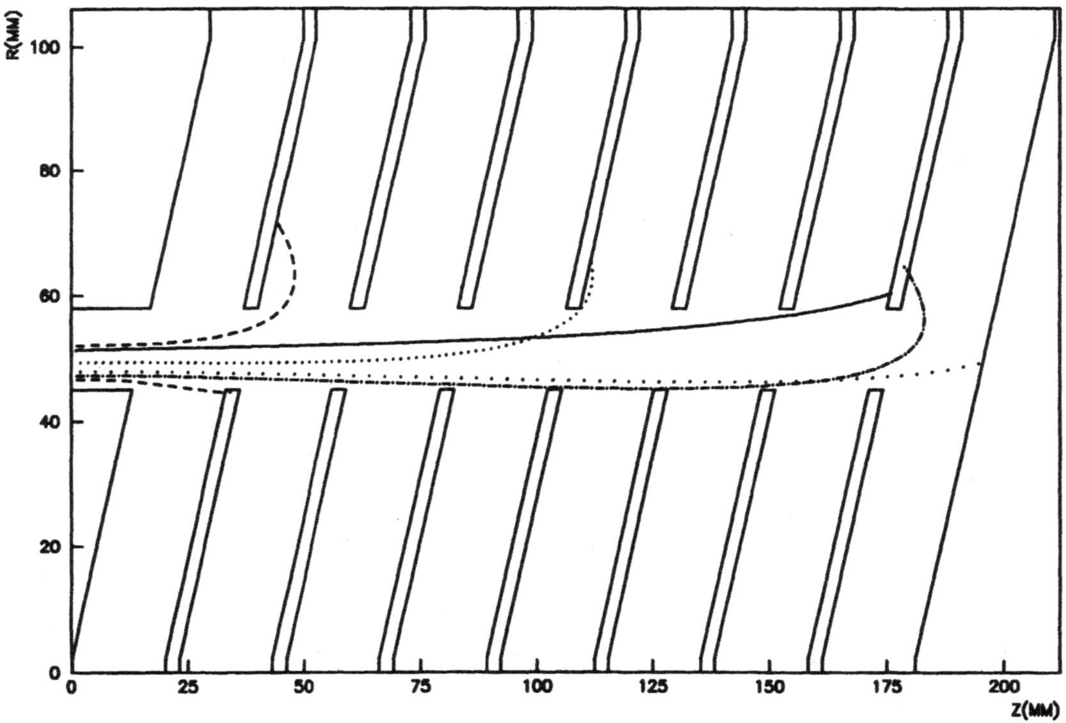

Fig.8b) Computer simulations of electron trajectories in the new multistage collector :

(b) with a transverse magnetic field of 5 Gauss.

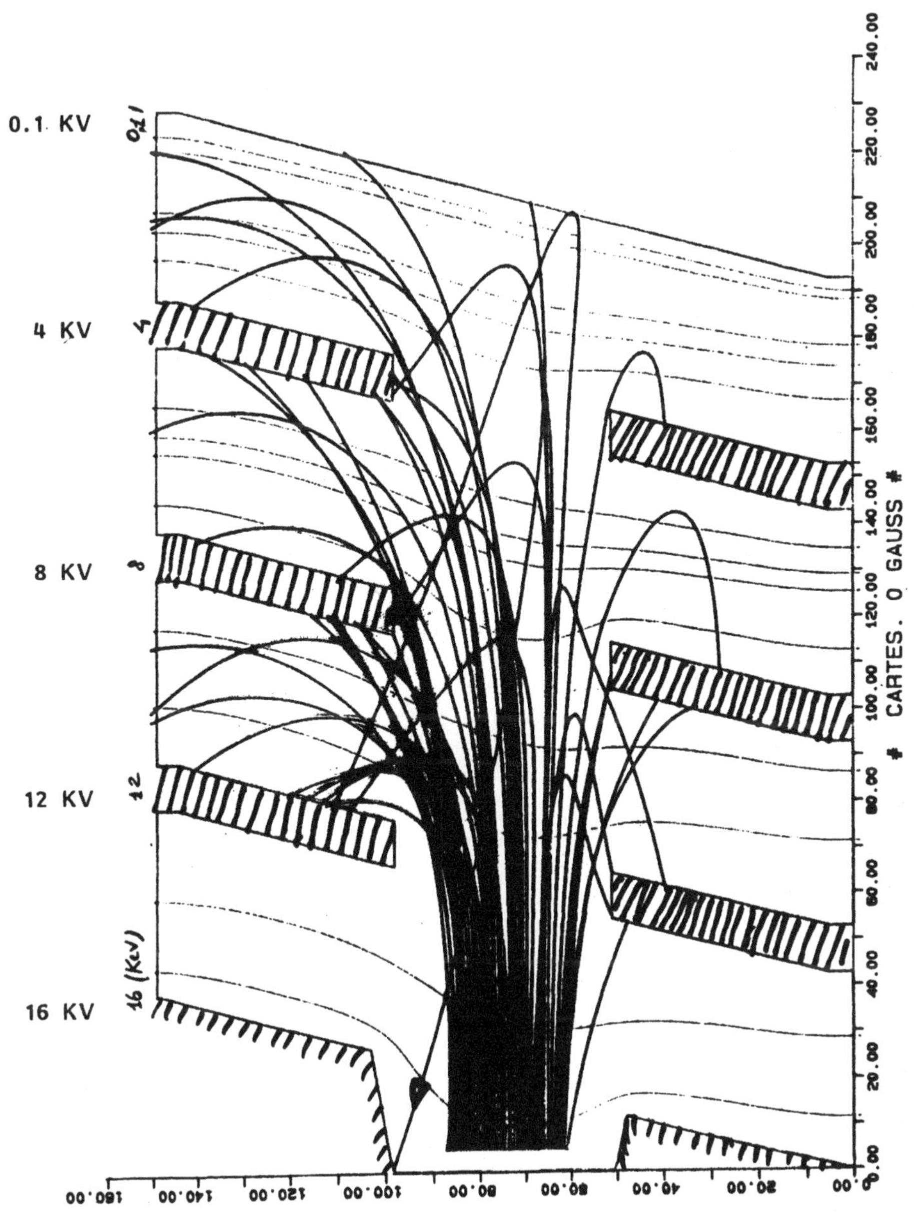

Fig.9a)The same simulations of fig.7 with more electron trajectories with space charge effect included.

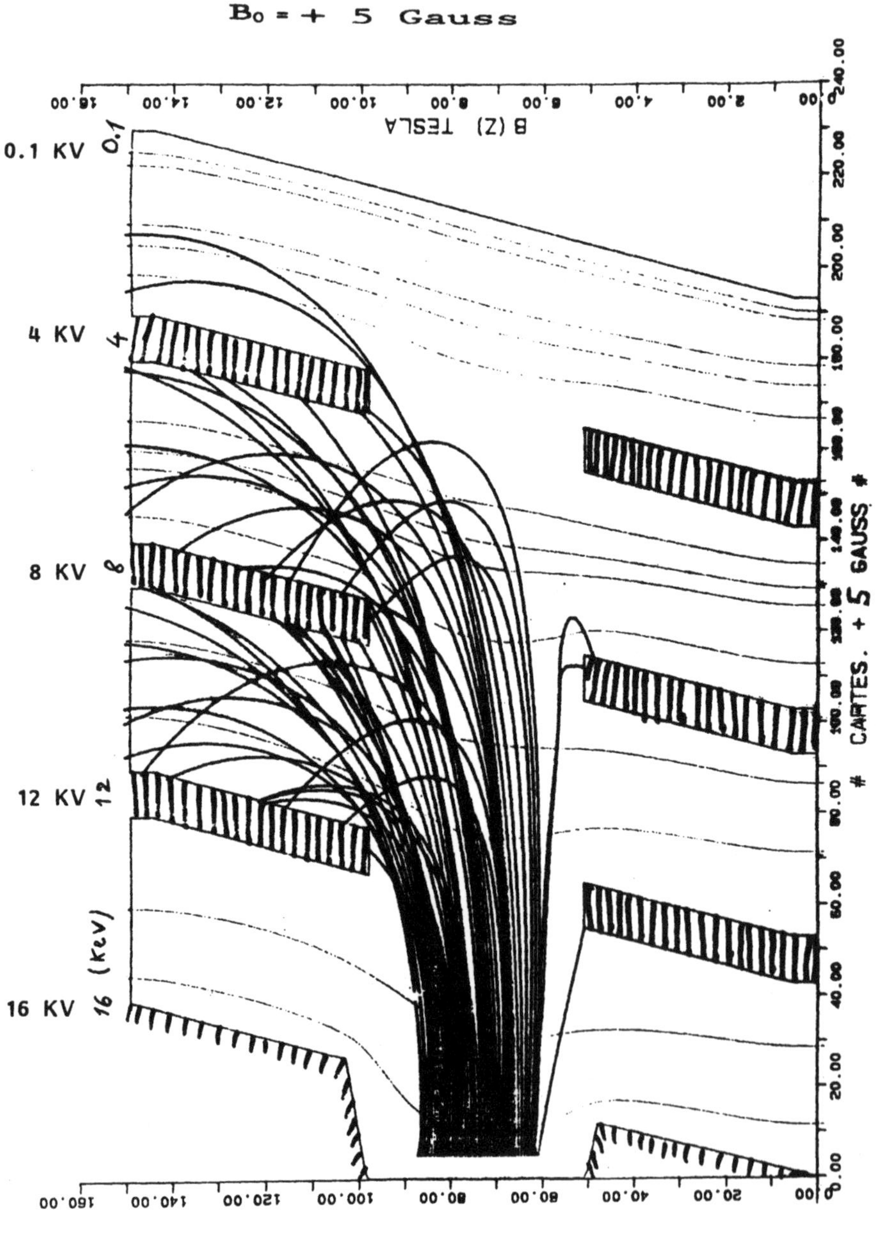

Fig.9b)The same simulations of fig.7 with more electron trajectories with space charge effect included.

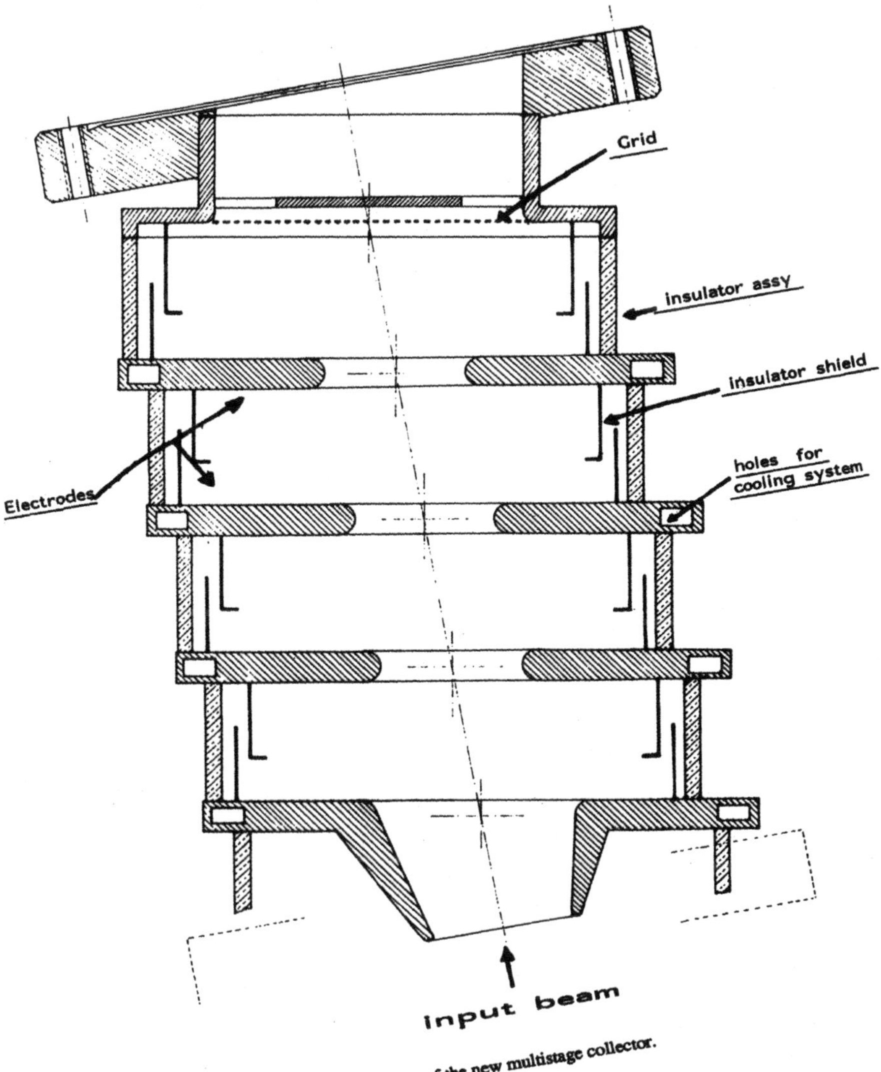

Fig.10 Mechanical design of the new multistage collector.

THE NEW ELECTRON BEAM COLLECTOR FOR LEAR

J. Bosser, R. Ley, and G. Tranquille

European Organization for Nuclear Research (CERN)
CH-1211 Geneva 23

V. Bikovsky, I. Meshkov, and A. Rogozin
V. Sinitsky, E. Syresin, and V. Funtikov

Centre of Applied Physics and Technology (CAPT)
Lipetsk, USSR
The Department of Novosibirsk Institute of Nuclear Physics
Novosibirsk, USSR

ABSTRACT

The present electron beam collector is derived from that used at ICE. It suffers from a poor electron recuperation efficiency and of a lack in reliability.
It has been decided to foresee the construction of a new collector which has been entrusted to CAPT at Lipetsk (USSR) under the authority of INP Novosibirsk.
This paper describes the desigh of the new collector taking into account the present environmental constraints and the need to insert a vacuum valve. First results of trajectory calculation are also presented.

1. INTRODUCTION

Since more than two years CERN has made the effective experience of an electron-cooler inserted in LEAR. The overall project, derived from ICE[1)], has been managed by a KfK. team with the collaboration of some CERN members.

Many reports have been published on the results obtained with this device [2,3,4)], they are promising.

Electron cooling is in itself a complicated process and implies much hardware and software development. To be fully used in LEAR, it cannot suffer from any lack in reliability. Amongst all the "sensitive parts" the collector is the most unreliable. Its sophistication and bad reliability during tests has not yet allowed the LEAR Electron-Cooler to be put into normal operation.

In consequence CERN, has decided to look for another collector. The design, construction and test has been entrusted to CAPT with the collaboration of CERN (essentially on mechanical and vacuum techniques and computer simulations).

2. SPECIFICATIONS FOR THE NEW COLLECTOR

Some specific requirements must be fulfilled for this new collector which has to be installed on the LEAR Electron-Cooler. First of all, the collector must provide an efficient collection of the electron beam on a surface with a potential close to that of the cathode. "Efficient" in this case means that potential difference between cathode and collector is as small as possible and simultaneously that the current loss due to secondary and reflected electrons from the collector do not cause an increase in vacuum pressure inside the storage ring due to the residual gas desorption under bombardment of the chamber walls by electrons.

The high level of losses occurring in the present electron cooling device at LEAR (see Table 1) produces monotonous worsening of the vacuum during the working cycle and it results in the instability of the parameters of the cooled antiproton beam and other undesirable effects. It seems true that the main reason of this high level of current loss is due to the specific arrangement of the electron collector[1,5], where the "reflecting electrostatic needle" is used. The other reason to reconsider the LEAR-Collector scheme is its bad behavior under high voltage, and therefore high current, environment [4,6]. Particularly, it is at present impossible to collect electron beams with an energy higher than 15 keV.

Table 1 - Main parameters of the electron-cooling device

PARAMETER		PRESENT SYSTEM	SYSTEM UNDER DEVELOPMENT
Electron energy	keV	15	35 (40)
Electron beam diameter	cm	5	5
Electron beam current	A	2.4	3.3 (4)
Collector potential	kV	1.5 to 2.5	up to 3.5
Repeller potential	kV	0.8 to 1.5	up to 3
Current losses	mA	7.5	0.3
Average vacuum at collector level	pTorr	50	$\leq$10

3. THE DESIGN

During the elaboration of the new collector design some practical considerations were taken into account. The main difficulty of the new collector construction is that it must be integrated to LEAR which is intensively used for physics experiments. This dictated some additional conditions for the design and construction. They are:

1) The overall time which can be devoted for design, construction and test of the new collector is very short -less than one year-.
2) The collector must be supplied together with a vacuum valve in order to dismount and remount it without any significant influence on LEAR. It is worth remembering that the baking procedure is tedious and risky; it happens no more than once a year.
3) Most of the power supplies of the "old" system should be re-used in order to simplify and reduce the necessary investment and work.
4) The present allowed "volume" is limited.

The first condition excludes any application of a sophisticated collector system like reviewed in Ref. 5 (pp. 37-38). As a result, the simplest collector scheme was chosen. It is similar to the one used for the first electron cooling device at NAP-M storage ring[7]. The water-cooled Faraday cup together with a suppressor (repeller) electrode form the main frame. Magnetic shielding to form magnetic field geometry is sufficient to obtain a homogeneous electron density distribution onto the collector surface.

To fulfill the second condition, and not to distort significantly the main solenoid longitudinal magnetic field, an additional coil is placed near to the collector entrance. It can be powered with the power supply already used for the present collector entrance coil, "BBC", at LEAR.

4. SIMULATIONS

The collector scheme is given in Fig. 1.

The computations have been done using a dedicated software package[8].

In Fig. 2 are presented:

- The electron trajectories. They take into account the effect of the space charge of the primary electron beam. The electron beam input energy is 20 keV.
- The equipotential lines. They show an electrostatic trap inside the collector Faraday cup, near to collector output for the secondary electrons. This should forbid the exit of secondary electrons with initial energy less than 350 eV.
- The corresponding magnetic field distribution on the axis (radius $r = 0$). One can see the existence of a magnetic mirror near to the collector entrance of the primary beam.

The magnetic field lines are plotted in Fig. 3. They take into account steel saturation effects. The magnetic shield is placed outside the vacuum chamber. The used software package named "Poisson" is standard at CERN. The entrance coil is fed by a current of 6.986 kA.

Many other computations have been made (and are underway) with other electrical potentials and magnetic fields.

If possible, one would try to simulate the trajectories of secondary electrons issued from the collector surface with different angles and energies.

A test bench is under construction at CAPT. It will be used later this year to test the performances of this project.

5. REFERENCES

1) Bell, M. et al., "Electron Cooling in ICE at CERN", Nucl. Instr. and Methods, Vol. 190, N2, 237-255 (1981).
2) Poth, H. et al., "First Results from Electron-Cooling Experiments at LEAR, Zeitschrift für Physik.
3) Wolf, A. et al., "Results from Electron Cooling Experiments at LEAR", Proc. of the 1988 EPAC Conference (Rome), Vol.1, 204-206 (1988).
4.) Bosser, J., Ley, R. and Tranquille, G., "Status of Electron Cooling Experiments at LEAR". This Workshop.
5) Poth, H., "Electron Cooling; Theory, Experiment, Application", CERN-EP/90-04, 36-38 and Fig. 22 (1990).
6) Poth, H. et al., "Further Results and Evaluation of Electron Cooling Experiments at LEAR", Nucl. Instr. and Methods, A287, 328-332 (1990).
7) Kudelainen, V., Meshkov, I. and Salimov, R., "The Electron Cooling Device", CERN 77-08, Part B, 63-64 (1977).
8) Herrmannsfeld, B., "Electron Trajectory Program", SLAC-226.UC-28.

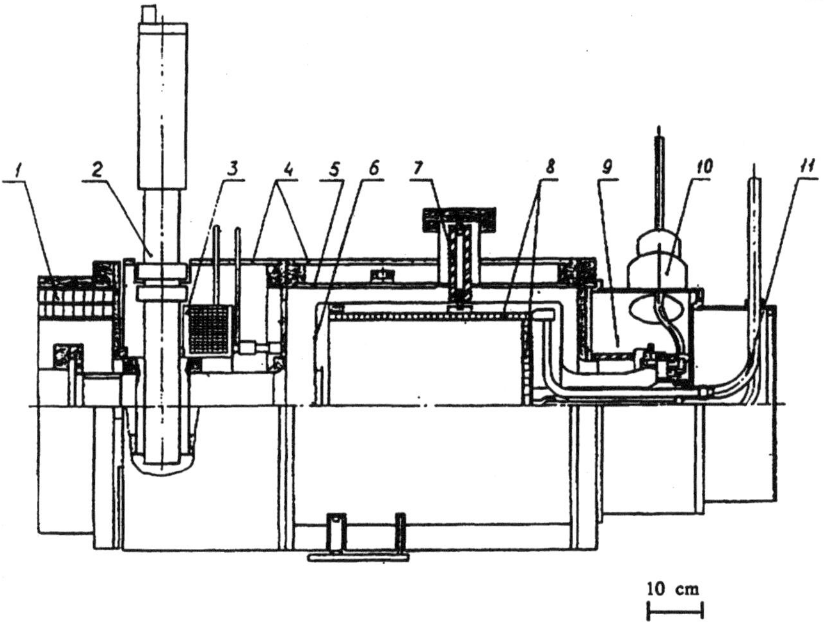

1 - solenoid
2 - vacuum valve
3 - compensation coil
4 - magnetic shielding
5 - vacuum chamber
6 - repeller
7 - supporting insulator
8 - collector
9 - vacuum insulator
10 - feed through
11 - water cooling tubes

Figure 1: The new collector scheme

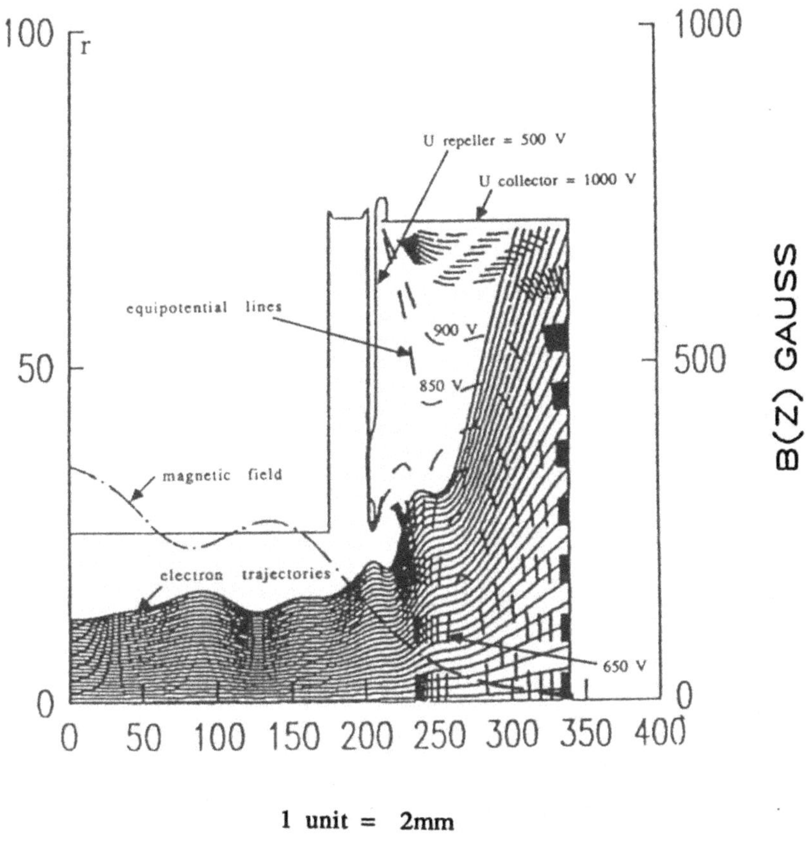

Figure 2: **The electron trajectories with the equipotential lines and the corresponding magnetic field on axis (I coil = 1500 A)**

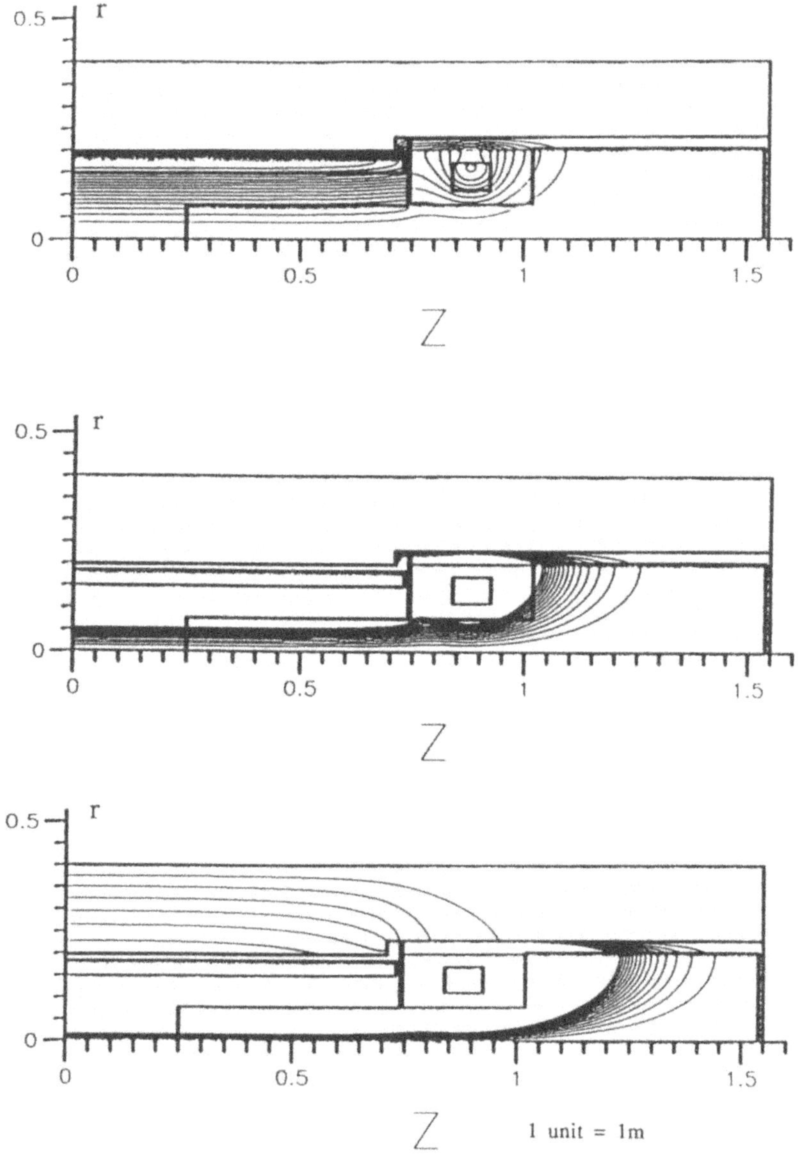

Figure 3: The magnetic field lines as a function of radius

The new Control System for the LEAR Electron Cooler

G. Tranquille
T. Pettersson

PS Division
CERN
CH 1211 Geneva 23
Switzerland

Abstract

The Low Energy Antiproton Ring (LEAR) electron cooler is designed to cool ion beams up to a momentum of 380 MeV/c. Until now it has been controlled via a standalone system comprising of a LSI 11/23 microcomputer with its associated CAMAC loop. For the device's full integration into LEAR and for future operational requirements, it has been decided to convert the whole control system to a LEAR based system. This paper describes the present system and the hardware and software changes necessary to integrate the cooler completely into LEAR. The operational aspects of running the cooler in the variable momentum range of the LEAR machine will also be discussed.

1. Introduction

The utility of electron cooling at LEAR has been demonstrated on a number of occasions [1] [2] [3], but until now its routine operation has been delayed for a number of reasons, one of them being the incompatibility of the present control system with that of LEAR. The present system (Figure 1) consists of a LSI 11/23 microcomputer which controls all the electron cooler parameters via a fibre-optic serial CAMAC loop. The fibre-optic link is necessary for the control and acquisition of parameters in the Faraday cage some 40 kV above ground potential.

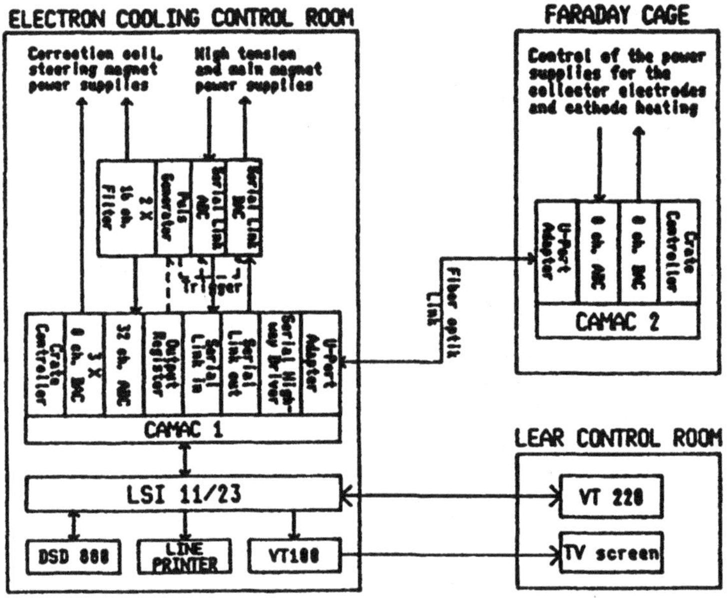

Figure 1: Hardware setup for the present control system.

To simplify the task of writing the controls software, it was decided to split the different tasks into separate programmes. The CAMAC interfacing is handled by two distinct tasks. The first acquires all the values, converts them into their appropriate format and updates a common partition in the computer's memory every second. These values are displayed on a VT220 terminal via a display programme (Figure 2 on page 3) which reads the common block acquisition values. A second task used for individual parameter setting has the highest priority of all the tasks and is invoked via an interrupt from an application programme. The programme waits for data concerning the parameter(s) to be modified to be transferred and then reads the appropriate new control values from the common partition.

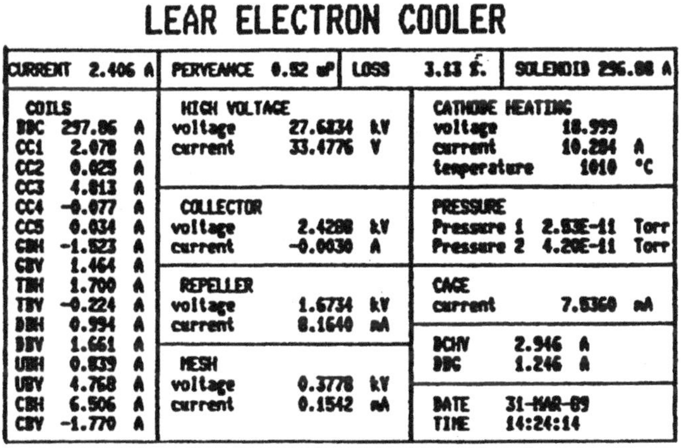

Figure 2: Display of the acqusition values on the present system.

Parameter control (Figure 3 on page 4) is handled by a task that enables the control at any time of up to four different parameters. All 'controlable' parameters can be hocked or unhocked from the control panel and can be changed directly, or in small increments or toggled with a previously stored reference value. Of course if the electron beam energy is changed or the main solenoid field is modified all the parameters scale automatically. A database editor is used for creating new parameters or modifying existing parameter characteristics (CAMAC address, LSB etc). A flow diagram showing how the different programmes interact with eachother is shown in Figure 4 on page 5.

The present system has performed well since the cooler's installation in the LEAR ring in 1987 but due to lack of computer memory the present system cannot be further extended. Any new application programmes can only be run with one or more of the previously mentioned tasks stopped. Moreover the fact that maintenance of the LSI/PDP series of computers has been discontinued has lead to a situation that any minor hardware problem on the computer could mean that cooler operations have to be stopped. For these reasons and also for reasons concerning the future operation of the cooler it was decided to convert the present system to a system compatible with the LEAR control system [4].

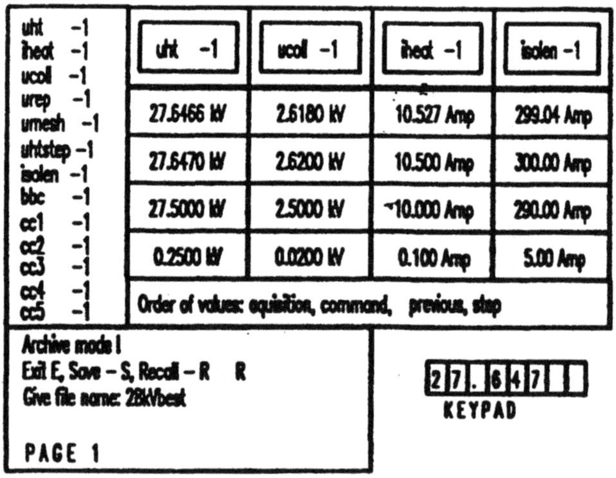

Figure 3: The panel for the parameter control programme.

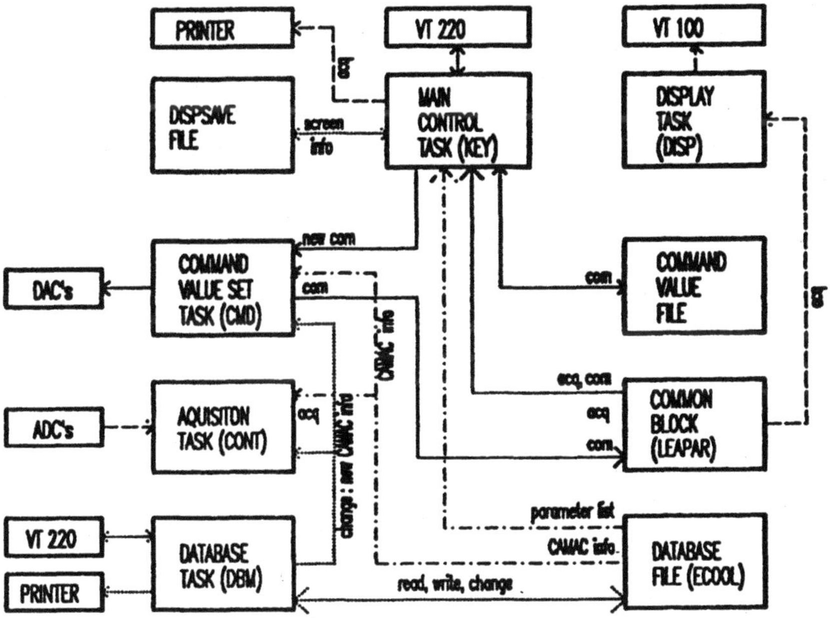

Figure 4: Flow diagram showing the programme interaction on the present control system.

2. The new control system

2.1 Hardware structure

The new system uses a workstation console and handles all CAMAC access through a dedicated microprocessor. For this reason an extra crate had to be inserted on the existing electron cooler CAMAC loop and at the same time a number of new modules were purchased for increased precision in both the control and the acquisition of the parameters.

Another crate was also added to the LEAR CAMAC loop for the installation of the new generation function generators known as GFD−GOCT [5],[6] (GOCT stands for Gfd OCtal Transceiver, GFD stands for Digital Function Generator) and a timing event decoder module. The GOCT device is the CAMAC interface module and the GFD the power supply interface module. The GOCT−GFD communicates over a serial link using industry standard communications integrated circuits. The GFD device contains a M68000 microprocessor that calculates the power supply ramp values as described by the downloaded function in real time. Two of the GFDs installed for the electron cooler are used to power the main solenoid and the two compensation solenoids while a third is loaded with the electron beam energy data on three different tables. All this is necessary for the 'pulsed mode' (Figure 5 on page 6) operation. The cooler solenoids are synchronised to the LEAR deceleration cycle and any orbit or tune corrections are applied at the standard LEAR flat–tops. When the beam has been decelerated onto a flat–top by the radio–frequency cavity and then debunched, the GFD containing the beam energy data is triggered on the appropriate table via the timing event decoder. This causes the high voltage power supply to ramp to the operational value in 100 msecs and 6 seconds later the cooling process is switched off by reducing the voltage to zero. The same process is applied at subsequent flat–tops until the machine reaches its final operational energy. It was shown in previous machine development sessions that some 15 minutes in cooling time could be gained in decelerating a beam to 105 MeV/c using this method instead of stochastic cooling.

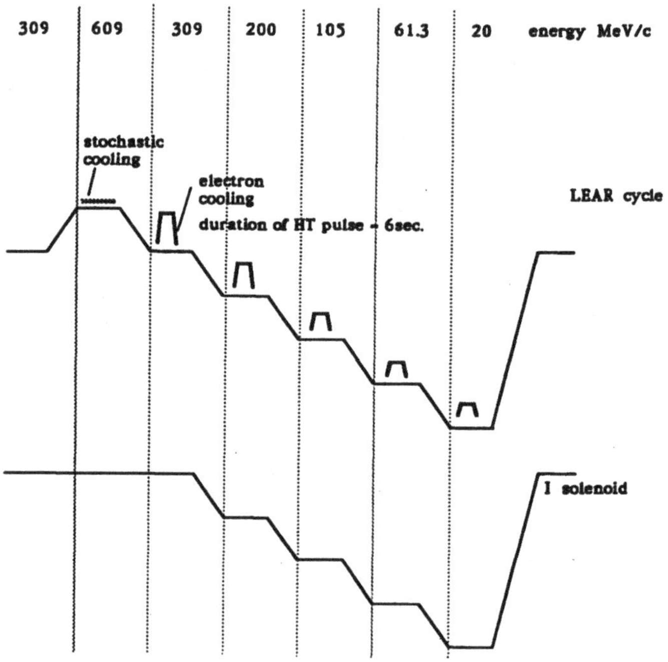

Figure 5: Principle of the pulsed mode of operation. The cooler solenoids are synchronised with the LEAR magnets throughout the whole deceleration process. On the flat–tops appropriate orbit and tune corrections are performed and the high voltage power supply is triggered to rise to the operational value and cool the circulating beam for 6 seconds.

2.2 Software structure

The control system is accessed via a limited number of subroutines which handle the interfacing to the different databases, the operating system, and the network. These environment routines cover the low level details of the operating system, details that the ordinary user does not wish to be confronted with when programming application.

The control system supports two types of process parameters :

- singled valued parameters that require a direct access for control and acqusition e.g. correction coil power supplies

- multiple valued parameters that are controlled from GFDs and can assume many different values in a given period of time e.g. the solenoid power supply when the cooler runs in the 'pulsed mode'.

All parameters of the cooler, be they single valued or multivalued parameters are handled in databases. Currently there is one database for single valued parameters, one for the GFDs and timing, and one for non–hardware related parameters such as pick–up calibration values or gun perveance data.

The serial CAMAC access is confined to a single board microprocessor acting as an intelligent controller with the single valued parameters as an intergrated live database. The access to any given parameter requires the execution of a series of CAMAC instructions followed by the conversion of the acquired data into engineering units. The normal user has no need to know how the data is transferred to/from his programme from/to the hardware as long as it is performed in the correct manner. The CAMAC controller is connected to a local area network, Ethernet, and any other computer in the same network may send requests to it.

2.3 Existing programmes

The acquisition and control of all the electron cooler's parameters are performed from one programme. As on the old control system, up to four parameters can be controlled simultaneously in exactly the same manner as before i.e. direct input, increment/decrement in steps, or toggle with a previously saved value. Scaling of all parameters is automatic but a provision has also been provided to bypass the scaling of parameters. This can be extremely useful if one wishes to step the electron beam energy away from the circulating ion beam without perturbing the magnetic settings, or if one wishes to see the effect of the longitudinal magnetic field on cooling. Figure 6 on page 8 shows the display screen for the control programme and Figure 7 on page 9 shows how the individual parameter control panel is set up.

In addition programmes have been written for the remote control of the security synoptics and for the inversion of some of the correction coils power supplies. All the programmes are activated via a touch panel on the workstation display using a mouse. As all the workstation consoles are also connected to the local area network, it is possible, from the electron cooling equipment building, to control and acquire any LEAR machine parameter. This is most useful in the setting–up phase where we can acquire parameters such as the beam intensity, the beam orbit or the Schottky scans Figure 8 on page 10 through the computer without having to run back and forth between the LEAR control room and our equipment building.

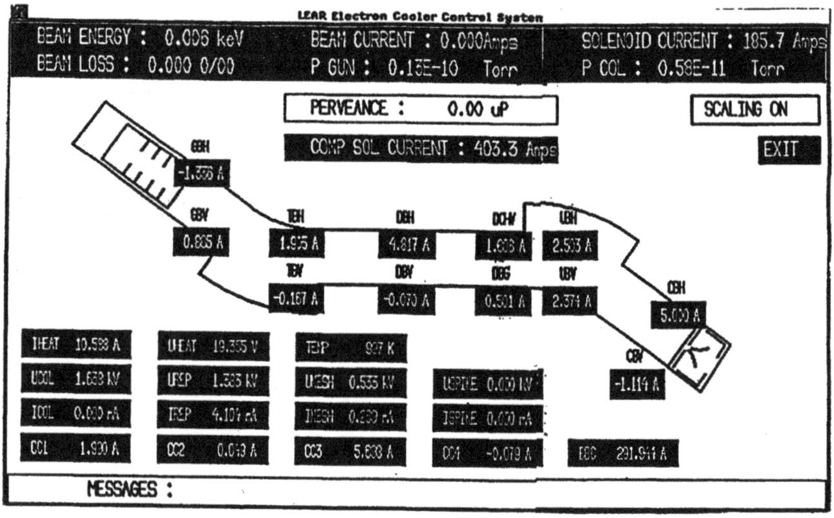

Figure 6: Display of the new control programme running on the workstation consoles. The parameters are represented in a synoptic form and can be activated by using the mouse device. Values are refreshed approximately every second.

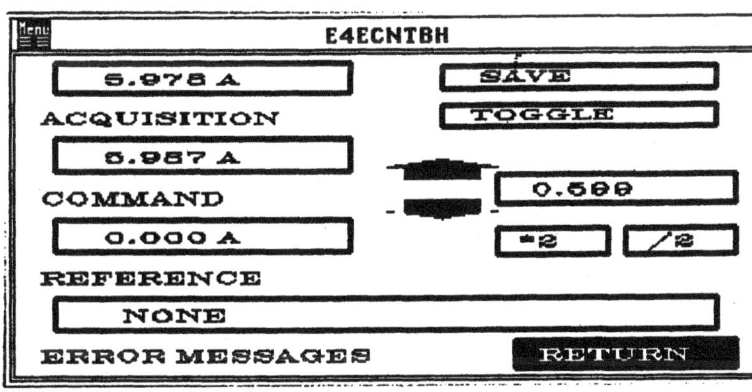

Figure 7: The parameter control panel. Values can be modified directly (by clicking in the command value box), or in increments (by clicking on the up/down arrows, or by toggling between the current command value and a saved reference value (by clicking in the toggle box). New increment values can be entered and command values can be saved to a reference buffer.

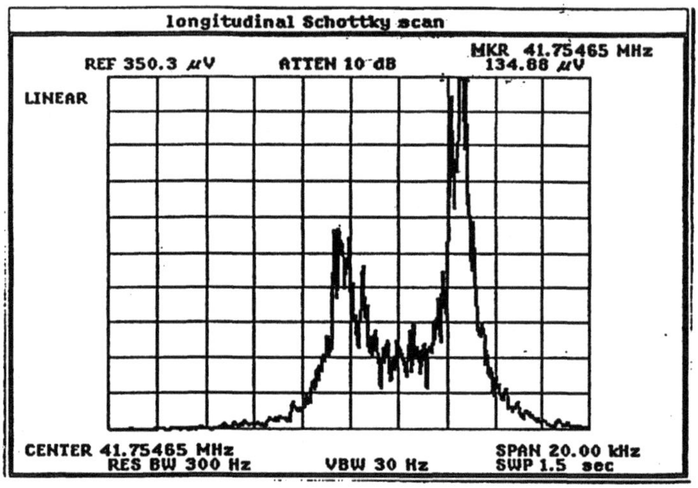

Figure 8: Computer display of the longitudinal Schottky scan of an electron cooled anti-proton beam. The spectrum analysers displaying the Schottky signals can be acquired via a CAMAC/HPIB interface module. Data transfer is normally of the order of a second and remote setting of individual analysers is also possible.

3. Conclusions

We have shown that the electron cooler control system can be completely modified using existing hardware and software structures. The new control system should be fully operational this Summer which will enable us to put the cooler into operation for momenta below 200 MeV/c. Due to the increased capabilities of the workstation based system we can now expect a variety of diagnostics programmes (electron trajectory measurements, longitudinal frition force measurements etc) to be written in a relatively short period of time in order to investigate the physical mechanisms of electron cooling in more depth.

4. Acknowledgements

We would like to thank J. Bosser and R. Ley for constructive discussions on the new control system, and to G. Molinari and F. Ollenhauer for their help in some of the hardware installations.

References

1. H. Poth et al., First Results of Electron Cooling Experiments at LEAR, Zeitschrift für Physik A332(1987) and A. Wolf et al., Results From Electron Cooling Experiments at LEAR, Proc. EPAC88, Rome 1988 and J. Bosser et al., Electron Cooling Tests at LEAR, CERN/PS/AR/ Note 88 – 21

2. H. Poth et al., Further Results and Evaluation of Electron Cooling Experiments at LEAR, NIM A287 (1990) and J. Bosser et al., Summary of the MD on Comparative Measurements of Electron Cooling with Protons and Antiprotons CERN/PS/AR note 89 – 6

3. J. Bosser et al., Status of Electron Cooling Experiments at LEAR, this workshop

4. T. Pettersson et al., The LEAR Control System – Phase III, Proc. IVth Workshop on Physics at LEAR with Cooled Low Energy Antiprotons, Villars – sur – Ollon 1987

5. P. Lienard, The LEAR GFD, CERN/PS (to be published)

6. E. Tanke, The LEAR GOCT, CERN/PS/LI/Note 87 – 04

AN ULTRA HIGH VACUUM SYSTEM FOR THE LEGNARO COOLER

A. Aragona, V. Chimenti, A. Clozza, V. Lollo, G. Serafini,
L. Tecchio*, G. Turchetti.

Istituto Nazionale di Fisica Nucleare
Laboratori Nazionali di Frascati
Via Enrico Fermi 40, 00044 Frascati
ROMA

* Istituto di Fisica Superiore, Università di Torino, Italy
* Istituto Nazionale di Fisica Nucleare, Sezione di Torino, Italy

ABSTRACT

An ultra high vacuum system, using a non conventional sputter ion pump, will be described. It was developed for the Legnaro cooler. The working pressure, that it is possible to reach, with this pumping system, in the interaction region, is of the order of 10^{-11} mbar. It will be assembled and tested at Legnaro laboratories.

1 INTRODUCTION

Ion accelerators are of increasing interest today in the world. Electron coolers [1)] are generally used to improve ion beam characteristics. As known a cooler device is essentially a monoenergetic electron beam interacting with an ion beam to compensate its blow-up [2)]. The main goals for these devices are: electron beams of very good quality and very low operating pressure in the interaction section.

Here we are going to describe an ultra high vacuum system developed for the Legnaro cooler.

2 UHV SYSTEM.

2.1 General Description.

The Legnaro cooler is shown in figure 1. The vacuum chamber is mainly composed by an electron gun, an accelerating section, an interaction chamber, a decelerating section and an electron collector.

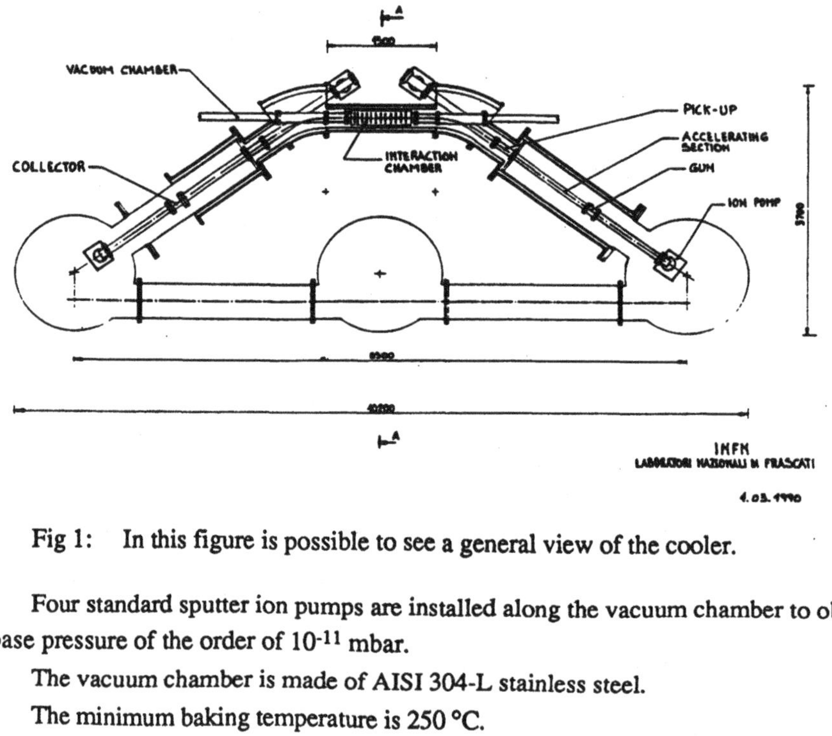

Fig 1: In this figure is possible to see a general view of the cooler.

Four standard sputter ion pumps are installed along the vacuum chamber to obtain a base pressure of the order of 10^{-11} mbar.

The vacuum chamber is made of AISI 304-L stainless steel.

The minimum baking temperature is 250 °C.

A cryopump will be used as fore-vacuum pump and during baking.

Finally a dry roughing pump is foreseen as primary vacuum pump.

Standard Conflat flanges are generally used. In some cases custom flanges with elicoflex gaskets are used.

The vacuum chamber will be placed inside a solenoid in order to reduce the electron beam transverse dimensions.

The main working parameters of the cooler are shown in table 1:

PARAMETER LIST

Beam energy	700 keV
Beam current	3.5 A
Beam diameter	3.5 cm
Working pressure	10^{-11} mbar
Outgassing rate of electron gun	$\leq 4\times10^{-7}$ mbar l s^{-1}

Table 1.

Our main goal, as said before, is to reach a pressure in the range of 10^{-11} mbar in the interaction section in working condition, i.e., when the electron gun is on and the electron beam circulates in the cooler.

For the base vacuum system we have two possible choices for the pumping device in order to obtain the required pressure value :

1) Sputter ion pump.
2) Getters:
   a) NEG (non evaporable getters)
   b) Titanium sublimation pumps.

In our apparatus we have chosen a base conventional vacuum system composed by 4 conventional ion pumps (400 l/s each) and 2 titanium sublimation pumps.

Now there is a problem: how is it possible to match together the high outgassing rate of the electron gun when it is working (about 4×10^{-7} mbar l s^{-1}) and the very low operating pressure value in the interaction section ($p=10^{-11}$ mbar)?

A possible solution for this question is the installation of an extra high speed pumping system very close or, better, inside the interaction chamber.

For this solution there are two possible choices: NEG or non conventional sputter ion pumps

There are advantages and disadvantages for both solutions of the problem. The NEG pumps are simple to install inside the interaction chamber but their working requires an activation at high temperature (about 700 °C in some cases). So, if the NEG pumps are not enough thermically shielded they can induce an high outgassing rate, from the vacuum chamber walls, that must be removed by the base vacuum system. The sputter ion pumps are mechanically more complicated and require an external magnetic field to operate, but they don't induce thermal load and the electric field can be easily shielded. Under these considerations, our choice was to design and to install a system of sputter ion pumps which works with the magnetic field generated by the central solenoid. In figure 2 is shown a drawing of a prototype of this kind of pump.

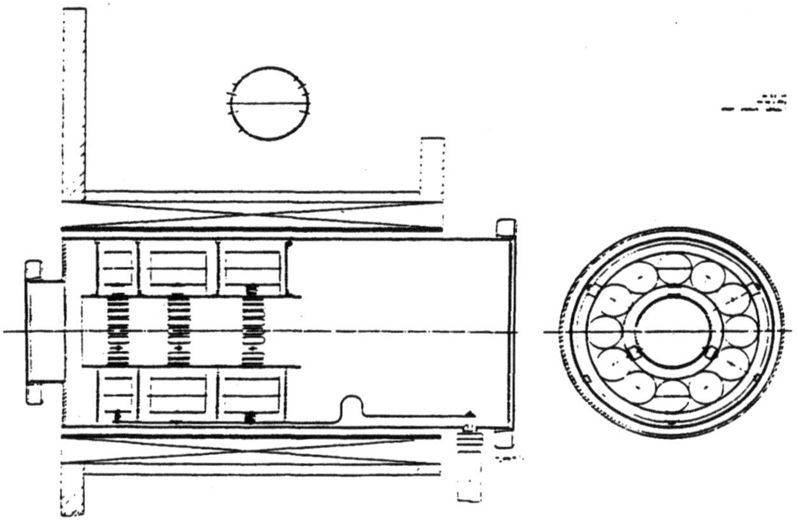

Fig 2: In this figure is shown a schematic drawing of the pump. that will be installed in the interaction section. The three elements forming the pump are, from left to rigth: A, B and C respectively.

Taking in mind that the outgassing rate of the electron gun, in working condition, is about 4×10^{-7} mbar l s^{-1} and that the conductance of the vacuum chamber between the gun and the interaction section is small (about 5 l/s), it is possible to reach the desired pressure value using, in the interaction region, a sputter ion pump which pumping speed is about 500 l/s.

This pumping speed must be valid for pressures down to: $p=10^{-11}$ mbar.

2.2 Dimensioning The System.

It is well known that a sputter ion pump works on the basis of the Penning discharge; see figure 3.

The parameters that influence ion pump behavior are: the magnetic field, the electric field and the geometry of the cell.

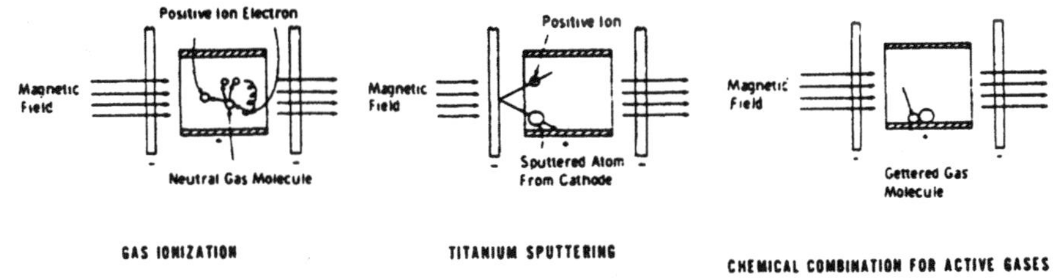

Fig 3: In this figure is outlined the principle of working of a sputter ion pump.

Once fixed the values for magnetic field B and electric field E, pumping speed depends on the diameter of the cell, taking in mind that the length of a cell is normally taken 1.5 times its diameter.

The following formulas 3) give the pumping speed and the minimum working pressure for an ion pump:

$$S = 2.5 \times 10^{-6} a\, V^{1/2} \left(B\phi - \frac{3.6 \times 10^{5}}{B\phi} \right) \left(1 - e^{-2.5\phi} \right) \qquad (1)$$

$$p_{min} = \frac{5 \times 10^{4}}{V \phi^{3} \left(B\phi - \dfrac{3.6 \times 10^{5}}{B\phi} \right)^{3}} \text{ torr} \qquad (2)$$

where: S = pumping speed (l/s), a = length of a cell (cm), V = anode potential (V), B = magnetic field (gauss) and ϕ = diameter of a cell (cm). In figure 4 are plotted the pumping speed for a single cell (fig. 4 left panel) and the minimum working pressure (fig. 4 rigth panel) versus the diameter of the cell.

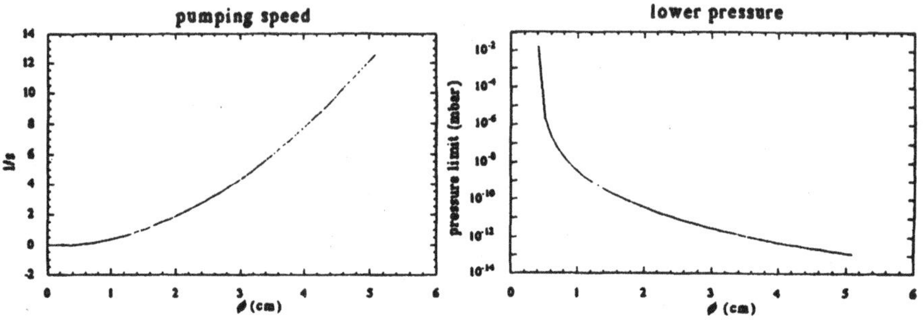

Fig 4: In this figure are plotted the pumping speed (left panel) and the lower pressure (rigth panel) versus the diameter of the cell. The values refere to a single cell.

According to previous formulas and experimental works [4], we have realized an apparatus for testing 3 kinds of cells with the same diameter and different length. The diameter of the cell was chosen so large in order to obtain an high pumping speed at pressures down to 10^{-11} mbar.

The table 2 shows the dimensions of the cells for the pumps A, B and C:

CELL DIMENSIONS (in mm)

	A	B	C
a	40	70	70
b	7	15	7
ϕ	45	45	45

Table 2

where: a is the length of a cell, ϕ is the diameter of a cell, b is the distance between the cell and the cathode; all dimensions are in mm.

2.3 Characterization.

We have measured the pumping speed, using the method of known conductance, as a function of magnetic field and pressure, having fixed high voltage supply of about 7 kV.

We have also paid attention to the behavior of the pumps after repeated vents of the vacuum system, with dry nitrogen or air, till atmospheric pressure.

In figure 5 are plotted the pumping speed of the pump A versus magnetic field strength (fig. 5 upper left panel), the pumping speed of pump A versus pressure at different values of the magnetic field (fig. 5 upper rigth panel) and the pumping speed of the pumps A, B and C versus pressure (fig. 5 lower panel).

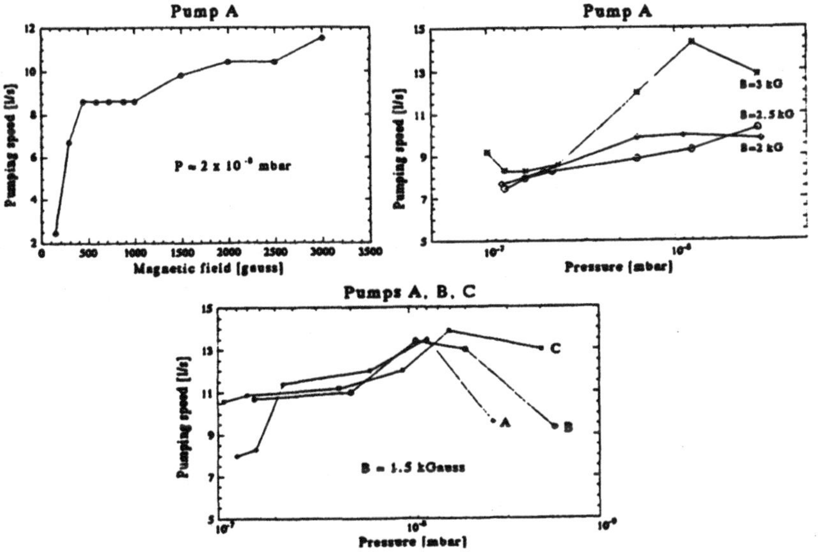

Fig 5: In this figure are plotted the pumping speed versus magnetic field, at a pressure of $2x10^{-8}$ mbar, of the pump A (upper left panel), the pumping speed of the pump A versus pressure for different values of the magnetic field (upper rigth panel) and the pumping speed of the pump A, B and C versus pressure (lower panel).

The behavior of the pumping speed versus magnetic field is characterized by an edge around the value of about 500 gauss. Below this limit the pumping speed falls suddenly to very low values, above the edge the pumping speed doesn't change significantly, maintaining a value, for unit cell, in the range of 10 l/s at a pressure of about $2x10^{-8}$ mbar.

The values of the pumping speed versus pressure, at different intensity of the magnetic field, show an almost constant trend for the values of magnetic field B=2kG and B=2.5kG, while for the value of B=3kG it is possible to see a raise of the pumping speed at a pressure of about 10^{-8} mbar. However, in all cases the pumping speed, for unit cell, at a pressure of 10^{-8} mbar, is maintained at a value of about 10 l/s or more.

Looking at the behavior of the three different type of cells it is possible to see that the pump A shows a non constant trend of the pumping speed. For the pumps B and C the pumping speed at pressures between 10^{-7} mbar and 10^{-8} mbar is similar, while for pressures down to 10^{-9} mbar it seems that the pump C shows a better value of the pumping speed.

Further tests, consisting in repeated venting of the vacuum system to atmospheric pressure both with dry nitrogen and with air, have pointed out the better behavior of the pump B both in starting and in limit pressure.

So we have chosen for assembling our pumping system the B geometry of the cells. Taking in mind that each segment of a pump is composed by 12 cells, it is possible to reach a pumping speed of the order of 500 l/s, at a pressure of 10^{-8} mbar, just assembling 5 segments together to make a pump.

3. CONCLUSIONS.

Pumping speed of our system seems to be more than sufficient to achieve electron cooler design requirements ($p = 10^{-11}$ mbar in the interaction region).

Moreover, pumping speed seems to depend significantly on the treatment done before starting the pump. Venting with dry nitrogen and argon Penning discharge seems to give better results. This work is now under development.

With our prototype of the pumping system, after an adequate baking to 250 °C, we have reached a limit pressure of: $p = 10^{-11}$ mbar.

REFERENCES

1 G. Y. Budker, At. Energy 22, 346 (1967)

2 H. Poth, "Proc. CERN Accelerators school, Advanced Accelerator Physics", Oxford 1985, ed. S. Turner, Geneva 1987

3 M.D. Malev and E.M. Trakhtenberg, Vacuum 23, 403 (1973)

4 S. L. Rutherford, Varian report VR-27, 1 (1964)

POLARIZATION OF PROTONS OR ANTIPROTONS USING A BEAM OF POLARIZED ELECTRONS

M. Conte
Dipartimento di Fisica-Uni. Genova and INFN-Genova,
A. Penzo
INFN-Trieste,
A. Pisent
INFN-Lab.Naz. Legnaro (Padova),
M. Pusterla
Dipartimento di Fisica-Uni. Padova and INFN-Padova

Abstract

The possibility of polarizing a stored beam of antiprotons by means of an intense beam of polarized electrons is here investigated. The polarizing mechanism is discussed and a numerical example for a proton energy of 92 MeV is produced.

1. Introduction

The production of a beam of polarized antiprotons is interesting to perform experiments in machines like LEAR or at higher energies, and different methods have been proposed to achieve this polarization[1]). In particular experimental tests on protons are now under way at the Indiana Cooler Ring based on the principle of the Spin-Splitter [2]).

In this paper we sketch an alternative idea, based on the developement of intense sources of polarized electrons [3]). Antiprotons circulating in a storage ring can be polarized longitudinally in the presence of a magnetic field of adequate intensity, parallel to their trajectories, if they move together with a beam of electrons of the same velocity and polarized in the same direction. All arguments developed in this note hold in the same way for protons that shoud be used for a test of the principle.

The lay-out of the system is shown in fig.1. In the ring is introduced a solenoid where protons and electrons together cover common straight paths along the axis. As it happens in the electron cooling devices both beams must converge to the entering point of the solenoid and separate from its end in such a manner that the electrons can be recovered on electrons beam line (and possibly recycled) whereas the protons continue their motion within the storage ring.

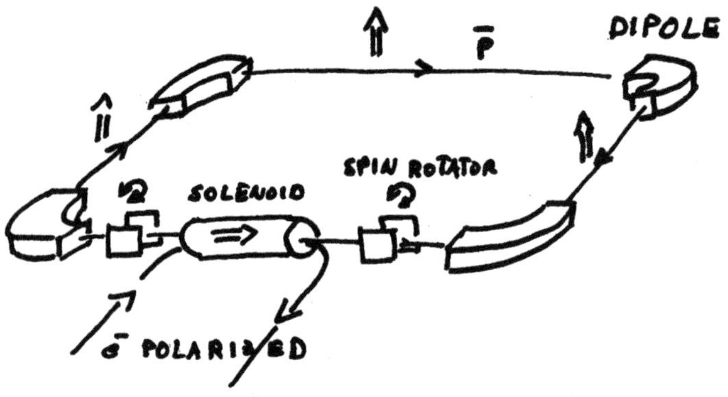

Figure 1: Lay-out of the system: ⇑ is the direction in which the polarization of p's is built up.

During the process of building up of the antiproton beam polarization the usual depolarizing effects that occur in a ring should be avoided. For this reason one requires to the spin equilibrium direction $\vec{n}$ to be axial in the solenoid. This condition is fulfilled by choosing an energy such that $a\gamma = 2$ ($a = 1.79$ is the proton syromagnetic anomaly, γ is the relativistic factor), or by introducing the two rotators shown in fig.1 to maintain free the choice of the electron energy.

The polarizing apparatus is completed by the presence of a RF system that sends monocromatic radiation of the desired frequency to a well defined region of the solenoid: this radiation is the necessary fundamental ingredient that allows the polarization of the beam by inducing magnetic dipole transitions on the electron-proton pairs (see below). A possible realization of this apparatus will be described later in this note.

2. Dynamical Polarization

To calculate the polarization of the protons (antiprotons) in a configuration of the type described in the previous section, one must take several parameters into account:

a) the temperature of the electrons;

b) the transition probability and the life-times of the coupled spin states of the system electron-proton as excitations occur by resonant absorption of a photon coming from a microwave cavity; this mechanism appears quite similar to the "dynamical polarization" that is used to produce polarized targets[4]): the main difference staying in the fact that one deals in one case with unbound state of electron proton system, while in polarized targets the electron proton pairs are always cosidered bound states. One necessarely requires that the electron and the proton are at atomic distances ($\simeq 10^{-8} cm$).

The physical mechamism of the "dynamical polarization" follows immediately. In fact we have to consider that each pair electron-proton with spin $\vec{S}$ and $\vec{I}$ respectively is subject to several interactions due to their spins:

a) the electron Zeeman's effect in the external field B_0,

$$H_e = -g\mu_e \vec{S} \cdot \vec{B}_0, \qquad [1]$$

with g=2 and $\mu_e = -0.93 \times 10^{-20}$ erg/gr. If we consider the Oz axis in the direction of the field $\vec{B}_0$ we have the two levels:

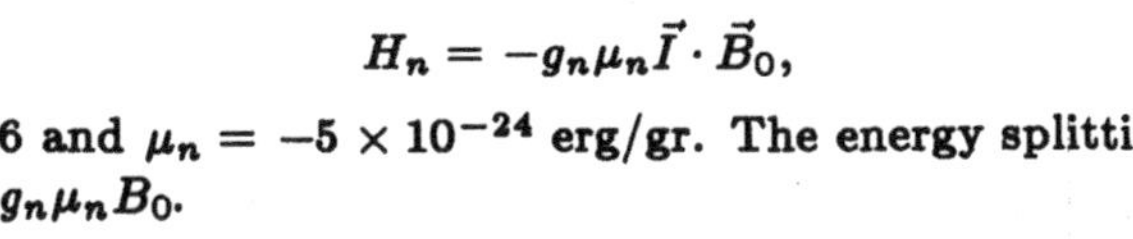

where the energy splitting is: $\Delta E_e = h\nu_e = g\mu_e B_0$ and $|+\rangle$ indicates the spin state $S_z = +\frac{1}{2}$.

b) The nuclear Zeeman interaction with the field:

$$H_n = -g_n\mu_n \vec{I} \cdot \vec{B}_0, \qquad [2]$$

with $g_n = 5.586$ and $\mu_n = -5 \times 10^{-24}$ erg/gr. The energy splitting is written as $\Delta E_n = h\nu_n = g_n\mu_n B_0$.

The eigenstates and the corresponding energy levels of the combined system of this pair of spins is obtained by the diagonalization of the Hamiltonian

$$H_0 = H_e + H_n, \qquad [3]$$

which is immediate from the previous analysis of eigenstates. We therefore may represent the whole system of eigenstates and levels in the following scheme:

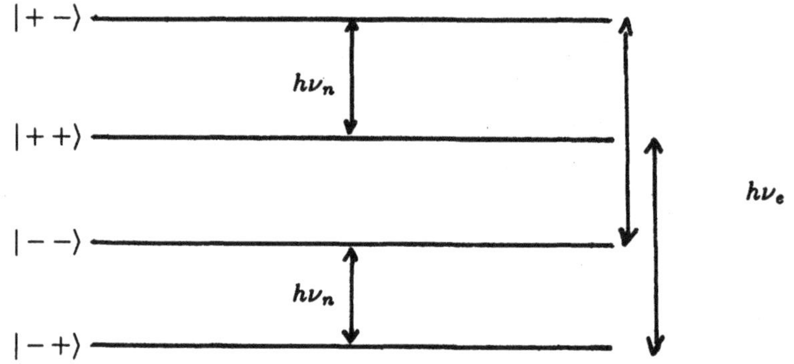

where for example $|+\,-\rangle$ indicates the spin state $S_z = +\frac{1}{2}$ (electron), $I_z = -\frac{1}{2}$ (nucleon).

c) the dipole-dipole interaction. This interaction between the two magnetic moments mixes the eigenstates just considered by a mixing parameter ϵ of the order of $\langle H'\rangle/\langle H_0\rangle$ where:

$$H' = \frac{g g_n \mu_e \mu_n}{r^3}\left[\vec{S}\cdot\vec{I} - \frac{3}{r^3}(\vec{I}\cdot\vec{r})(\vec{S}\cdot\vec{r})\right]; \qquad [4]$$

more precisely we obtain:

$$\begin{aligned} \psi_{+-} &\simeq |+\,-\rangle + \epsilon|+\,+\rangle \\ \psi_{++} &\simeq |+\,+\rangle - \epsilon|+\,-\rangle \\ \psi_{--} &\simeq |-\,-\rangle - \epsilon|-\,+\rangle \\ \psi_{-+} &\simeq |-\,+\rangle + \epsilon|-\,-\rangle. \end{aligned} \qquad [5]$$

One can induce resonant transitions between the states ψ_{++} and ψ_{--}, ψ_{+-} and ψ_{-+}, with RF fields of frequencies $\nu_e - \nu_n$ and $\nu_e + \nu_n$ respectively; although the probabilities of such transitions are ϵ^2 times smaller than the corresponding ones for processes $\psi_{++} \to \psi_{--}$, $\psi_{+-} \to \psi_{--}, \psi_{++} \to \psi_{+-}, \psi_{-+} \to \psi_{--}$, an appreciable change of the population of spin states may be achieved, having the advantage that the life-times of the levels so reached are much longer ("forbidden transitions"). So the electron proton pair drives the nucleon polarization by means of the forbidden transitions. The mechanism is identical, as metioned already, to the one of the bound states electron-proton and electron-nucleus that one uses for polarized targets, except for several physical phenomena such as spin diffusion and others that enhance the effect quantitatevely in solid cristals and are absent in our case. However we achieve a reasonable polarizarion of protons (antiprotons) by repeating the process several times $\simeq 10^9$ with the help of a storage ring, as it is demonstrated in the following section.

3. A Possible Set of Parameters

Typically $B_0 \simeq 2.5T$, then ν_e=70 GHz and ν_n=100 MHz. The RF frequency required is then $\nu_e + \nu_n$=70.1 GHz and the spectral purity required is of the order of 10^{-3}. With those values the levels are split enough to speak about "forbidden transition".

To give a numerical example we then assume as electron current density $J_e = 150A/cm^2$ with a kinetic energy in the laboratory system $W_e = 50KeV$ ($\gamma = 1.098; \beta = 0.413$). The electron density is $n_e = J_e/(e\beta c) = 7.56 \times 10^{10}/cm^3$. The protons (or antiprotons) energy must be $W_p = 92MeV$ to have the same velocity.

We consider the electron beam totally polarized and the geometrical sizes of the reagion of overlapping of the two beams to be $1cm^2$ base and $1m$ axis length. In the overlap reagion we assume an average minimum distance between electrons and protons the inverse of the cubic root of the electron density defined above, i.e. about $10^{-4}cm$. Under the assumption that each electron drives one proton to polarization we obtain that the number of proton prolarized per revolution is:

$$7.56 \times 10^{10} \times 10^2 \times 10^{-12} \times \simeq 10.$$

Therefore after 10^9 revolutions the number of polarized protons in the ring is of the order of 10^{10}.

To conclude we just mention that the requirements imposed to the electron source are extremely demanding, particularly in terms of average current, and quite away from the performances of the present laser-driven photocathodes. Nevertheless refinements in the conception are possible and future developements of the photocathode technology should lead to a feasable proposal.

References

[1]) Particles and Fields, Series 34, (Polarized Beams at SSC-Polarized Antiprotons) Editors A.D.Krish and A.M.T.Lin A.I.P.-Conf. Proc. N^{0}145, New York 1986.

[2]) Split Collaboration "The Spin-Splitter: Study of a Method for In Situ Polarization in Storage Rings" Proposal No. 89-110 (88-104) IUCF.

[3]) R.Calabrese, G.Carugno and L.Tecchio, Nuclear Instr. and Methods, A292(1990) 728-730.

[4]) M.Borghini "Proton Spin Orientation" CERN 68-32 Geneva 1968.

A DENSE ELECTRON TARGET FOR THE MEASUREMENT OF CROSS SECTIONS OF DIELECTRONIC RECOMBINATION (DR) AND RADIATIVE ELECTRON CAPTURE (REC)*

M. Kleinod, R. Becker, E. Jennewein, U. Pröbstel
Institut für Angewandte Physik der Universität Frankfurt am Main
Robert-Mayer-Str. 2-4, D-6000 Frankfurt am Main 1, FRG

A. Müller, S. Schennach, J. Haselbauer, W. Spies, O. Uwira, M. Wagner
Institut für Kernphysik der Universität Gießen
Leihgesterner Weg 217, D-6300 Gießen, FRG

N. Angert, J. Klabunde, P. Mokler, P. Spädtke, B. Wolf
Gesellschaft für Schwerionenforschung, Postfach 110 552, D-6100 Darmstadt, FRG

ABSTRACT

An electron target similar to electron cooling devices has been designed and constructed to measure the recombination of highly charged heavy ions with free electrons. The high density electron beam (up to 10 A/cm^2) merges with the UNILAC heavy ion beam in a single-path arrangement. First results on radiative capture measurements are presented.

INTRODUCTION

When cooling partially stripped ions in heavy ion storage rings by the use of electron coolers, losses of these ions by recombination with electrons have to be considered. In contrast to bare nuclei, where radiative capture is the main loss mechanism, the remaining electrons of the partially stripped ions can also be employed in dielectronic recombination (DR) processes. There the incoming free electron loses energy by exciting a core

*Supported by BMFT under contract no. 060F656

electron, thus being captured in an excited state depending on the kinetic energy of the free electron and the excitation energy. The doubly excited state will then decay by the emission of a photon. Since the cross sections for the DR processes are expected to be as large as 10^{-17} cm^2, the loss rates can introduce severe problems in the cooling of heavy ion beams. Therefore an electron target has been constructed and built for the study of these recombination processes for various ions and energies up to 15 MeV/u delivered by the UNILAC. The corresponding electron energy to observe capture near zero relative velocity for the colliding partners will then ask for values of up to 8 keV (s. fig. 1)[1].

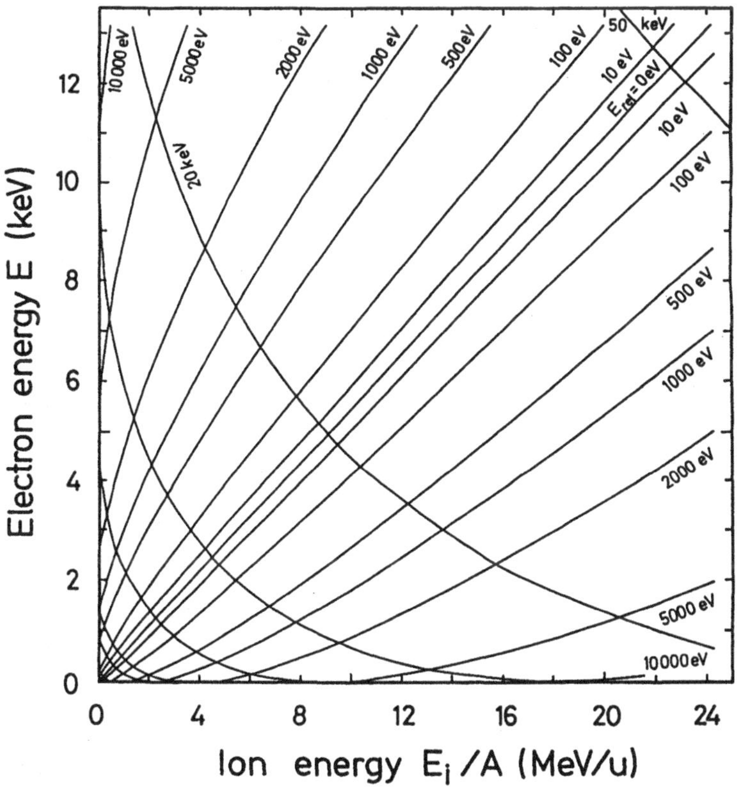

Fig. 1 Accessible relative energies for given electron and ion laboratory energies in a merged beams experiment

Charge transfer collisions of the highly charged ions with the residual gas will also produce ions which have decreased their charge state. This competing process prevents experiments at low laboratory velocities, only with high-β beams the cross section for charge transfer are low enough to allow for distinctive results, still demanding residual gas pressures below 10^{-10} mbar. On the other hand, the kinematics result in very low longitudinal velocity spreads by the transformation into the center-of-mass system, resulting in high resolution necessary for the study of such resonant processes as DR.

ELECTRON TARGET

Electron-ion recombination processes can be studied by using an electron cooler as an electron target[2,3]. This involves the use of complicated toroidal magnetic guiding fields to bend the electron beam in and out of the ion beam line. A different approach uses a convergent electron gun, where the ion beam is fed into the electron target through a hole in the cathode[4]. This design is very similar to an electron beam ion source (EBIS) for the production of highly charged ions at low velocities[5]. In both cases, the reaction rate is proportional to the electron density and the interaction length. Without beam compression in a convergent gun, which otherwise increases the transverse temperature of the electron beam[4], the achievable cathode saturation current density sets a limit at about 10 A/cm^2. For this high value the necessary magnetic focusing field strength has to be 1 Tesla homogeneous to 10^{-4} to conserve the low radial energy spread of about 0.1 eV by the cathode temperature. This demand can easily be accomplished by the use of a superconducting magnet. Further, the vacuum requirements are fulfilled at the same time by the cryopumping action using the cold bore of the magnet as the vacuum vessel for the electron beam like in an EBIS.

To avoid toroidal fields for the superconducting design we decided in contrast to the typical electron cooler set-ups to use only one straight solenoid (s. fig. 2). The electron gun and collector are placed 22.5 mm off-axis fully immersed in the high field region. It takes 45 cm to merge the electron beam of 3 mm diameter into the ion beam; together with an interaction length of 45 cm this asks for a homogeneous region of 1 m length. By the use of distributed single current loops to extend the homogeneity[6], the overall length of the solenoid is only 1.9 m.

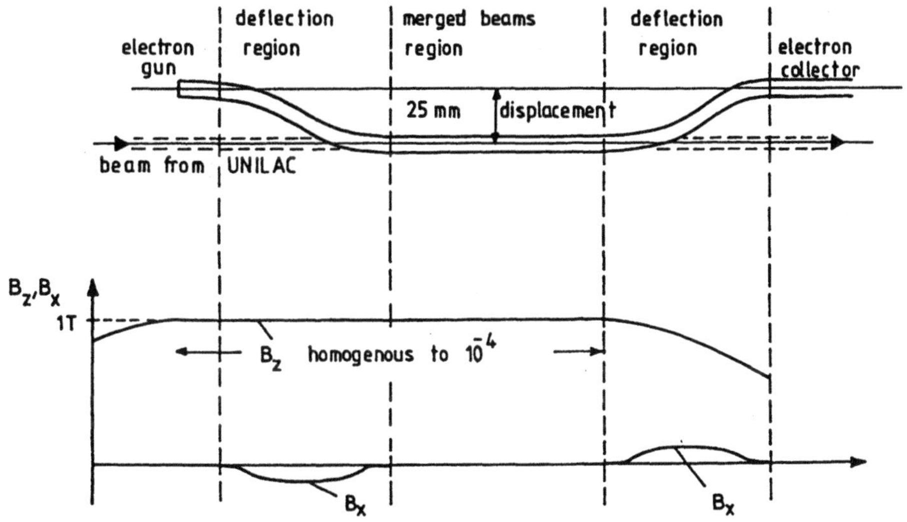

Fig. 2 Merging of electron beam and ion beam by appropriate flux lines

Elliptical windings produce the necessary bend of flux lines to merge and demerge the electron beam to and from the ion beam ($B_\perp = 0.15\,B_0$). Their return flux is being compensated by additional elliptical windings, the unwanted longitudinal field being removed again by appropriate annular current loops (s. fig. 3).

The electron gun uses resonant focusing thus reducing the transverse energy to < 0.5 eV and operates up to 800 mA at 8 keV, delivering 10 A/cm^2 (i.e. current density $n_e \sim 10^{10}$ cm^{-3}) in a high perveance mode[7)]. The design was performed using the SLAC electron optics program[8)]. The low transverse energy remains unchanged, when the anode voltages are lowered in proportion while decreasing the magnetic field proportional to the square root of the energy.

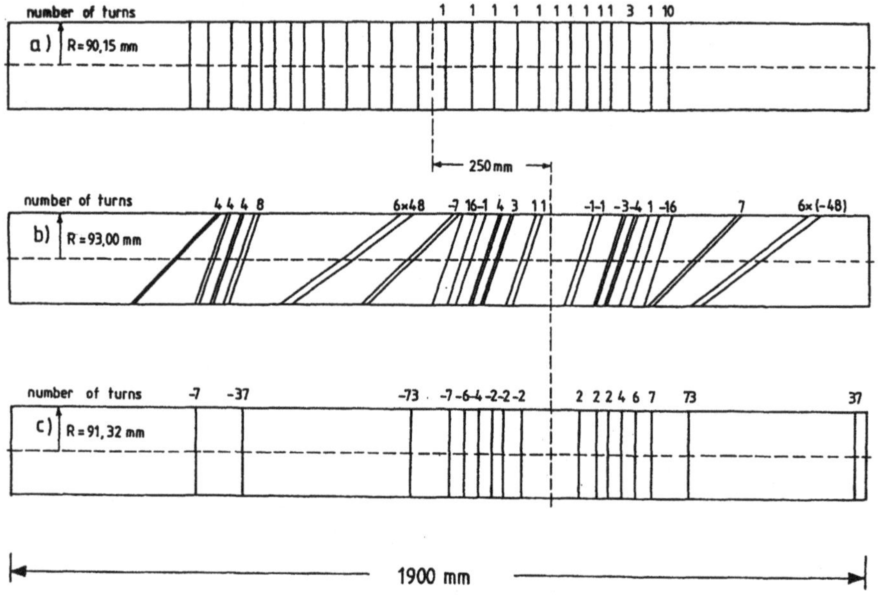

Fig. 3 Winding scheme of the magnet (main windings not shown) for
a) extension of axial homogeneity to 1 m length
b) transversal field regions by ellipses
c) compensation of the axial field due to the ellipses

As mentioned before, the kinematics lower the longitudinal energy spread considerably by transformation into the center-of-mass frame. The potential drop across the electron beam of about 100 V will be reduced to be less than 0.4 eV near zero relative velocity (s. fig. 4). When measuring DR resonances at higher relative energy the longitudinal energy will be increased. Therefore the electron gun can be operated in a low perveance mode with only 1 A/cm^2 ($n_e \sim 10^9$ cm^{-3}) at maximum, thus decreasing the beam depression also by a factor of 10. Furthermore, the potential drop across the electron beam could be lowered by using space charge compensation by protons in the interaction region when applying appropriate electrode potentials to form a trap and feeding hydrogen gas to this region.

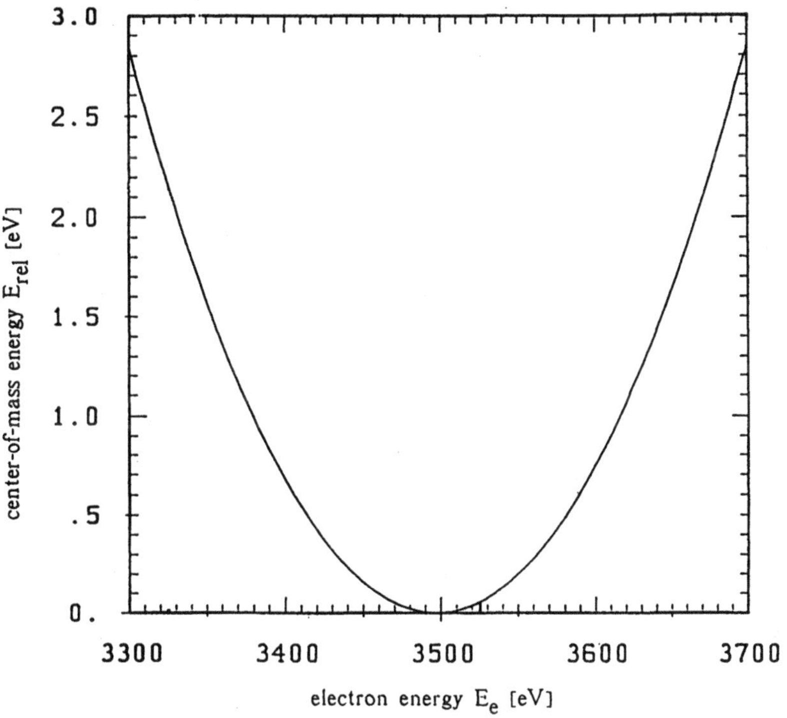

Fig. 4 Center-of-mass energy vs. electron laboratory energy for a fixed ion beam energy of 6.3 MeV/u in a merged beams experiment

BEAM LINE SET-UP

After collimation of the UNILAC ion beam, which also serves for differential pumping of the beam line to better than 10^{-10} mbar, the ions are focused into the interaction region by magnetic quadrupole lenses (s. fig. 5). The two dipole magnets are used for charge state definition and analysis respectively. Two sets of three rotatable steerers serve for the counteraction against the dipol fields of the elliptical windings being used to merge the electron beam. The beam positions - and thus the complete overlap - are controlled by the use of a pair of precision movable slits in front of and behind the interaction region. After ion beam analysis, the parent ion beam is collected in a Faraday cup, while the ions having undergone a capture process are detected by a position-sensitive multi-channelplate detector.

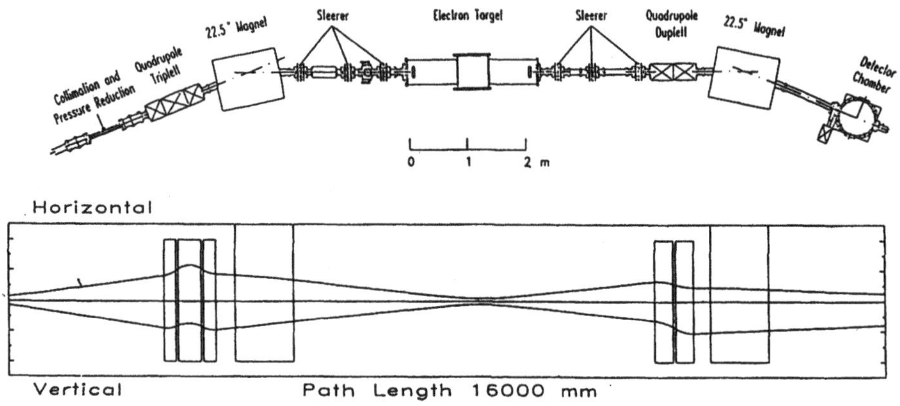

Fig. 5 Set-up of the beam line and its optics at the UNILAC

FIRST MEASUREMENTS

In a first experiment[9] a beam of 65 nA 6.3 MeV/u U^{28+} ions was merged with a 20 mA electron beam ($n_e = 4 \cdot 10^8$ cm^{-3}) whose energy was changed from 3325 to 3525 eV in multiple scans of 200 steps of 1 V each, recording the number of U^{27+} at each step (s. fig. 6). The total recombination rate shows a pronounced peak at about 3500 eV, which corresponds to the match of both longitudinal velocities. Then only the velocity spreads in both beams define the center-of-mass energy, leading to a finite rate for the diverging cross section at zero relative energy.

Since the longitudinal energy spread is vanishing by the kinematics (5.3 eV -> 1 meV), the assumption of a transverse energy spread of 0.2 eV leads to a calculated recombination rate of $2 \cdot 10^{-8}$ $cm^3 sec^{-1}$ for U^{28+}. After subtraction of a background of about $0.5 \cdot 10^{-7}$, the peak rate is 5 times higher than expected from calculations[10]. Unfortunately there is no direct experimental information about the velocity distribution of the electron beam, which would confirm the assumed transverse energy spread, but even a lower spread of 0.1 eV (corresponding to the cathode temperature) results in a discrepancy factor of still 2.5.During the next beam time we hope to determine the radial energy spread by measuring DR resonances at non-zero energies.

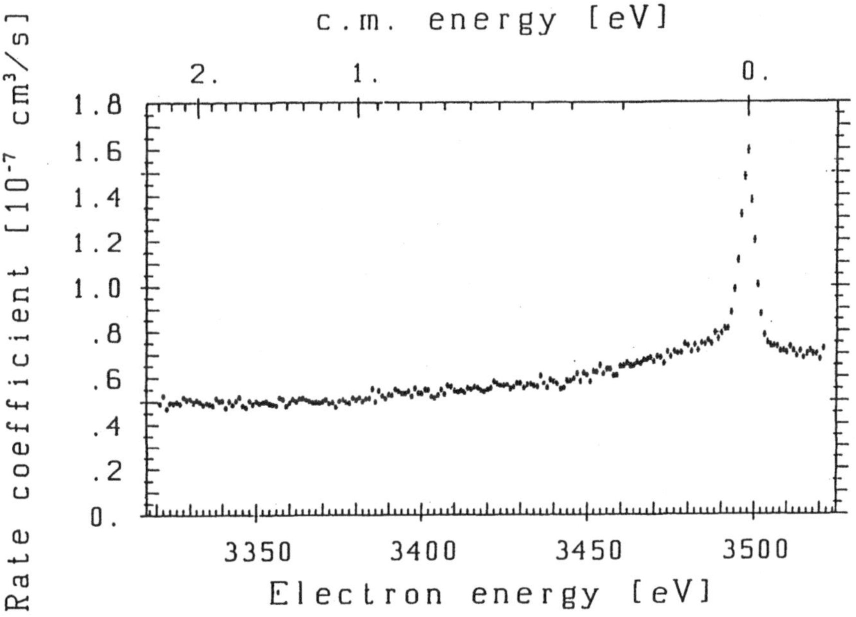

Fig. 6 Experimental recombination rate of 6.3 MeV/u U^{28+} ions with free electrons vs. electron laboratory energy

SUMMARY

In a first experiment the electron target was used to measure radiative recombination rates for a U^{28+} beam. The observed enhancement might be the result of additional DR resonances near zero relative velocity, which has to be checked in a further experiment. But if there is confirmation of the results on radiative capture, the extrapolation from completely stripped low-Z-ions, where theory was stated recently in a similar experiment[11], to partially stripped ions with high nuclear charge can only be used with caution. Especially the electron cooling of such ions in a storage ring may become very difficult because of unexpected recombination losses.

REFERENCES

1) Becker, R., et al, GSI Sci. Report 1985, GSI-86-1 (1986) 378
2) Andersen, L.H., et al., Phys. Rev. Lettrs. 62 (1989) 2656
3) Kilgus, G., et al., Phys. Rev. Lettrs. 64 (1990) 737
4) Dittner, P.F., Phys. Rev. Lettrs. 51 (1983) 31
5) see, e.g., Int. Symp. on Electron Beam Ion Sources and Their Applications, Upton, N.Y., 1988, ed. A. Hershcovitch, AIP Conf. Proc. 188
6) Becker, R., et al., EPAC 88, Proc. Europ. Part. Acc. Conf., Rome, 1988 (ed. S. Tazzari), World Sci. Publ., (1989) p. 607
7) Pröbstel, U., et al., GSI Sci. Rep. 1986, GSI-87-1 (1987) 271
8) Herrmannsfeldt, W.B., SLAC 166 (1973)
9) Schennach, S., et al., GSI Sci. Rep. 1989, GSI-90-1 (1990) 148
10) Müller, A., et al., to be publ.
11) Andersen, L.H., et al., Phys. Rev. Lettrs. 64 (1990) 729

Cooling by inelastic scattering of partially ionized ions

K. Kilian, Institut für Kernphysik, Forschungszentrum Jülich GmbH,
Postfach 1913, D–5170 Jülich

Intrabeam scattering (IBS) is one of the processes which have to be considered if one wants to store beams with highest phase space density in a synchrotron. IBS couples the longitudinal and transverse degrees of freedom of the stored beam. Strong bunching for example which introduces a large longitudinal momentum spread Δp_z in the beam usually leads to an increase of the transverse temperatures. IBS even couples longitudinal beam energy into the internal temperatures of the beam if one is above transition energy.

One has considered for IBS only elastic processes so far. The intrabeam collisions of partially ionized ions $A^{\xi+}$, however, can in addition be inelastic with two other types of collisions:

- charge changing which lead to unwanted beamlosses
- charge conserving which excite one of the ions.

The latter could provide a new possibility for phase space cooling[1)] of heavy ion beams.

$$A^{\xi+}+A^{\xi+} \rightarrow \left\{ \begin{array}{lll} A^{\xi+}+A^{(\xi+1)+}+e & \text{only if } E_c\text{>ionization threshold } E_i & \multirow{2}{*}{\text{charge changing losses}} \\ A^{(\xi+1)+}+A^{(\xi-1)+} & \text{charge exchange} & \\ \hline A^{\xi+}+A^{\xi+} & \text{elastic, mixes } \Delta p_{x,y,z} & \\ A^{\xi+}+A^{*\xi+} & \sigma_{inel} \text{ can be - } 10^{-16}\ \text{cm}^2 & \text{no losses} \\ \quad A^{\xi+}+h\nu & \text{photon emission for cooling} & \end{array} \right.$$

An excitation process will take out part of the ion–ion intrabeam collision energy E_c and the excited ion $A^{*\xi+}$ radiates its excitation energy E_γ away in a photon emission with a time constant smaller than the typical intrabeam collision time. In a focused and bunched beam this process cools the ensemble <u>transversely and longitudinally</u> towards p_o, the center momentum.

For the intrabeam (center of mass) momentum deviation p^* in the moving ion–beam system we can write

$$p^*_{x,y,z} \simeq \begin{cases} \Delta p_z/\gamma & \text{(longitudinal)} \\ \Delta p_{x,y} = p_o\,\theta_{x,y} = p_o\sqrt{\epsilon_{x,y}/\pi\beta^*_{x,y}} & \text{(transversal)} \end{cases}$$

where $\Delta p_{x,y,z}$ are the typical deviations from the average laboratory beam momentum p_o. γ is the Lorentz factor, θ the angular deviation from the beam axis, β^* the lattice function, and ϵ the beam emittance.

The intrabeam collision energy which is relevant in all cross section considerations is:

$$E_c = \frac{(p^*)^2}{M} \approx \frac{1}{M}\left[\frac{(\Delta p_z)^2}{\gamma^2} + \frac{p_o{}^2}{\pi}\left(\frac{\epsilon_x}{\beta^*_x} + \frac{\epsilon_y}{\beta^*_y}\right)\right]$$

One sees that a strongly focussing lattice (low β^*) provides large collision energies. In fig. 1 the intrabeam energy is shown for ^{20}Ne+^{20}Ne intrabeam collisions as a function of the beam momentum. There is a threshold energy E_{th} for the lowest ion excitation and a threshold for ionization E_i at higher energy. It would be ideal to find ions with a ionization energy E_i large as compared to the collision energy E_c and an E_{th} very small as compared to E_i in order to reduce ionization losses.

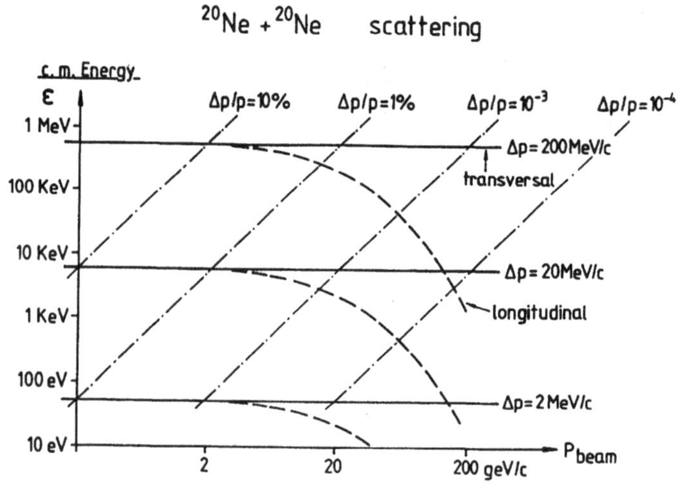

Fig. 1: Schematic plot of the intrabeam energy for the case of a ^{20}Ne beam, as function of the beam momentum and the absolute and relative beam momentum spread.

Note that for relativistic beams there is a factor γ between the longitudinal laboratory momentum spread Δp_z and the smaller center of mass momentum spread $p^*_z = \Delta p_z/\gamma$. In the case of a Ne^{7+} beam there are inelastic excitations only above $E_{th} = 16$ eV, charge exchange collisions above $E_{CEX} = 32$ eV and ionization above at $E_i = 239$ eV.

A ring like COSY–Jülich with > 3 GeV/c per charge could store $^{20}Ne^{7+}$ at about 20 GeV/c and provide sufficiently high intrabeam energies of some 100 eV with a moderate momentum spread of a few 10^{-4}.

A small absolute E_{th} helps in so far as it allows for good equilibrium emittances (Fig. 2). As shown in fig. 2, the Δp values approach each other after inelastic collisions. After repeated inelastic scattering the collision energy E_c approaches the threshold for inelastic excitation E_{th}. Radiative ion cooling then stops.

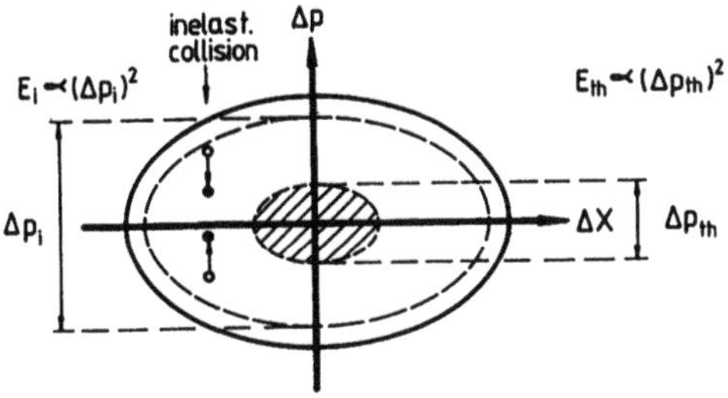

Fig. 2: Schematic plot of the phase space occupied by ions. The outer ellipse indicates the ring acceptance. If momentum differences are smaller than Δp_i then there are no ionization losses. Cooling stops when momentum differences approach Δp_{th} because the collision energies E_c reach E_{th}, the lowest inelastic threshold.

In the longitudinal phase space one can reduce the bunch length and increase again Δp_z by increasing adiabatically the RF voltage ($\Delta p_z^{max} \propto \sqrt{U_{RF}}$). This allows to go on in cooling with the attractive result of shortening the bunch length. Adiabatic reduction of the RF voltage afterwards gives an adiabatic increase of bunch length and allows finally for a further reduction of the momentum spread Δp_z.

Relevant for the cooling time constant is the time between inelastic collisions τ_{inel} for a single ion in the beam. It holds

$$\tau_{inel}^{-1} \approx \sigma_{inel} \cdot (c \cdot \Delta\beta) \cdot N/(L \cdot A) = \sigma_{inel} \cdot v \cdot \rho$$

with an average effective excitation cross section σ_{inel} and an average velocity spread $\Delta\beta$ in the ion–ion system.

As an example take $N \simeq 6 \cdot 10^{10}$ ions in a bunch with $L \simeq 10$ m bunch length and $A \simeq 1$ cm^2 average beam cross section. For $\Delta\beta \simeq 1.4 \cdot 10^{-4}$ (in a ^{20}Ne beam of 2 GeV/c this corresponds to $\Delta p/p \sim 10^{-3}$ and a relative ion ion energy of $E_c \simeq 50$ eV) and $\sigma_{in} \simeq 3 \cdot 10^{-17}$ cm^2 one finds: $\tau_{in}^{-1} \simeq 0.8 \cdot 10^{-2}$ s^{-1} or a collision time $\tau_{in} \sim 2$ minutes. The cooling time constant will be longer by a factor $<E_c/E_\gamma>$, the average ratio of the collision energy E_c over the photon energy $E_\gamma = h\nu$. This may be an order of magnitude. The cross sections σ_{in} for inelastic processes in symmetric $A^{\xi+}$ ion collisions are not well known.

A similar estimate can be made for beam losses if one inserts the cross section σ_L for charge changing processes. There data are known e.g. for $Bi^+ + Bi^+$ scattering[2]: $\sigma_L(E) = 3.14 \cdot 10^{-17}\ cm^2 \cdot E^{0.81}$, with E in keV. In practice one has to average $\sigma \cdot v$ over the energy spectrum f(E) in the beam $< \sigma \cdot v > = \int_0^{E_{max}} \sigma(E) \cdot v \cdot f(E)\, dE$.

The intrabeam cooling has to be faster than the intrabeam losses. Charge changing collisions will be negligible only if we can find a working point where $\sigma_{in} \cdot (E_\gamma/E_c) >> \sigma_L$. Then the photon emission after inelastic intrabeam scattering may be a useful new way of phase space cooling. Some features of this cooling would be:

+ it works best for dense beams where other cooling methods run into problems
+ it works with partially stripped heavy ions
+ each ion species has its optimal intrabeam collision energy E_c
+ the optimum E_c can be adjusted by the choice of: beam momentum, momentum spread, focussing strength, RF amplitude
+ bunching is needed and short bunches are provided
+ no extra effort is needed at the correct working region. A ion storage ring with RF is sufficient.
+ From atomic physics one can expect the most attractive ratio between inelastic and ionization cross–section (smallest intrabeam losses) for heavy singly ionized ions[3].

It is a pleasure to thank G. Baur and D. Möhl for exciting discussions.

References

1. K. Kilian, CERN–EP internal report 84–05, Oct. 1984
2. F. Melchert et al., Nucl. Instr. Meth. A278 (1989) 65
3. G. Bauer, IKP – KFA Jülich, priv. comm.

LASER COOLING AT THE HEIDELBERG TEST STORAGE RING (TSR)

W.Petrich, V.Balykin[1], M.Bock, C.Ellert, M.Grieser, D.Habs, E.Jaeschke, H.Miesner, M.Music, D.Schwalm, P.Sigray[2], M.Steck, B.Wanner, A.Wolf
Physikalisches Institut der Universität Heidelberg and
Max-Planck-Institut für Kernphysik, D-6900 Heidelberg, F.R.G.

R.Grieser, G.Huber, R.Klein, S.Schröder
Institut für Physik der Universität Mainz, D-6500 Mainz, F.R.G.

T.Kühl, R.Neumann
Gesellschaft für Schwerionenforschung, D-6100 Darmstadt, F.R.G.

[1] on leave from Academy of Sciences, Troitzk, Moskow Region, USSR

[2] on leave from Manne - Siegbahn - Institut, Stockholm, Sweden

For the first time, successful laser cooling of ions at relativistic velocities has been observed at the Test Storage Ring. Longitudinal temperatures of a 7 MeV $^{9}Be^{+}$ and a 13 MeV $^{7}Li^{+}$ beam were reduced by three orders of magnitude down to 6 K and 180 mK, respectively. Different cooling schemes have been investigated in recent experiments. The combination of laser cooling with electron precooling offers the possibility to further decrease the longitudinal temperature.

Laser cooling has been performed in traps[1)] and in single-pass experiments at low velocities[2)]. However, a storage ring is a fertile tool to combine the high velocity of the ions and the precision of laser techniques with long interaction times. Indeed, one of the aims of laser cooling at the Heidelberg Test Storage Ring (TSR) (fig.1) is a precision measurement related to the theory of special relativity. Also, a planned Ramsey-type experiment will use a laser cooled ion beam[3)]. As compared to electron cooling, lower temperatures in the millikelvin region can be reached by laser cooling within shorter cooling times. In this temperature regime, the movement of ions is dominated by their Coulomb repulsion, which can lead to the occurrence of ordering, i.e. crystallisation[4)]. However, by collective effects such as the creation of plasma waves, an ultracold ion beam exhibits a considerably different behavior as compared to a hot beam.

	$^7Li^+$	$^9Be^{2+}$	$^9Be^+$
lower level	2s $^3S_1(F{=}5/2)$	2s $^3S_1(F{=}5/2)$	2s $^2S_{1/2}$
upper level	2p $^3P_2(F{=}7/2)$	2p $^3P_2(F{=}7/2)$	2p $^2P_{3/2}$
lifetime of lower level	50 sec	2 sec	stable
lifetime of upper level	43 nsec	29.2 nsec	8.7 nsec
transition wavelength	548 nm	372.2 nm	313 nm
saturation intensity	9 mW/cm^2	40 mW/cm^2	467 mW/cm^2
theor. cooling force[1,2]	44 meV/m	95 meV/m	380 meV/m
theor. cooling time[2]	4.8 μsec	2.8 μsec	2.0 μsec
theor. temperature limit[2]	84 μK	123 μK	413 μK
participating ions	10% - 30%		100% [3]
Ar^+ ion laser line	514.4 nm	334.6 nm	300.2 nm
Ar^+ ion laser power	1200 mW	700 mW	220 mW
velocity β	6.4%	10.6%	4.2%
energy	13.36 MeV	47.67 MeV	7.29 MeV

Table 1: Parameters for laser cooling of $^7Li^+$,$^9Be^+$ and $^9Be^{2+}$

The principle of laser cooling is based on the fact, that – averaged over many cycles of (directed) absorption and (symmetric) spontaneuos emission – momentum is transferred to the ion[5]. Out of a broad velocity distribution , a narrow-band laser can resonantly excite a specific 'velocity class' v^*, which fulfills the Doppler condition $\nu_l = \nu_0\gamma(1+\beta)$, where ν_l is the laser frequency , ν_0 is the transition frequency in the ion rest frame, $\beta = v^*/c$ is the ion velocity in units of c and $\gamma = (1-\beta^2)^{-1/2}$. Hence, interaction with e.g. a copropagating laser beam accelerates ions of a certain velocity v^*. This results in a dip and a corresponding peak in the velocity distribution. If now the laser frequency is scanned in the correct direction, ions are successively accelerated and thus collected and pile up in a narrow velocity distribution . As an alternative to sweeping the laser frequency, the velocity distribution can be shifted across the Doppler-shifted resonance condition between ions and a laser field of fixed frequency (see below).

As strong laser cooling requires high transition probabilities, i.e. electric dipole transitions, the number of appropriate ions is limited. At the TSR, two ion species

[1]on resonance
[2]at saturation intensity
[3]interaction can be reduced by optical pumping

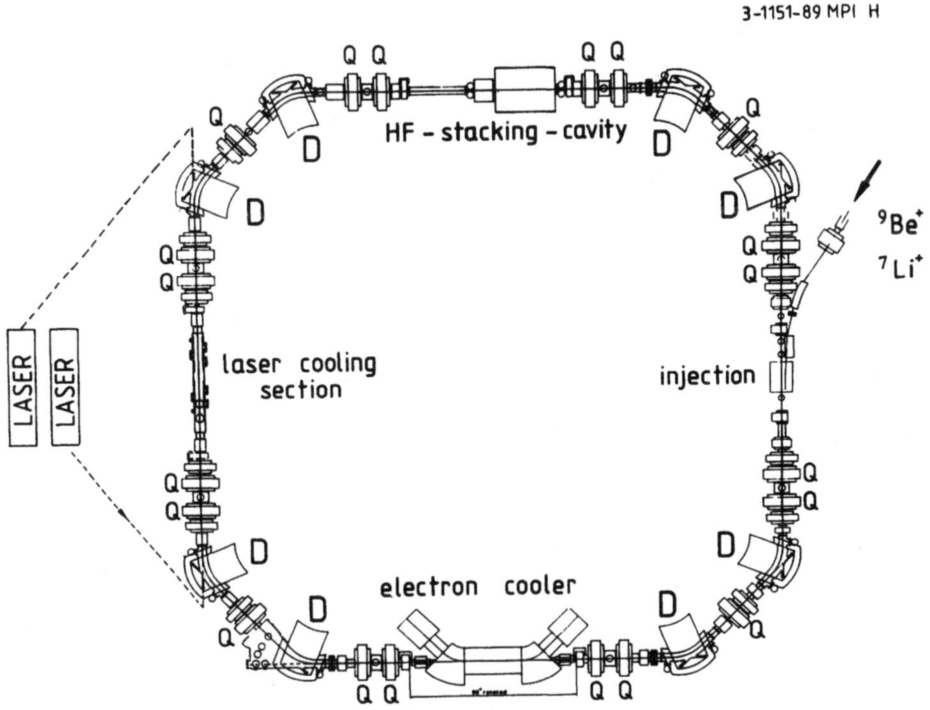

Figure 1: The Heidelberg Test Storage Ring (TSR).

In the second straight section after the injection, laser light is overlapped with the coasting beam. Fluorescent light caused by the interaction between the laser field and the ions can be monitored by photomultiplier in the middle of the cooling section. For the various purposes described in the text, two high-voltage tubes are installed in the middle of this straight section. For further description of the TSR see [8].

have been investigated, namely $^7Li^+$ [6] and $^9Be^+$, having transition frequencies in the visible and near ultraviolet region. Characteristic features are summarized in table 1 (see also below) together with the parameters of the proposed experiments on $^9Be^{2+}$. In the He-like $^7Li^+$ and $^9Be^{2+}$ systems, only the metastable triplet state fraction of the coasting beam is able to interact with the laser radiation. On the other hand, the $^3S_1 \longrightarrow {}^3P_2$ transition forms a closed two-level system with transition wavelengths in the visible and near-ultraviolet region, respectively. In the case of $^9Be^+$, *all* ions can be cooled in principle, but one has to handle the difficulties of optical pumping and ultraviolet laser light techniques. For all of the three ion species, sub-millikelvin temperatures can theoretically be achieved with cooling times in the microsecond region. In the described experiments, the ion energy is always chosen such, that an intense laser line of an Ar^+ ion laser can be used for cooling.
However, magnetic dipole transitions may also be used for cooling. The fine-structure of e.g. $^{35}Cl^{12+}$ or $^{40}Ca^{14+}$ and the hyperfine-structure of $^{209}Bi^{82+}$ or $^{235}U^{92+}$ exhibit an appropriate splitting of energy levels, where conventional laser technique can excite the corresponding transition [7].

LASER COOLING OF $^7Li^+$

A 13.3 MeV $^7Li^+$ ion beam is delivered by the accelerator facility of the Max - Planck - Institut für Kernphysik. During the charge-change process in the stripper gas of the tandem accelerator, 10% to 30% of the $^7Li^+$ ions are excited into the metastable triplet state (τ_r = 50 s), used for laser cooling. A 2 μA $^7Li^+$ beam (β=6.4%) is stored in the TSR. Due to collisions with residual gas, the beam life-time was limited to 2 s. Fig. 2 shows the relevant part of the level scheme. The $^2S_1(F=5/2)\rightarrow {}^3P_2(F=7/2)$ transition in the metastable triplet system forms a closed two-level system, having a transition wavelength of 548.5 nm. The lifetime of the excited state is 43 ns, corresponding to a natural linewidth of 3.7 MHz.
For laser cooling, a copropagating Ar^+ ion laser beam of a fixed wavelength (λ_1=514 nm) is merged with the ion beam. In the ion rest frame, the mean Ar^+ ion laser detuning was set to approx. 8 GHz with respect to the transition frequency. Simultaneously, counterpropagating light of a tunable ring dye laser (λ_2=585 nm) is overlapped with the ion beam. Its Doppler-shifted frequency is set to excite the high-energy end of the ion sample. Scanning the dye laser frequency from higher to lower values avoids the above described successive collection of ions. Monitoring the fluorescent light as a function of laser frequency thus yields information of the original momentum spread $(\Delta p/p)_{\|}=6\cdot 10^{-5}$ and the absolute velocity (fig. 3a). Reverse scanning collects

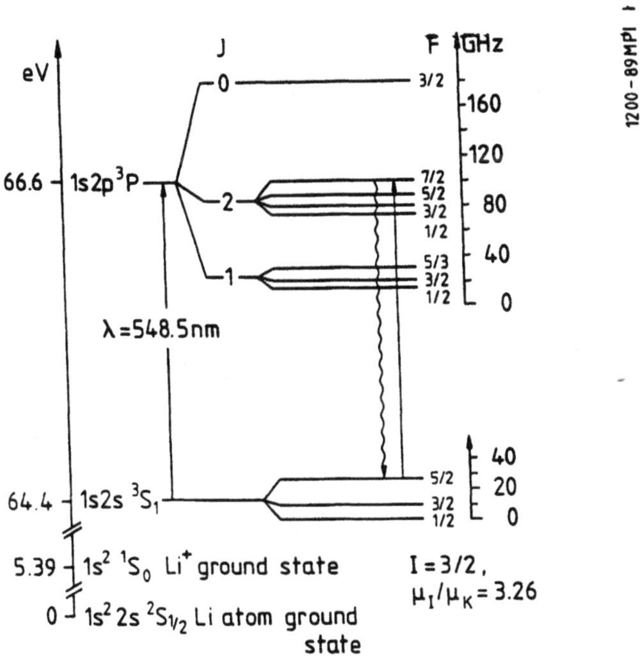

Figure 2: Part of the level scheme of $^7Li^+$, which is relevant for laser cooling

the ions and thus cools the velocity distribution. Ions which are then continuously shifted to lower energy by means of the decelerating laser-cooling force of the dye laser light constantly emit fluorescent light. This fact results in the (slightly decreasing) plateau in the spectra shown in fig. 3b. When the ions are shifted approx. 5 keV, they start to experience the strong accelerating light-pressure force of the copropagating Ar^+ ion laser light. Here, they are brought to bright fluorescence in the combined field of the two lasers. The width of the strongly increasing signal in fig. 3b gives an upper limit for the width of the cooled momentum distribution $(\Delta p/p)_{||}=7\cdot 10^{-6}$. However, the longitudinal temperature is decreased from 260 K (fig. 3a) to below 3 K. In a recent experiment, an upper limit of 180 mK could be measured, using a more dedicated way of probing the velocity distribution after cooling.

An independent way of diagnostics is given by measuring the Schottky noise signal of the coasting beam[9]. It should be pointed out, that in contrast to the fluorescent light spectra, the Schottky pick up of course does not distinguish between singlet- and triplet-state ions. Fig. 4a shows the 30 th harmonic of the Schottky noise spectrum

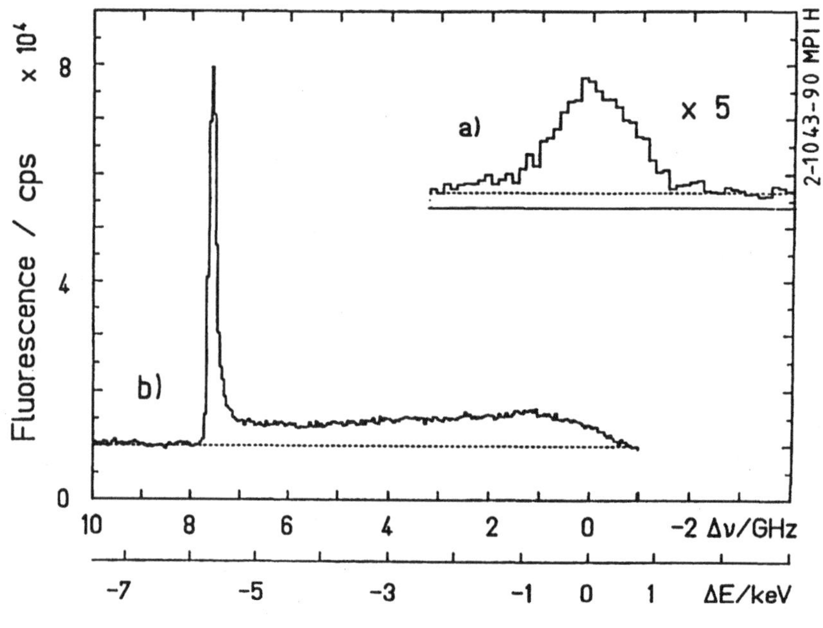

Figure 3: Fluorescent light spectra of $^{7}Li^{+}$, which are emitted when scanning the dye laser frequency. The frequency is swept (a) against or (b) in the cooling direction [6].

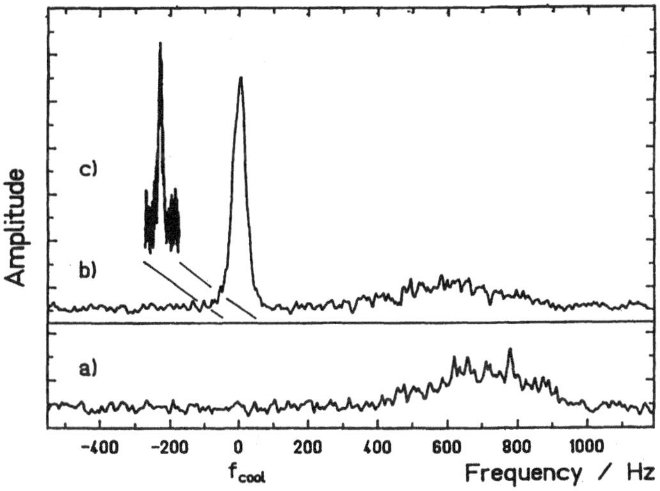

Figure 4: Schottky spectrum of (a) an uncooled and (b,c) a cooled $^7Li^+$ beam [6].

of the uncooled distribution, taken 2.5 s after injection. For the measurement in fig. 4b, this delay time is used for a laser-cooling sweep, which now stops before the ions are shifted across resonance condition for the Ar^+ ion laser frequency . In this way, the cooled triplet state ions are confined in the combined field of the two lasers. In fig. 4c, the same procedure was applied while measuring the Schottky noise with a higher resolution of the spectrum analyser. Assuming incoherent particle motion, interpretation of the shown spectra would mean an increase of the number of particles by a factor of 4. Moreover, the number of triplet state ions seems to rise by two orders of magnitude, what can be ruled out by theoretical as well as experimental investigations. However, a more detailed look onto the interpretation of Schottky spectra yields a strong influence of collective effects[10]. From this point, signal enhancement due to collective phenomena such as the creation of plasma waves seems to be very likely. The observed integral signal enhancement is just the opposite of what has been measured in electron cooling experiments, where an integral decrease of noise intensity was found[11],[12].

LASER COOLING OF $^9Be^+$

In order to avoid the disadvantage of only cooling a small fraction of the stored beam (like the metastable triplet state fraction in the $^7Li^+$ case), laser cooling

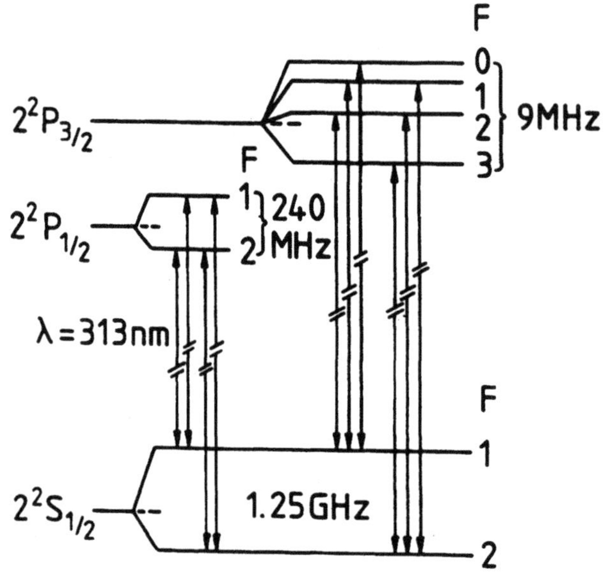

Figure 5: Part of the level scheme of $^9Be^+$, which is relevant for laser cooling.

of a 7.29 MeV $^9Be^+$ beam (β=4.2%, I=2 μA) is investigated. The ion velocity is chosen such, that the strong 300 nm line of an Ar^+ ion laser can excite the $^2S_{1/2}$(F=1,2) $\leftrightarrow$ $^2P_{3/2}$(F=0,1,2,3) transition (λ_0=313 nm), which has a mean radiative lifetime of 8.7 ns. Fig. 5 shows the relevant part of the level scheme. As the hyperfine splitting of the $^2P_{3/2}$(F=0,1,2,3) state of 9 MHz is smaller than the natural linewidth of 18 MHz, no *particular* transition can be selected by the laser frequency. If the electron populates the $^2P_{3/2}$(F=1) or the $^2P_{3/2}$(F=2) substate, the ion can decay into *any* of the two groundstate hyperfine sublevels. The large hyperfine splitting of these two substates of 1.25 GHz can thus cause the ion to come out of resonance with the narrow-band laser field. This process is well known as 'optical pumping'.

However, the effect of optical pumping can be suppressed by the use of a high voltage tube, where the ion velocity can be changed locally. By that, the ion experiences different Doppler-shifted laser frequencies inside and outside the tubes. The voltage can be chosen in such a way that the two Doppler-shifted frequencies match with the two possible transition frequencies in the ion rest frame. In this way, the $^2S_{1/2}$(F=1) $\rightarrow$ $^2P_{3/2}$(F=0,1,2) and the $^2S_{1/2}$(F=2) $\rightarrow$ $^2P_{3/2}$(F=1,2,3) transition can

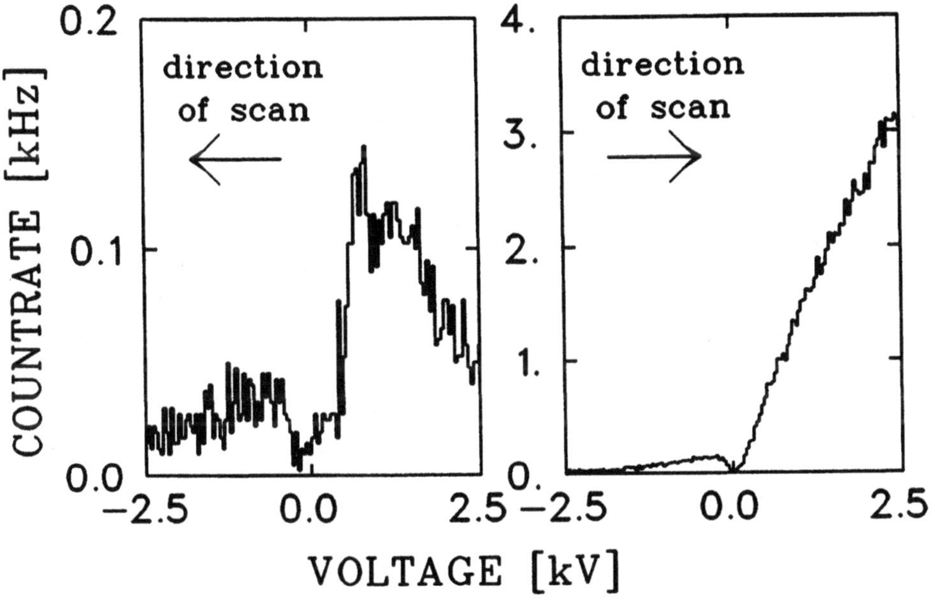

Figure 6: Fluorescent light spectrum of laser cooling of $^9Be^+$ with the high voltage scan method. Sweeping the voltage in the cooling direction yields fluorescent light spectra, which are dominated by momentum transfer rather than optical pumping.

be excited simultaneously, when averaging over one round trip. Hence, the effect of optical pumping can be suppressed.

For probing the velocity profile a second HV-tube was locally scanning the velocity distribution across the laser resonance in a very short time. The fluorescence signal is monitored perpendicular to the beam axis as a function of the applied voltage. Fig. 6a shows the fluorescence spectrum taken directly after injection, where the width yields an initial temperature of $T_{||} \approx 3700$ K, corresponding to $(\Delta p/p)_{||}=3\cdot 10^{-4}$. The dip at 0 V and the peak at 456 V are caused by optical pumping as well as by momentum transfer through the light pressure force[13)].

A stable condition for laser cooling requires a point, where the total force F equals zero and where the slope of the force dF/dv is negative. For an accelerating laser cooling force, this can be achieved by an auxiliary deceleration. For laser cooling of $^9Be^+$, mainly two methods were applied to oppose the accelerating laser cooling force: the so-called 'high voltage scan method' and artificial deceleration with an induction accelerator.

a) high-voltage scan method

In this experimental procedure, the laser frequency is detuned in such a way that the ions are out of resonance outside the two tubes. By applying a voltage, ions can be brought into resonance inside the tubes. When this voltage is swept from -5 kV to +5 kV, the velocity distribution can be compressed by the laser cooling force of the copropagating Ar^+ ion laser. A constant offset of 456 V was kept between the two scanning tubes in order to suppress optical pumping. Fig. 6b displays the fluorescent light, which is emitted during the scan. The increase in signal clearly shows, that ions continuously can be kept in resonance and thus the velocity distribution can be cooled. The rising countrate is comparable with the situation in the $^7Li^+$ experiments, when the ions are collected by sweeping the dye laser (see fig.3b).

b) The induction accelerator

offers another possibility to shift the velocity distribution across the resonance condition. The accelerator acts like a transformer, where a current change in the primary winding causes a magnetic flux change and thus an induced voltage in the secondary winding. Here, the secondary winding is the ion beam itself, which in this way can be accelerated or decelerated[14)]. After shifting the ions with the induction accelerator by an amount of 7.5 keV within 1.85 s, a quick probe scan is done with one high voltage tube. From the fluorescent light spectrum emitted inside this tube (fig. 7), a relative momentum spread of $(\Delta p/p)_{\parallel} \leq 1 \cdot 10^{-5}$ and a corresponding temperature of $T_{\parallel} \leq 6$ K can be derived. Hence, the temperature of the $^9Be^+$ ion beam is reduced by almost three orders of magnitude. The variation of longitudinal velocity due to betatron oscillation $((\delta\beta/\beta)_{b.o.} \approx 10^{-5})$ as well as the frequency jitter of the unstabilized Ar^+ ion laser of $\Delta\nu \approx 100$ MHz (corresponding to $(\delta\beta/\beta)_{f.j.} \approx 10^{-5}$) are very likely to be the main reasons preventing a further reduction of the longitudinal temperature.

However, the transverse temperature still is in the order of $T_{\perp} \approx 10^6$ K. Theoretical considerations show, that electron precooling offers the possibility to drastically improve the laser cooling procedure. First investigations are done on the 7.29 MeV $^9Be^+$ ion beam. After 8.3 s of electron cooling, the above described diagnostic scan yields a longitudinal temperature of $T_{\parallel} \approx 170$ K, which has to be compared to 3700 K before electron cooling. The transverse beam profile (fig.8) shows the expected reduction of the beam size due to electron cooling of the transverse component. The measured beam diameter of 2.6 mm after 8.5 s of electron cooling corresponds

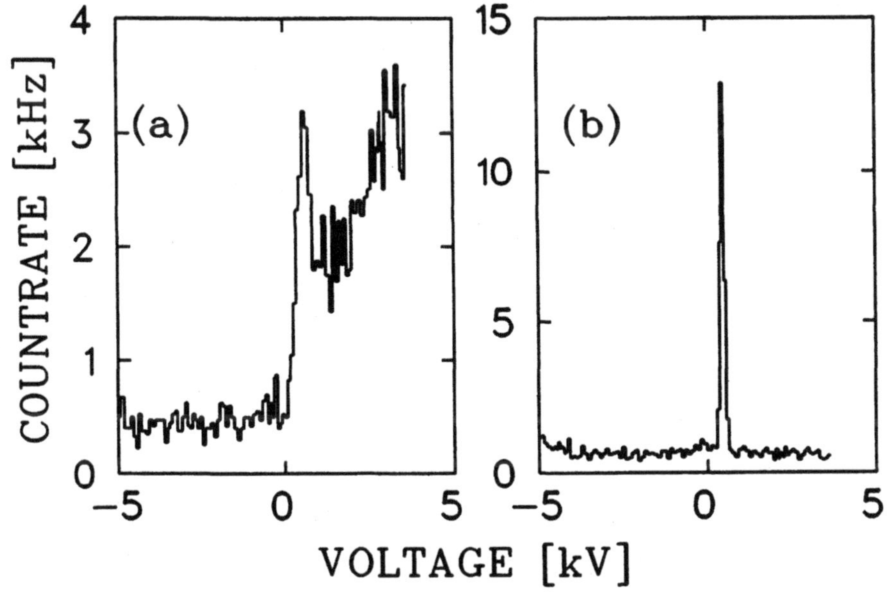

Figure 7: Laser cooling of $^9Be^+$ with the induction accelerator. The longitudinal temperature of the uncooled distribution (a) of $T_{||}^i$=3700 K is reduced down to $T_{||}^f \leq 6$ K (b).

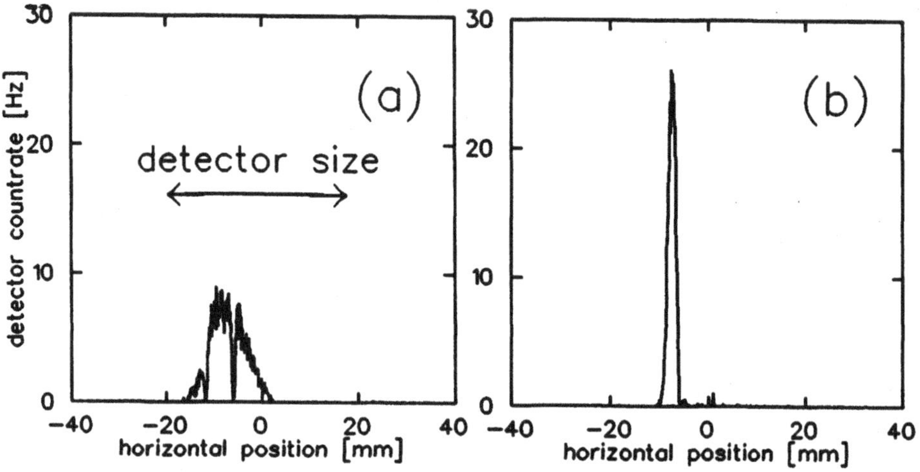

Figure 8: Transverse profile of a coasting $^9Be^+$ beam measured with the beam profile monitor[16]. Spectrum (a) is taken after 1.5 s of electron cooling, while (b) shows a considerable reduction of the beam size after 8.5 s of electron cooling.

to a transverse velocity spread of $\delta\beta_\perp \approx 10^{-5}$ ($T_\perp \approx 4\cdot10^3$ K). The beam diameter is further decreased to below 1 mm for longer cooling times[16]. As transverse cooling also means a decrease of the beam divergence, intra-beam scattering starts to play an important role. This effect seems to be a major reason for the limited reduction of longitudinal temperature [15].

However, the velocity distribution is further compressed by simultaneous laser and electron cooling[15] to $(\Delta p/p)_\parallel \leq 1\cdot10^{-5}$ ($T_\parallel \leq 6$ K). In this first measurement, the frequency jitter of the laser has been as well in the order of ± 100 MHz, which again strongly hampers the cooling as well as it influences the detected signal. Thus, part of the present activities concentrate on the stabilization of the laser equipment.

However, at very low transverse temperatures, the coupling between the longitudinal and transverse movement via Coulomb collisions is enhanced. Thus, *longitudinal* laser cooling may also lead to a further reduction of the *transverse* momentum spread by means of intra-beam scattering. In this way, the creation of ultracold ion beams by combined electron and laser cooling seems feasible.

During the described cooling experiments on $^9Be^+$, the Schottky signal is strongly enhanced, similar to what has been observed in the $^7Li^+$ case. Preliminary analysis of some spectra show an increase of the integrated squared signal of more than a factor of 5. In contrast to the $^7Li^+$ case, an increase of the total number of ions, which can be cooled, is of course excluded. Even more, for $^9Be^+$ it is possible to measure the *complete* velocity distribution via the fluorescent light and to simultaneously detect the collective beam response with the Schottky pick-up electrodes.

In summary, it was shown, that the longitudinal temperature of a $^7Li^+$ beam and a $^9Be^+$ beam was reduced by almost three orders of magnitude down to 180 mK and 6 K, respectively. Three different cooling schemes were described, namely sweeping the laser frequency, the so-called high-voltage-scan method and laser cooling with an induction accelerator. The Schottky-noise signal showed a strong contribution from collective phenomena such as the creation of plasma waves. First simultaneous laser end electron cooling was investigated.

REFERENCES:

1) D.Wineland, R.E.Drullinger, F.L.Walls, Phys. Rev. Lett. **40** (1978) 1639 and W.Neuhauser, M.Hohenstatt, P.Toschek, H.G.Dehmelt, Phys.Rev.Lett. **41** (1978) 233

2) S.V.Andreev, V.I.Balykin, V.S.Lethokov, V.G.Minogin Pis'ma Zh. Eksp. Teor. Fiz. **34** (1981) 463 (JETP Letters **34** (1981) 442) and W.D.Phillips, H.Metcalf, Phys.Rev.Lett. **48** (1982) 596

3) D.Habs, this conference

4) D. Habs in: 'Frontiers in Particle Beams', Lecture Notes in Physics **296** (1986) 310

5) R.J.Cook, Phys. Rev. **A 20** (1979) 224

6) S.Schröder et al., Phys. Rev. Lett. **64** (1990) 2901

7) T.Kühl et al. in: 'Proceedings of the Workshop on Crystalline Ion Beams', Wertheim F.R.G, (1988), p.294

8) D.Habs et al., Nucl.Instr.Meth. **B 43** (1989) 390

9) D.Pestrikov, this conference

10) S.Cocher,I.Hoffmann, submitted to Part. Acc., preprint GSI-89-61 (1989)

11) V.Parkhomchuk,D.V.Pestrikov, Sov.Phys.Tech.Phys. **25** (1980) 7

12) M.Steck et al., Nucl. Inst. Meth. **A 287** (1989) 324

13) W.Petrich, MPI-report MPI H - 1989 - V42

14) C.Ellert et al., in preparation for Nucl. Inst. Meth.

15) A.Wolf et al., this conference

16) B. Hochadel et al., to be published

FIRST RESULTS ON COMBINED LASER AND ELECTRON COOLING OF $^9Be^+$ IONS AT THE TSR

A. Wolf, V. Balykin[1], C. Ellert, M. Grieser, D. Habs, B. Hochadel, E. Jaeschke, D. Krämer, H.J. Miesner, M. Music, W. Petrich, D. Schwalm, P. Sigray[2], M. Steck, and B. Wanner

Physikalisches Institut der Universität Heidelberg and
Max-Planck-Institut für Kernphysik
D-6900 Heidelberg, Federal Republic of Germany

T. Kühl and R. Neumann

Gesellschaft für Schwerionenforschung (GSI)
D-6100 Darmstadt, Federal Republic of Germany

R. Grieser, G. Huber, R. Klein, and S. Schröder

Institut für Physik der Universität Mainz
D-6500 Mainz, Federal Republic of Germany

ABSTRACT

Electron cooling of ions with low charge-to-mass ratio, such as $^9Be^+$ and $^7Li^+$, was performed in the heavy ion Test Storage Ring (TSR) in Heidelberg. The low-emittance ion beams were laser cooled simultaneously to electron cooling, and the fluorescence of the ions induced by the laser was used for beam diagnostics. For $^9Be^+$ ions, preliminary results of electron cooling and its combination with laser cooling are presented, and the possiblity of indirect transverse laser cooling of dense ion beams by intrabeam scattering is discussed.

1. INTRODUCTION

Since several years laser radiation is applied to manipulate and cool neutral atoms and ions nearly at rest in the laboratory. For fast ions in a storage ring laser cooling has been demonstrated recently at the TSR[1,2] using 13-MeV $^7Li^+$ ions and 7-MeV $^9Be^+$ ions, and at the ASTRID ring[3] using 100-keV $^7Li^+$ ions. After these initial experiments the cooling of ions by laser radiation presents itself as a promising method for obtaining very cold ion beams in a storage ring. It would be of particular interest[4,5] to reach low rest-frame temperatures in all degrees of freedom since the ion beam is then expected to show properties of a strongly coupled one-component plasma, such as spatial ordering of the ions.

---

[1]On leave from Inst. of Spectroscopy, Academy of Sciences, Troitzk, Moscow region, USSR.
[2]On leave from Manne Siegbahn Institute, Stockholm, Sweden.

Table 1: Ion and electron beam parameters for electron cooling of $^9Be^+$ at the TSR.

Ion energy	7.3 MeV
Ion momentum p_0	352 MeV/c
Beam velocity v_0	1.26×10^9 cm/s (0.042c)
Stored ion number[a]	1×10^8; 1×10^7 [b]
Ion beam mean lifetime	5...15 s
Ring circumference	55.4 m
Electron energy	445 eV
Electron current	15 mA
Electron density n_e	3.7×10^6 cm^{-3}
Transverse electron velocity	1.3×10^7 cm/s
Magnetic field strength	14 mT
Overlap factor[c] η_e	0.022

[a] Approximate values at injection
[b] Value if ring aperture reduced by pinholes
[c] Ratio of effective electron beam length to ring circumference

Very low temperatures indeed are predicted for laser cooling. Considering only the *longitudinal* degree of freedom and taking into account the friction and the diffusion due to the laser interaction, equilibrium temperatures below 1 mK in the beam rest-frame can be envisaged, which correspond to a relative velocity spread below 10^{-8}. On the other hand, for the *transverse* motion direct laser cooling appears very difficult to achieve in fast ion beams. Thus, for transverse cooling of slow atomic beams[6] laser radiation directed transversely to the particle beam is used. For fast beams, the problem mainly consists in obtaining sufficient laser intensities during the short interaction time of the ions with such fields.

It was suggested earlier[7] that the combination of laser cooling and electron cooling could help to obtain very low beam temperatures in *all* degrees of freedom. The main advantage of this combination consists in the transverse cooling provided by the electrons. Thus the beam size and the transverse ion velocities can be strongly reduced. This helps to optimize the longitudinal cooling and also, as the ion density is increased, enhances the coupling between the longitudinal and the transverse ion motion due to intrabeam scattering. The heat transfer due to this coupling could finally lead to further reduction of the transverse beam temperature by indirect ('sympathetic') laser cooling. Moreover, an auxiliary acceleration of the ions is provided by the longitudinal friction force in the electron beam so that steady-state laser cooling can be obtained as explained below. Concerning the possible interference of laser and electron cooling theoretical estimates[7] indicate that the heating of the ions in the electron beam should not significantly deteriorate the

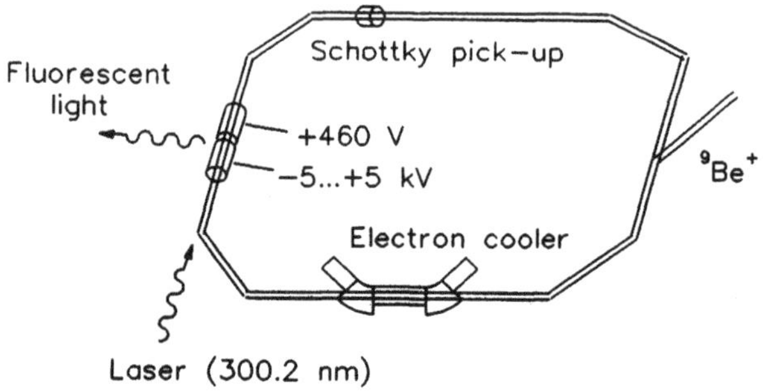

Fig. 1: Set-up for laser and electron cooling of $^9Be^+$ ions at the TSR. Drift tube potentials are shown for simultaneous laser/electron cooling yielding the signals shown in Fig. 8.

equilibrium temperature of laser cooling.

The experimental investigation of combined laser and electron cooling started recently at the TSR for both $^9Be^+$ and $^7Li^+$ ions. First results of beam diagnostic measurements are discussed below starting with electron cooling of $^9Be^+$ ions and continuing with the combined cooling of these ions.

2. ELECTRON COOLING OF $^9Be^+$ IONS

In comparison to the usual situation of electron cooling in storage rings, the singly charged ions $^9Be^+$ and $^7Li^+$ have a particularly low charge-to-mass ratio. Thus, as listed in Table 1, their beam velocity is relatively low even if the magnetic rigidity is close to the maximum possible at the TSR. The electron energy in the cooling device is well below 1 kV, and the electron current correspondingly small. Under these conditions magnetized electron cooling[8] is essential for cooling down the ion beam within its lifetime in the storage ring. In this section, we first present the available diagnostic measurements and then discuss the basic parameters describing the electron cooling performance.

2.1 Diagnostic Measurements

2.1.1 Principles of laser fluorescence diagnostic

The cooled $^9Be^+$ beam was monitored as usual by Schottky noise spectra and, additionally, by measurements of the fluorescence intensity in the laser cooling section. In particular, the laser fluorescence proved to be convenient for determining the longitudinal cooling time and the equilibrium momentum spread.

Laser fluorescence diagnostic is based on the concept that the fluorescent-light intensity is proportional to the number of ions in a narrow velocity class reso-

Table 2: Parameters for laser diagnostic and cooling of $^9Be^+$ at the TSR.

Laser wavelength	300.2 nm
Laser intensity	≈ 100 mW/cm^2
Laser beam size	± 4 mm
Transition	$1s^2 2s\,^2S_{1/2} - 1s^2 2p\,^2P_{3/2}$
Saturation intensity	230 mW/cm^2
Energy shift for Doppler re-pumping[a]	460 eV
Overlap factor[b] η_L	0.09

[a] See text
[b] Ratio of geometrical laser/ion overlap length to ring circumference

nant with the laser. As a function of the longitudinal ion velocity, the strength of the laser interaction is described by a resonance curve resulting from the laser frequency, the absorption profile of the ions, and the Doppler formula. As long as no laser cooling is performed the width of this resonance profile is always small compared to the ion velocity spread. Measuring the fluorescence intensity as a function of the laser frequency yields the Doppler profile of the ion beam, which is proportional to the ion velocity distribution.

In the experiments with $^9Be^+$ a fixed-frequency argon-ion laser was used. For diagnostic measurements, instead of varying the laser frequency, a variable *local* change of the ion velocity was applied in that part of the laser–ion interaction region where also the fluorescent light was observed (Fig. 1). This was achieved by an external potential applied to a drift tube ≈ 0.7 m long. The acceleration of the ions when they enter the tube brings different velocity classes of the ion beam in resonance with the laser, depending on the drift tube potential. The laser–ion interaction region outside of the drift tube provided the laser cooling resonance. In order to activate laser cooling an auxiliary acceleration of the ions is required[2]; in the present case, this is accomplished by tuning the electron cooling energy accordingly, cf. Section 3.1.1.

The principal parameters relevant for the laser diagnostic and cooling in the $^9Be^+$ experiment are listed in Table 2. The laser transition has two hyperfine components separated by 1.2 GHz; if only one of these components is driven by the laser optical pumping occurs, which leads to a fast reduction of the absorption rate of the ions. The optical pumping due to the cooling resonance was counteracted by driving the second hyperfine components simultaneuosly, using a second drift tube (Fig. 1) and applying a constant potential in this 'Doppler re-pumping' zone. A single velocity class strongly interacting with the laser, as required for laser cooling, is generated by this procedure.

The above basic concept of fluorescence diagnostic does not always apply

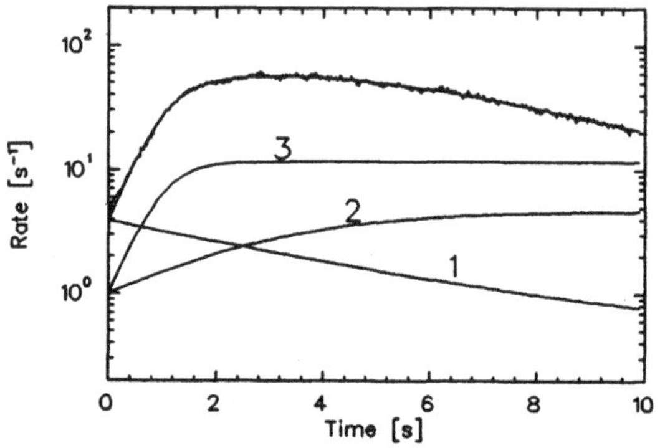

Fig. 2: Fluorescence rate as a function of time after the injection of $^9Be^+$ ions. A function explained in the text is fitted to the signal. With the fitted parameters theoretical curves are shown for: (1) decay of fluorescence without cooling, (2) increase of fluorescence due to transverse cooling, (3) increase due to longitudinal cooling. Curves 2 and 3 normalized to an initial rate of 1 s^{-1}.

to the $^9Be^+$ Doppler profiles obtained by sweeping the drift tube potential. Thus, not only the number of resonant ions, but also the population of the hyperfine sublevels in the steady state of optical pumping must be considered. Relatively easy interpretation is possible if the ions are normally off resonance and interact with the laser only within the drift tube at a suitable external potential. Then, re-pumping by the laser does not occur; re-pumping at a much lower rate is caused by the passage of the ions through the storage-ring magnets, which is non-adiabatic with respect to the hyperfine transition frequency. The fluorescence intensity becomes proportional to the number of ions in *two* velocity classes spaced corresponding to the energy shift listed in Table 2. The problems related to optical pumping are discussed in more detail elsewhere[2)].

2.1.2 Time depencence of fluorescence

The effect of electron cooling on the ion velocity distribution was observed as an increase of the fluorescence intensity within the first seconds after the injection (Fig. 2). This increase occurred if the laser resonance profile was situated close to the maximum of the cooled velocity distribution. Initially, the ion beam is larger in diameter than the laser beam. Hence, the increase of the fluorescence intensity is attributed both to the reduction of the longitudinal velocity spread of the ions and to the increase of the overlap area of the two beams. Information about longitudinal and transverse cooling can be extracted from the measurement by describing the

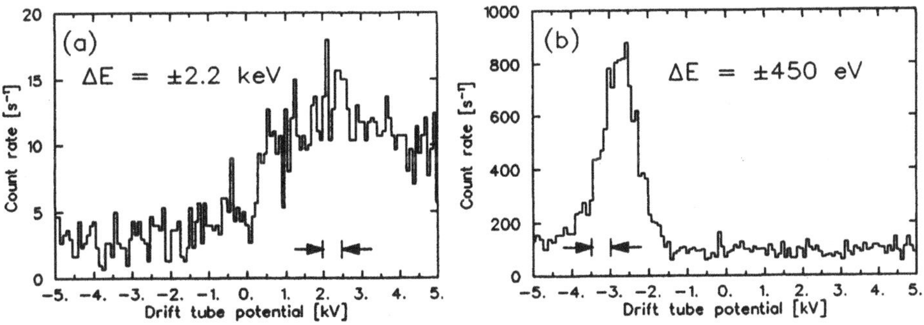

Fig. 3: Fluorescence rate as a function of the drift tube potential close to the fluorescence detector, representing the energy distribution of the ions. (a) Initial distribution obtained directly after injection at an injection rate of 5 s^{-1}. Average of 200 scans of 2.5 s each, laser intensity only 0.5 mW in order to avoid optical pumping effects; 5×10^7 stored ions. (b) Distribution after longitudinal electron cooling; average of 20 scans of 1.3 s taken at a delay of 8.3 s after each injection. 1×10^7 stored ions; beam diameter at profile monitor 2 mm.

fluorescence intensity $I(t)$ by the simplified formula

$$I(t) = \frac{a e^{-t/\tau_0}}{(e^{-t/\tau_\parallel} + w_1)(e^{-2t/\tau_\perp} + w_2)} + b \quad . \tag{1}$$

The following parameters were determined by fitting this expression to the signal: the beam lifetime $\tau_0 = 4.4$ s, the longitudinal and transverse cooling times $\tau_\parallel = 0.39$ s and $\tau_\perp = 3.7$ s, the ratio of the final to the initial momentum spread $w_1 = 0.1$, the ratio of the laser beam cross-section to the initial ion beam cross-section $w_2 = 0.26$, and the amplitude a. A fixed background count rate b, determined independently, was used. The longitudinal cooling time $\tau_\parallel$ was determined from this measurement only; the given fit results for the other parameters are in reasonable agreement with independent measurements discussed in the following. These measurements also show that the transverse cooling time constant decreases as the beam cools down (cf. Fig. 4).

2.1.3 Doppler profiles

The Doppler profiles of the ion beam, representing the longitudinal velocity distribution, were obtained by recording the fluorescence intensity as a function of the potential applied to the drift tube close to the photon detector. The velocity class of ions shifted into the laser resonance inside the drift tube varied during the scan according to the applied voltage. The spectra shown in Fig. 3 represent the ion

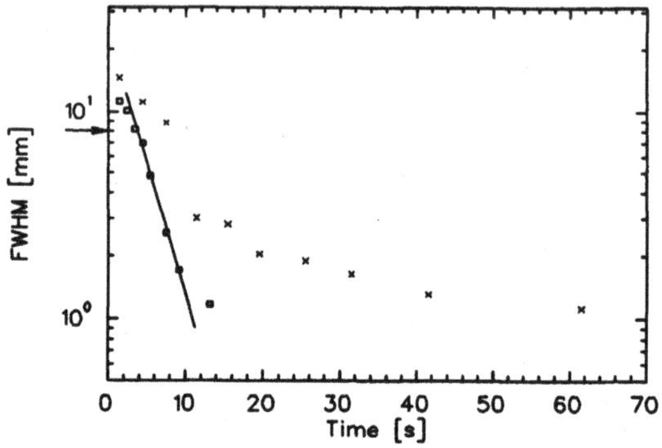

Fig. 4: Decrease of the horizontal beam diameter (full width at half maximum) as a function of time after the injection. Crosses for 1×10^8 stored ions at injection, without pinholes; squares for 1×10^7 stored ions at injection, with pinholes. Line indicates the highest cooling rate observed; corresponding time constant: 2.6 s. Arrow indicates the critical diameter d_{cr}.

energy distribution with the restriction that in fact two closely spaced resonances contribute to the signal (see Section 2.1.1).

First, a Doppler profile was measured at a high injection rate and an integration time long compared to the delay between the injections (Fig. 3a). In this initial Doppler profile the width is large compared to the two hyperfine resonances and the initial energy deviation of the ions can be extracted in a straightforward manner. Later (Fig. 3b), scans were performed at a delay of 8 s after injection, when the beam had been cooled down to small transverse dimensions and had reached the longitudinal equilibrium. The energy deviation ΔE was determined assuming two resonances of equal height, spaced by the energy shift listed in Table 2. The initial and final momentum deviation of the ions obtained from these measurements will be used below to deduce the longitudinal cooling force and the equilibrium temperature.

Schottky spectra measured at a number of stored ions around 2×10^7 yielded compatible results. In comparison to beam noise spectra, Doppler profiles can be useful at high ion beam intensity or at the low beam temperature obtained by laser cooling. In these cases the relation between the velocity distribution and the Schottky spectrum is obscured because of collective density fluctuations in the ion beam.

2.1.4 Transverse ion beam profiles

Important complementary information could be obtained in a later run, when horizontal profiles of the ion beam were recorded using a residual-gas ionization

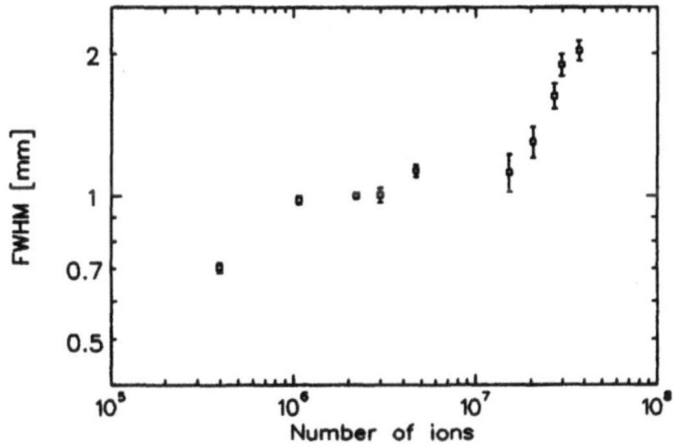

Fig. 5: Horizontal beam size (FWHM) as a function of the number of stored ions obtained by two different methods described in the text.

beam profile monitor[9]. The resolution of this device was determined to ±0.25 mm. From the beam diameter d (FWHM) at the profile monitor the average transverse momentum (divergence) $p_\perp/p_0$ at any ring position s can be obtained using the relation

$$p_\perp/p_0 = \frac{d}{2\sqrt{\beta_x(s_m)\beta_x(s)}} \quad , \tag{2}$$

where $\beta_x(s_m) = 10$ m is the lattice function at the profile monitor. The values of the lattice function at the electron cooler and in the middle of the laser cooling section were 3.5 m and 6 m, respectively. The ring-averaged value was $\overline{\beta_x} = 3.3$ m.

In order to obtain the variation of the beam size during electron cooling of the ion beam, horizontal profiles were integrated over 1 s intervals at varying delay after injection. In general, the profiles were averaged over a number of injection cycles of the order of 10. The FWHM beam size as a function of time after injection is shown in Fig. 4 for two situations. For the lower points the ring aperture was limited by three centered pinholes of 20 mm diameter in the laser cooling section, which are routinely used to ensure the overlap of laser and ion beam. For the higher data points these pinholes were removed, such that the initial ion beam diameter was larger by about 30% in this particular measurement.

In both series of data the highest transverse cooling rate, corresponding to a time constant of 2.6 s, is reached at a critical beam diameter $d_{cr} \approx 8$ mm discussed below. The difference in the time dependence of the beam diameter for longer delay after injection probably can be explained by the difference in beam intensity between both cases. The data from Fig. 4 obtained at times > 18 s are shown as a function of the beam intensity in Fig. 5 (right points). For lower beam intensity more data, also shown in Fig. 5, were obtained by varying the injected current and measuring

Table 3: Electron cooling times, cooling forces, and equilibrium ion beam properties deduced from diagnostic measurements.

Longitudinal cooling:	
Time constant $\tau_{\parallel}$	0.39 s
Initial momentum deviation $\Delta p_{\parallel}/p_0$	$\pm 1.5 \times 10^{-4}$
Normalized cooling force $F_{\parallel}^{*}(v_{\parallel}, v_{\perp})$	0.55 eV/m
Longitudinal ion velocity[a] $v_{\parallel}$	1.9×10^{5} cm/s
Transverse ion velocity[a] $v_{\perp}$	2×10^{6} cm/s
Transverse cooling:	
Time constant $\tau_{\perp}$	2.6 s
Critical beam diameter[b] d_{cr}	8 mm
Critical divergence[c] $p_{\perp}(d_{cr})/p_0$	0.7×10^{-3}
Normalized cooling force $F_{\perp}^{*}(v_{\parallel}, v_{\perp})$	0.6 eV/m
Longitudinal ion velocity[a] $v_{\parallel}$	$\approx 0.5 \times 10^{5}$ cm/s
Transverse ion velocity[a] $v_{\perp}$	0.9×10^{6} cm/s
Longitudinal equilibrium:	
Momentum deviation $\Delta p_{\parallel}/p_0$	$\pm 3.1 \times 10^{-5}$
Ion temperature	1.4×10^{-2} eV/k_B = 160 K
Transverse equilibrium:	
Beam size[b]	± 0.5 mm
Divergence[c] $p_{\perp}/p_0$	0.9×10^{-4}
Ion temperature	1.1×10^{-1} eV/k_B = 1230 K

[a] In beam rest-frame
[b] At profile monitor
[c] At electron cooler

the equilibrium after 15 s (pinholes inserted). It appears that the beam diameter as a function of the ion current increases in steps rather than smoothly.

2.2 Discussion of Electron Cooling Results

2.2.1 Cooling time and cooling force

The different diagnostic measurements performed with the laser and with the transverse beam profile monitor yield quite a consistent picture of the cooling-down process. Thus the horizontal profiles show that the size of the ion beam reaches that of the laser beam at a delay of about 5 s after injection, as described by the slowly rising factor of the intensity function [Eq. (1)]. In particular, this justifies the interpretation of the initial rise in Fig. 2 being due to longitudinal cooling.

The results deduced from the diagnostic measurements for the electron cooling performance with $^{9}Be^{+}$ ions are summarized in Table 3. The measurements yield the *longitudinal* cooling force at the initial longitudinal ion velocity $v_{\parallel}$ in the beam rest-frame. The force is obtained from the longitudinal cooling time constant, the initial momentum deviation $\Delta p_{\parallel}$, and the overlap factor η_e according to

$$F_{\parallel}(v_{\parallel}, v_{\perp}) = \frac{\Delta p_{\parallel}}{\tau_{\parallel}\eta_e} \quad . \tag{3}$$

Note that the listed transverse velocity in the rest frame, $v_{\perp}$, is much higher than $v_{\parallel}$ and was determined from an estimated intial beam diameter during the measurement of Fig. 2. For comparison to theory the result is given in Table 3 as a normalized force defined by

$$F_{\parallel}^{*} = \frac{F_{\parallel}}{n_e} \cdot 1 \times 10^{8} \text{ cm}^{-3} \quad , \tag{4}$$

where n_e is the electron density. The maximum longitudinal cooling force *without* magnetization effects, calculated for vanishing longitudinal electron temperature and the transverse electron velocity according to Table 1, amounts[8] to 0.22 eV/m (normalized to $n_e = 1 \times 10^{8}$ cm^{-3} for a Coulomb logarithm of 8). The observed force of 0.55 eV/m clearly exceeds this value. It is in fact in reasonable agreement with measurements for other ions (mainly protons)[10], which yield a longitudinal force of ≈ 0.5 eV/m at an ion velocity in the rest frame of 2×10^{6} cm/s.

For the *transverse* cooling, the cooling time constant reaches its highest value at the critical beam diameter $d_{cr} = 8$ mm and then remains constant during the approach to the transverse equilibrium. The cooling force, which has its maximum at this moment, is determined from the time constant and the average transverse momentum $p_{\perp}(d_{cr})$ using

$$F_{\perp}(v_{\parallel}, v_{\perp}) = \frac{2p_{\perp}}{\tau_{\perp}\eta_e} \quad , \tag{5}$$

where the factor of 2 is due to the oscillatory character of the ion motion in transverse direction[11]. The transverse force is normalized in the same way as the longitudinal component and also listed in Table 3. From theory[12,13] it is expected for the present conditions that the longitudinal and transverse cooling forces should be roughly of the same size. [Although the angle of the ion velocity vector in the rest frame to the beam axis is $\approx 85°$, the angle of the cooling force vector is predicted to be $\approx 55°$ in the initial cooling phase; the force vector should turn to $\approx 80°$ in equilibrium. As a function of the velocity v in the rest frame the cooling force scales $\propto v^{n}$ with $1 < n < 2$; at 1×10^{6} cm/s the predicted cooling force, for which neither the longitudinal velocity spread of the electrons nor magnetic field errors are considered, is about a factor of 2 higher than the observed force.][12,13]

Presumably, the reason for the occurrence of the critical beam diameter is the effective electron temperature. From $p_{\perp}(d_{cr})$ an effective electron temperature of $\approx 10^{-3}$ eV/$k_B \approx 10$ K can be estimated. It lies about a factor of 10 above the optimum ($\approx 10^{-4}$ eV/k_B) that can be expected for magnetized electron cooling.

2.2.2 Equilibrium beam properties

The available data concerning the equilibrium momentum deviation of the ion beam, listed in Table 3, indicate a longitudinal temperature of the order of 10^{-2} eV/k_B, about a factor of 10 above the effective electron temperature. The measurement was performed at a time after injection, at which the ion beam was not yet cooled down completely in transverse direction. In order to decide if the equilibrium temperature was determined by intrabeam scattering (IBS), an expression given earlier[14,7] for the IBS heating rate is used. For the special case of $^9Be^+$ in the TSR, the following equilibrium temperature $T_{\|}$ as a function of the ion number N in the presence of longitudinal cooling with a time constant $\tau_{\|}$ is calculated:

$$T_{\|} = \frac{(\tau_{\|}/s)N}{(p_{\perp}/p_0)^3} \cdot 5.92 \times 10^{-20}\ \mathrm{eV}/k_B \quad . \tag{6}$$

This yields $T_{\|} = 2.9 \times 10^{-2}$ eV/k_B for the divergence of 2×10^{-4} and 1×10^7 stored ions, corresponding to the conditions of Fig. 3b. Since the rough estimate agrees with the observed equilibrium temperature within a factor of 2, IBS indeed appears to be the dominant heating mechanism. However, further measurements, in particular at longer delays after injection and for varying ion beam intensities, are desirable.

The importance of IBS for the longitudinal equilibrium during electron cooling at the TSR also became clear in measurements with other ions. Thus, for C^{6+} beams the longitudinal ion velocity spread clearly increased when the electron cooling was optimized to yield the smallest ion beam diameter.

The transverse equilibrium temperature of the $^9Be^+$ beam was about a factor of 10 higher than the longitudinal one, whereas the cooling times in both direction differed by a factor of about 6. This indicates that the transverse heating rate was at least as strong as the longitudinal one. The measured equilibrium beam size as a function of the ion beam intensity (Fig. 5) shows an irregular increase of the beam size as a function of the ion current, which could indicate transverse heating due to higher-order resonances crossed as the space-charge tune shift varies. Thus, an incoherent space-charge tune shift[15] of 0.01 is estimated for $^9Be^+$ in the TSR at an intensity of 1×10^7 stored ions and an average beam diameter of 0.6 mm, corresponding to the equilibrium size of 1 mm at the profile monitor. On the other hand, depending on details of the TSR lattice, also transverse IBS heating might occur at a rate comparable to that implied by Eq. (6) because of the coupling of longitudinal and transverse motion in the ring[14]. For the realization of indirect transverse cooling by IBS it would be essential to avoid this heating. Numerical IBS calculations for the TSR lattice and further measurements, in particular as a function of the machine working point, will be needed in order to identify the dominant heating mechanism.

3. COMBINED LASER AND ELECTRON COOLING

3.1 Study of the Combined Cooling Force

In the first investigation of combined laser and electron cooling, the laser resonance velocity was kept fixed and the electron acceleration voltage was varied in small steps. The velocity distribution of the ion beam was monitored at a delay of 3 s after injection by recording the Schottky spectrum. The Schottky spectra are shown in Fig. 6 and a sketch of the combined force of electron beam and laser is shown for corresponding situations in Fig. 7. In the combined force, *one* stable point exists if the electron velocity is higher than the laser resonance, and *two* stable points can appear if the electron velocity lies below the laser resonance. The longitudinal cooling rate around the stable points is proportional to the slope of the force at the zero crossing.

When the laser was turned off (Fig. 6f), the ion signal appeared at the frequency given by the electron acceleration voltage (arrow). With the laser turned on, the situation was practically not modified as long as the electron velocity was higher than the laser resonance (Fig. 6a,b); this is expected since the force at the stable point is only very weakly influenced by the laser. On the other hand, when the electron velocity was very close to the resonance, the cooling rate was essentially determined by the laser; similar to earlier observations of laser cooling, the strong cooling was accompanied by an enhancement of the beam noise power (Fig. 6c). According to Fig. 7, setting the electron velocity slightly *below* the resonance should yield even stronger cooling than a very close matching; correspondingly, particularly strong noise enhancement was observed for the next-lower electron acceleration voltage (Fig. 6d). Also, the ion noise signal clearly appeared at the frequency defined by the laser. For even lower electron velocities, the ions were again attracted to the stable point of the electron cooler (Fig. 6e).

In this series of measurements the laser cooling force was reduced as compared to the usual situation because the optical re-pumping described in Section 2.1.1 (Table 2) was not applied; even though, the principal operation of simultaneous laser and electron cooling is demonstrated. It is seen that efficient steady-state cooling can be obtained and that a small detuning between the electron velocity and the laser resonance presents the most favourable situation. An appropriate detuning also reduces the sensibility of the ion beam to small fluctuations of the electron acceleration voltage.

3.2 Longitudinal Equilibrium

We now discuss first measurements of the equilibrium beam properties during combined laser and electron cooling, noting that the diagnostic methods, in particular for the longitudinal temperature, are still under development. Contrary to the measurement described in the last section, optical re-pumping was now applied. The longitudinal velocity distribution was analyzed by both fluorescence and Schottky noise spectra. During the recording of a fluorescence spectrum, the ions

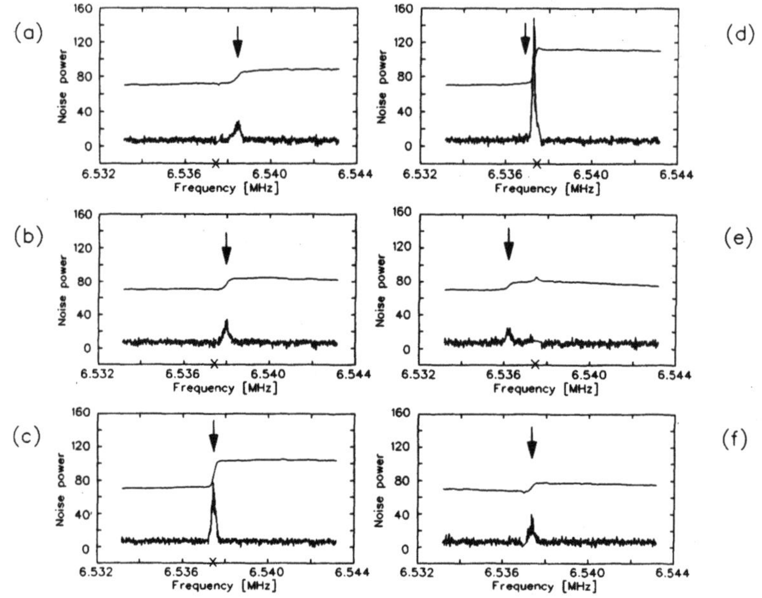

Fig. 6: Beam noise spectra of ions under the influence of the combined force of electron and laser cooling. Frequency corresponding to the laser resonance kept fix (cross). Frequency corresponding to the electron velocity varied (arrow). (a)–(e): Laser on; (f): laser off. Average of 10 scans of 5 s taken at a delay of 3 s after each injection; 2×10^7 stored ions. Noise power given in arbitrary units suitable for comparison to Fig. 8. Upper traces show integral of noise power.

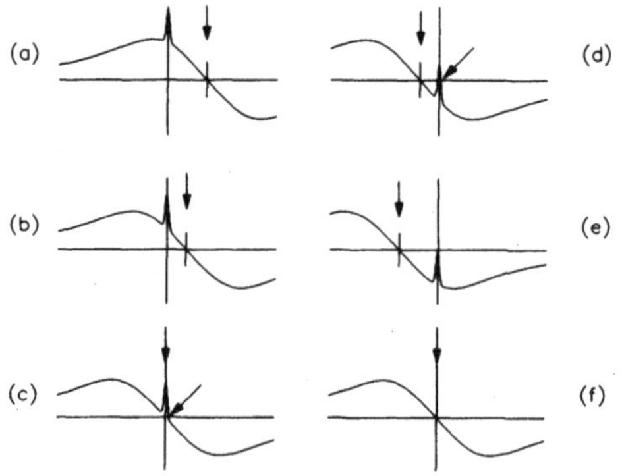

Fig. 7: Sketch of the combined force of electron and laser cooling for relative detuning simulating the experimental situation (Fig. 6a–e).

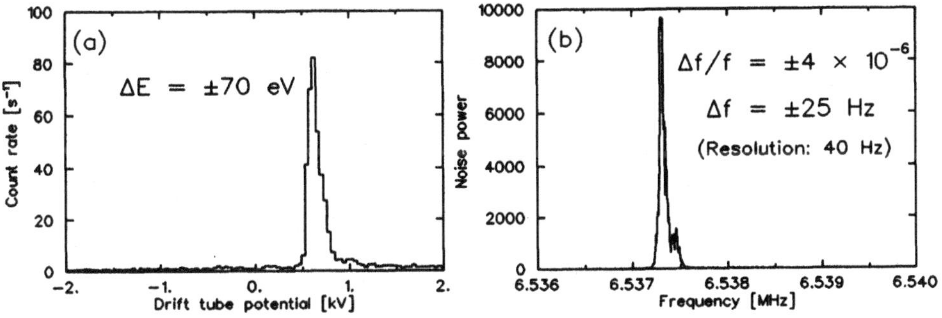

Fig. 8: Longitudinal beam diagnostic during simultaneous laser and electron cooling for 2×10^7 ions at a delay of 5.6 s after injection. (a) Doppler profile obtained by varying the drift tube potential close to the fluorescence detector (electron velocity tuned according to Fig. 6d). (b) Beam noise spectrum obtained simultaneously; noise power given in arbitrary units suitable for comparison to Fig. 6f.

interacted with a fixed laser resonance over most of the interaction length; however, close to the photon detector the resonant velocity class was varied by sweeping the drift tube potential.

The structure observed in the the Doppler profile (Fig. 8a) is interpreted as the energy distribution of the ions close to the cooling resonance, corresponding to a voltage of +460 V. The fact that a single peak dominates indicates that most of the ions are in a single hyperfine state, determined by the the steady-state conditions of optical pumping. Note that a higher potential applied to the drift tube leads to stronger deceleration and brings ions of higher energy into resonance with the laser. Hence, the cooled distribution is situated on the high-energy side of the cooling resonance. The ion energy deviation given in Fig. 8a yields a momentum deviation of $\pm4\times10^{-6}$ and correspondingly a longitudinal temperature of 2.7 K. The effective width of the laser resonance was subtracted quadratically from the observed signal width, cf. Table 4.

The Schottky noise spectrum recorded under the same conditions is shown in Fig. 8b. It can be seen that the integrated beam-noise power recorded by the pick-up increased by a factor of ≈ 500 as compared to the situation with electron cooling alone at a momentum deviation of $\pm3\times10^{-5}$ (Fig. 6f). The relative width of the beam noise spectrum is $\pm4\times10^{-6}$ and corresponds to the resolution of the spec-

Table 4: Laser performance during the $^9Be^+$ cooling experiment and estimated equilibrium temperature for laser cooling.

	This exp.	Optimum[a)]
Laser frequency jitter	± 100 MHz	—
Velocity deviation of resonance	$\pm 3 \times 10^{-6}$	$\pm 3 \times 10^{-7}$
Cooling time	4.4 ms	22 μs
Equilibrium temperature[b)]	0.6 K	0.4 mK
Equilibrium momentum spread[b)]	$\pm 2 \times 10^{-6}$	$\pm 4 \times 10^{-8}$

[a)] From Table 2 of Ref. 7
[b)] Calculated for IBS heating rate of 12 meV/s (see text)

trum analyser; because of the observed noise enhancement, this relative linewidth probably cannot, as in usual Schottky spectra, be interpreted as the ion revolution frequency spread.

It appears interesting to compare the preliminary result for the longitudinal equilibrium temperature to a realistic estimate of the laser cooling performance during the present experiment (Table 4). Compared to the optimum conditions[7)] of $^9Be^+$ laser cooling, the width of the laser resonance as a function of the ion velocity was higher by a factor of about 10 because of a jitter of the laser frequency occuring on a short time scale ($\ll 1$ s). This will be improved in the near future by stabilizing the argon ion laser. In addition, the laser intensity was only half of the saturation intensity (see Table 2), and the interaction length of the ions with the laser might have been reduced with respect to the geometrical overlap by optical pumping. Considering that the square of the resonance width enters into the expression for the cooling time, a reduction factor of 200 with respect to the optimum is estimated.

The heating due to IBS is taken into account for 2×10^7 stored ions and a beam divergence of 7×10^{-4}. (In Fig. 8, obtained at a delay of 5 s after injection, the ion beam was not yet cooled down transversely.) The calculation using Eq. (6) shows that an equilibrium temperature of the order of 1 K can be expected in this case. Comparing this value with the experimental result of 2.7 K and considering that the estimate of the cooling time in Table 4 might still be optimistic, the equilibrium temperature observed for laser cooling could very well be determined by IBS, as was concluded above also for electron cooling.

3.3 Search for Indirect Transverse Cooling

During a recent run, a short investigation was done concerning the influence of laser cooling on the transverse beam size during simultaneous electron cooling. The number of stored ions was 4×10^6 and sufficient time after injection was allowed in order to reach the transverse equilibrium at a divergence of 1×10^{-4}. No significant change of the beam size was observed at the profile monitor depending on the laser being turned on or off. The equilibrium properties of the $^9Be^+$ beam are in the range where transverse–longitudinal heat transfer could lead to transverse cooling times below 1 s. The heat transfer rate can be obtained from Eq. (6) as the ratio $k_B T_{\|}/\tau_{\perp} = 0.24$ eV/s. At the transverse ion temperature of 0.15 eV/k_B this represents a cooling time constant of 0.6 s.

Indirect transverse cooling seems to depend most critically on efficient laser cooling and on a careful choice of the ion beam intensity, adapted to the longitudinal cooling rate. Therefore, in further studies the longitudinal beam properties should be monitored and the dependence of the longitudinal temperature during laser cooling on the number of stored ions should be determined. Moreover, as was discussed in Section 2.2.2, a better understanding of the dominant transverse heating process in the electron cooling equilibrium and numerical IBS calculations for the TSR lattice are needed.

4. CONCLUSIONS

The first results on electron cooling of $^9Be^+$ ions at the TSR show that the cooling rates are high enough to obtain longitudinally and transversely cold, dense beams of these ions. Magnetized electron cooling, required for cooling the relatively slow and heavy ions within a few seconds, was clearly observed. Diagnostics using laser fluorescence, Schottky noise, and a transverse beam profile monitor yielded detailed, consistent information about cooling rates and the equilibrium beam properties.

The mechanism of simultaneous laser and electron cooling worked according to the expectations; in the steady state, a strong longitudinal velocity compression was caused by the laser. Intrabeam scattering seems to determine the longitudinal equilibrium temperature both for electron and for combined laser/electron cooling; however, more measurements and better diagnostic methods are required in order to analyze in detail the equilibrium velocity distribution during laser cooling.

The ion beam densities obtained by the transverse electron cooling are high enough to enable fast transverse–longitudinal heat exchange. Although the indirect transverse laser cooling of ion beams, which might be caused by this exchange, could not be observed in an improvised first attempt, diagnostic measurement have been demonstrated, which improve the chances to produce this effect in a more systematic approach.

Further studies will focus on the improvement of the diagnostic methods for laser-cooled beams and on the optimization of laser cooling using the advantages offered by its combination with electron cooling.

REFERENCES

1. S. Schröder *et al.*, Phys. Rev. Lett. 64, 2901 (1990).
2. W. Petrich *et al.*, these Proceedings.
3. Aarhus group, these Proceedings (paper presented by J. Hangst).
4. J.P. Schiffer and P. Kienle, Z. Phys. A 321, 181 (1985).
5. R.W. Hasse, I. Hofmann, and D. Liesen *(edts.)*, "Proc. Workshop on Crystalline Ion Beams", Wertheim, Fed. Rep. of Germany, Oct. 4-7, 1988 (GSI-89-10, Darmstadt, 1989).
6. V.I. Balykin *et al.*, J. Opt. Soc. Am. B 2, 1776 (1985); A. Aspect *et al.*, Phys. Rev. Lett. 57, 1688 (1986); A. Aspect *et al.*, Phys. Rev. Lett. 61, 826 (1988).
7. A. Wolf *in* Ref. 5, p. 307-320.
8. Ya.S. Derbenev and A.N. Skrinsky *in* "Physics Reviews", vol. 3, ed. by I.M. Khalatnikov (Harwood Acad. Publ., Chur, 1981), p. 165-238, and references therein.
9. B. Hochadel *et al.*, to be published.
10. H. Poth, Nature 345, 399 (1990).
11. Ya.S. Derbenev and A.N. Skrinsky, Part. Accel. 8, 1 (1977).
12. A.H. Sørensen and H. Bonderup, Nucl. Instrum. Methods 215, 27 (1983).
13. A. Wolf, Thesis (KfK 4023, Karlsruhe, 1986).
14. A.H. Sørensen *in* CERN 87-10, ed. by S. Turner (CERN, Geneva, 1987).
15. H. Bruck, "Circular Particle Accelerators" (LA-TR-72-10, Los Alamos, 1972), p. 283*f.*

PHOTOCATHODE STUDIES FOR AN ULTRACOLD ELECTRON BEAM DEVICE

S. Zwickler, D. Habs, P. Krause, R. Neumann, D. Schwalm, and A. Wolf
Physikalisches Institut der Universität Heidelberg and
Max-Planck-Institut für Kernphysik,
D-6900 Heidelberg, Federal Republic of Germany

ABSTRACT

An additional electron cooling device will be constructed for the Heidelberg Test Storage Ring TSR. The electron temperatures will be two orders of magnitude lower than those achieved with a conventional electron cooler. First experimental results concerning the GaAs cathode and calculations of the acceleration section are presented.

1. INTRODUCTION

By using a cold GaAs photocathode instead of a thermocathode, an ultracold electron beam with a rest-frame energy spread of about two orders of magnitude smaller than in present electron cooling devices may be obtained [1]. At heavy ion storage rings like the TSR or the ESR such an electron beam device will be of great use both for electron cooling of ion beams to even lower temperatures and for the study of electron-ion interactions, where narrow resonances can be scanned with high resolution.

The work on the ultracold electron beam has been directed towards the development of the photocathode mounted in a special ultrahigh-vacuum setup, and towards the design of an adequate beam transport system which preserves the low energy spread of the electrons.

2. PRINCIPLES OF OPERATION

The circular GaAs crystal of 2 mm diameter is activated by covering it with cesium and oxygen/NF_3 layers leading to a negative electron affinity. Electrons lifted from the valence to the conduction band by infrared laser light ($\lambda = 800$ nm) can leave this cold cathode and have a narrow energy distribution. They are accelerated by two electrodes and guided in a magnetic field of 300 Gauss to form a parallel beam. Several crystals have been installed in the UHV test setup and gave quantum yields between 0.1 and 1% for infrared laser light and 3% for HeNe-laser light. So, with the 400 mW laser diode used until now, it was possible to produce the projected current of 1 mA already with 130 mW incident light. In order to achieve higher quantum yields, a very careful cleaning of the crystal in UHV is necessary. Besides the conventional methods of heat cleaning, it was also observed that surface contaminants such as CO and CH_4 were flash-desorbed by intense infrared laser light (350 mW/mm^2 at 800 nm). This will be further investigated with higher light intensities produced by a titanium sapphire laser and also by an argon ion laser with a power of up to 40 W.

3. LIFETIME OF THE QUANTUM YIELD

Another important point concerns the lifetime of the activated surface. In contrast to conventional thermal cathodes, the activated GaAs-cathode requires a vacuum of less than 10^{-10} mbar for stable operation. Even then the lifetime is limited by desorbed ions that contaminate the surface. Therefore the electron beam must be well aligned in the vacuum chamber to prevent the electrons from hitting any of the electrodes.

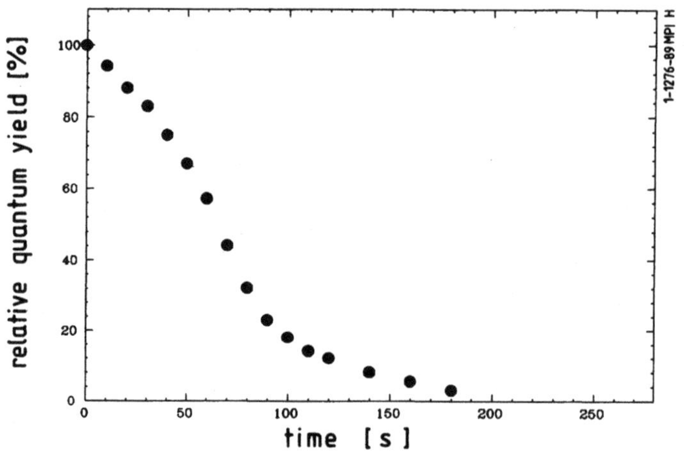

Fig. 1: Relative decrease of the quantum yield as a function of the illumination time at a laser power of 100 mW and the initial current of 0.8 mA.

An important observation was made when the activated surface was irradiated by a high laser intensity from the GaAlAs laser diode and no voltage was applied to extract the electrons. After each activation the lifetime of the photocathode showed a strong dependence on the light intensity, independent of the original quantum yield or the integrated extracted current. Figure 1 displays the fast decrease of the quantum yield at a laser power of 100 mW which corresponds to a current of 0.8 mA. In this case a contamination of the surface could be excluded because the electrons did not leave the cathode except during the short period when the quantum yield was measured. Hence, we conclude that it is the laser-induced cesium desorption which limits the stable operation of the photocathode at high current densities. A solution to this problem was found by continuous cesiation: at a low laser intensity and a Cs partial pressure of $6 \cdot 10^{-11}$ mbar in the test setup, an equilibrium between cesium desorption and adsorption was maintained so that a current of 0.1 mA could be extracted during several days.

4. THE ELECTRON ENERGY SPREAD

For determining the energy spread of the electron beam a retarding-field energy analyzer is situated at a distance of 4 cm in front of the cathode. By a small (25 μm) pinhole only a 10^{-4} fraction of the beam, originating from the cathode, is measured. In contrast to large installations intrabeam relaxation can be neglected at low electron energies (about 10 eV) and it is possible to measure the real temperature in longitudinal direction of the electrons emerging from the cathode.

The temperature of a conventional heated cathode installed in the test setup was determined to be 140 meV [FWHM of the energy distribution curve (EDC)] for electron currents in the range of 0.5 and 40 μA.

First measurements with the photocathode have been completed recently. After installing the GaAs cathode in the test setup it was activated by illuminating it with a 5 mW HeNe Laser (photon energy of 1.96 eV). It was observed that the energy distribution of the electron beam changed with the steps of activation. In the beginning, when the crystal was coated only with cesium, the electron beam had an energy spread of 120 meV. After additional NF_3 and Cs layers the energy spread changed to 130 meV and after ten activation steps even for small currents an energy spread of already 150 meV was obtained. Using the laser diode (photon energy 1.55 eV, laser wavelength 800 nm) smaller temperatures could be obtained, but this situation was comparable to the former one: at the beginning of activation when the quantum yield was about 0.1% an energy spread of only 85 meV was measured. After 30 activation steps following the so called YoYo technique a quantum yield of 1% was reached, but the EDC had a FWHM of 130 meV. This can be explained by the fact that for small values of negative electron affinity (NEA) the vacuum level acts as a low-energy cutoff on the electron distribution, thus producing narrowed EDC's. A higher value of NEA gives higher quantum yields, but also broader energy distributions. In contrast to other measurements [2,3] it could be demonstrated that the EDC is not dependent on the electron current between 0.5 and 30 μA. The beam energy was 12 eV. The EDC could not yet be determined for higher currents because the cathode power supply for higher voltage with appropriate accuracy is presently under construction.

Figure 2 demonstrates the dependence of EDC on the laser wavelength. Illuminating the crystal with light from the Ar^+ laser gave a FWHM of 220 meV, the HeNe laser gave 150 meV and the infrared light from the TiSa laser (same wavelength as laser diode) 120 meV of energy width. The fluctuations of current with Ar^+ and TiSa laser light (which is pumped by the Ar^+ laser) are directly correlated to power fluctuations of the Ar^+ laser. The TiSa laser has better optical characteristics than the laser diode: the beam is not so divergent, it has a Gaussian profile and illuminates the crystal homogenously. Furthermore the wavelength can be tuned easily.

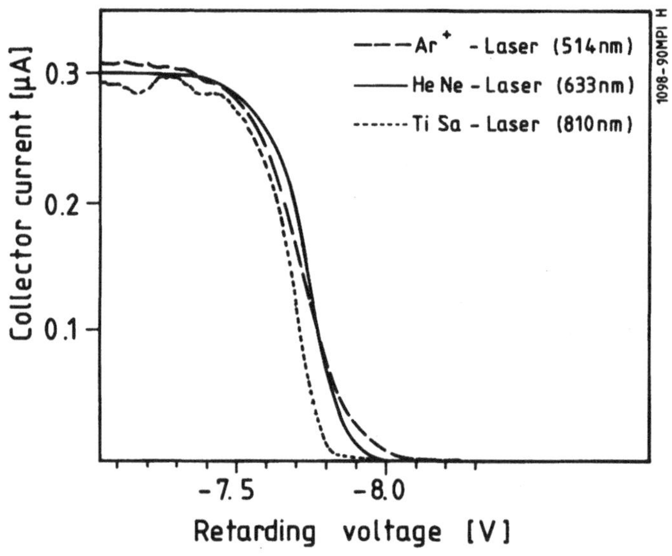

Fig. 2: The current of the collector (after the retarding field diaphragm) as a function of retarding voltage gives the integrated EDC. The total photocurrent was 1 μA.

Measurements of other groups [4] in the nA region show that energy spreads of 30 meV can be obtained with infrared light by cooling the GaAs to liquid nitrogen temperature. The task of further experiments will be to reach lower temperatures of the electron beam by cooling the cathode, to measure the temperature even at higher currents and higher energies, and finally to find the right wavelength and activation method to get an electron beam with high current and low temperature. It could be demonstrated that permanent cesiation of the crystal at constant electron current did not change the width of EDC.

5. OUTLOOK: THE BEAM TRANSPORT SYSTEM

In the beam transport system, a low energy spread of the electrons will be maintained by accelerating the electrons adiabatically with respect to the longitudinal plasma oscillations in the beam [5]. In the final design the electrons are accelerated non-adiabatically to 130 eV at the anode and will be decelerated again to 5 eV. This is the lowest energy possible in view of the space-charge forces in the beam.

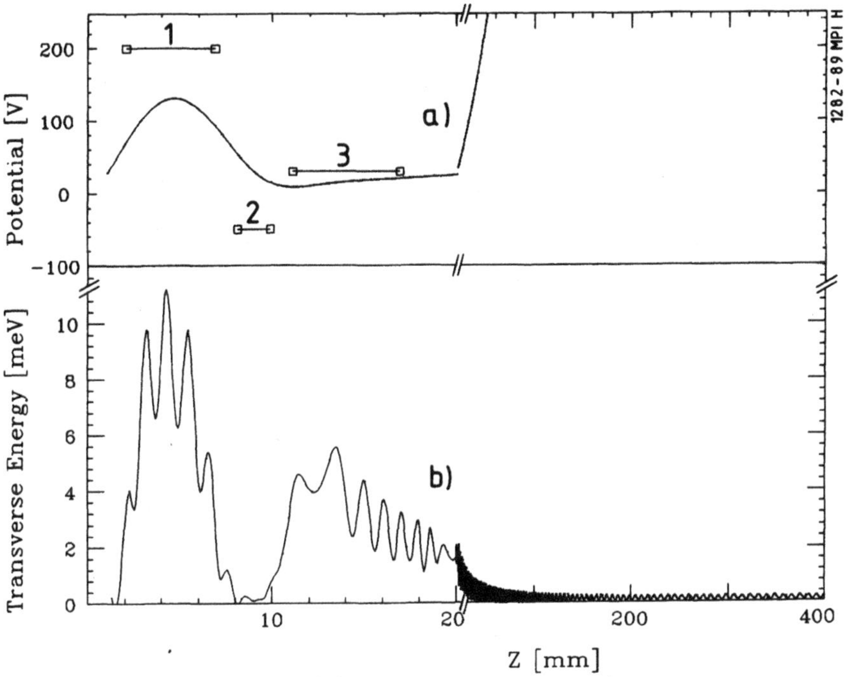

Fig. 3: Computer modelling of an electron gun with smooth optics. (a) Potential on the axis, where the bars numbered 1-3 indicate the length and potential of the first three electrodes. (b) Transverse energy at the boundary of the beam, $I_e = 1$ mA, $E_{kin} = 7$ keV, $B = 2$ kG.

Adiabatic acceleration then will take place in a NEC acceleration tube. Based on the computer simulations it was decided to use a large diameter of 18 mm for all acceleration electrodes. This results in a very smooth electric field gradient along the beam axis except for the deceleration section (Figure 3a). In a magnetic field of 2 kG, the transverse electric field in the acceleration region creates a very small additional transverse electron energy of 3 meV at the entrance which relaxes to below 0.2 meV at the end of the acceleration section (Figure 3b). For comparison, the transverse temperature of the electrons at the cathode is expected to be 1 meV. The transverse energy already includes the space-charge effect. Therefore the decrease of the transverse energy during the acceleration is caused by the lowering of the electron density. Since the low transverse energy of this smooth optic is achieved without using resonant focusing, all parameters connected with electric and magnetic fields such as the final beam energy can be chosen independently from each other.
The beam transport and acceleration system will soon be tested in a linear test bench using a thermoionic cathode, which will later be replaced by the GaAs cathode. The longitudinal temperature will be measured by an energy analyzer similar to the one used in the photocathode studies.

ACKNOWLEDGEMENT

Special thanks are due to Valerie Lebedev and Vaino Kudelaien for the help in the development of the energy analyzer.

REFERENCES

[1] D. Habs, J. Kramp, P. Krause, K. Matl, R. Neumann, and D. Schwalm, Physica Scripta T22, 269 (1988).
[2] T. Dodt, Dissertation, Kernforschungsanlage Jülich (1988).
[3] K. Ertl et al., Rev. Sci. Instrum. 59, 1938 (1988).
[4] H.J. Drouhin et al., Phys. Rev. B 31, 3859 (1985).
[5] V.A. Lebedev and A.N. Sharapa, Sov. Phys. Tech. Phys. 32(5), 594 (1987).

TEMPERATURE RELAXATION AND ADIABATIC ACCELERATION AT MAGNETIZED ELECTRON FLUX

A.V. Aleksandrov, N.S. Dikansky, N. Ch. Kot., V.I. Kudelainen, V.A. Lebedev, P.V. Logachov.

Institute of Nuclear Physics, Novosibirsk, URSS.

1. INTRODUCTION

The development of electron cooling method proposed by G. I. Budker (1) has required the elaboration of the methods of electron beam production with the minimal possible velocity spread. The reduction of electron temperature in the electron flux allows to increase the time decrements and to decrease the equilibrium temperature. Both aspects are very important inthe applications. The discovery of the longitudinal order in a motion of the cooled protons on the NAP-M installation was the new step on principle (2). This discovered a new direction in the acceleration physics of ultracooled beams that is characterized by a small value of the kinetic energy of electrons as compared with the average potential energy of their interaction (3.11). In the absance of intrabeam scattering the electron's acceleration in electron gun results in a decreasing of longitudinal temperature (4):

$$T_{\parallel} = \frac{T^2k}{2W} \qquad (1)$$

At the same time the transverse temperature (is assumpted the absence of the transverse velocities perturbation) does not change:$T_{\perp}$ = Tk. Here Tk is the cathode temperature, W is the electron energy.If the accompanyning magnetic field is sufficiently strong the transverse movement of electrons is magnetized and their effective temperature is defined by the longitudinal temperature, which value can be extremely small. At W = 500 eV Tk = 0.1 eV we get $T_{\parallel} \cong 10^{-5}$ eV $\cong$ 0.1 K.

The intrabeam scattering leads to an increasing of the longitudinal temperature. We can distinguish two effects: transverse-longitudinal relaxation when the temperature increases due to the transfer of the transverse motion energy into the longitudinal one (Boersch effect) (5), and the longitudinal-longitudinal relaxation when the temperature increases due to the thermodynamic relaxation only in the longitudinal direction (6).

The transverse-longitudinal relaxation is easily suppressed by sufficiently strong of suppressing of the longitudinal-longitudinal relaxation have strong confinements. For sufficiently small electron density the particular suppress can be received by use of more slowly (adiabatic) acceleration than in the Pierce gun (7). This paper is devoted to the expirimental study of the longitudinal temperature relaxation in electron beam and the study of opportunities of the adiabatic acceleration use for the supercooled electron beam preparation.

2. TRANSVERSE - LONGITUDINAL RELAXATION.

Immediately after electrostatic acceleration the longitudinal temperature in the rest frame is much lower than the transverse one. In further way of the electron beam its longitudinal temperature increases due to the transverse- longitudinal relaxation up to the equality of both temperatures. If one can neglect the influence of the magnetic field at $T_{\|} \ll T_{\perp}$ the relaxation can easily be calculed (7):

$$\frac{dT_{\|}}{dz} = \frac{\pi e^3 j L_c k}{W} \sqrt{\frac{m}{T_{\perp}}} \tag{2}$$

where $z = v_e \cdot t$ is a longitudinal coordinate, L_c ia a Coulomb logarithm, k is a numerical coefficient depending on the kind of the distribution function of electrons over transverse velocities (for the Maxwell distribution $k \cong 0{,}87$), j is a current density, W is an electron energy and $T_{\perp}$ is its transverse temperature. The increase of the longitudinal temperature goes quite rapidly, so at W = 450 eV, j = 0,5A / cm^2 both temperatures become equal through 3m. This effect can be suppressed by sufficiently strong accompanying magnetic field, so that an average Larmour radius of the electron transverse rotation $\rho_{\perp} = \sqrt{T_{\perp}^2}\, mc/eH$ is much lower than the distance between them:

$$\rho_{\perp} = \ll n^{1/3} \tag{3}$$

and the longitudinal electron temperature in sufficiently small so that Larmour radius is much less than the minimal probable distance between electrons:

$$\rho_{\perp} \ll e^2 / T_{\|}$$

Experimental data on the study of the transverse-longitudinal relaxation on the NAP-M (6) and " Solenoid model" (7) installations have been united by an empirical formula:

$$\frac{dT_{\|}}{dz} = \left(\frac{dT_{\|}}{dz}\right)_0 \mathrm{EXP}\left\{-\frac{2{,}8\, e^2}{\rho_{\perp}\,(e^2 n^{1/3} + T_{\|})}\right\} \tag{4}$$

where $(dT/dz)_0$ is a rate of the transverse - longitudinal relaxation without magnetic field (2).This expression has a relative accuracy about 30% for the next region of parameters: a magnetic field - 1-3 kG, a current density - 3-300 mA / cm^2, an electron energy -100-600eV.

3. LONGITUDINAL-LONGITUDINAL RELAXATION.

If the transverse-longitudinal relaxation is suppressed by a magnetic field then the main contribution to an increasing of the longitudinal temperature is given by the longitudinal-longitudinal relaxation.

In an ordinary electron gun the acceleration of electrons proceeds quickly as compared to the period of the plasma oscillations of the electrons. Therefore for the time of acceleration the relative location of the electrons in the beam practically does not change and their initial location with the chaotic distribution is preserved. Since the longitudinal temperature after acceleration is small (see (1)), the absence of correlations in positions of electrons leads to its increase by the value $\cong e^2 \cdot n^{1/3}$ (due to the electrostatic repulsion). As a result we get the estimation of the longitudinal temperature:

$$T_{\parallel} = \frac{Tk^2}{2W} + C e^2 n^{1/3} \tag{5}$$

where the constant C is about one. We can build a simple model to estimate the constant C. We guess $U_0 (n,T_{\parallel})$ is the total internal energy of the thermodynamically equilibrium (in the longitudinal degree of freedom) magnetized electron gas. By equaling the internal energy of electrons just after fast acceleration to the one after establishing the thermodynamic equilibrium we get:

$$\frac{Tk^2}{2W} + U_c = U_0 (n, T_{\parallel}) \tag{6}$$

where U_c is the correlative energy of nonequilibrium state of an electron beam just after fast acceleration. In the high temperature region $T \gg e^2 n^{1/3}$ the expression for the correlation energy of equilibrium state is well know (8):

$$U_c = -\frac{e^2}{2r_D} = -\frac{e^2}{2}\sqrt{\frac{4\pi e^2 n}{T}} \tag{7}$$

where r_D is Debye radius. In the opposite case $T_{\parallel} \ll e^2 n^{1/3}$ the electron beam is crystallized and the correlative energy is equal to:

$$U_c = T_{||}/2 - Ae^2 n^{1/3} \tag{8}$$

The coefficient 1/2 in the first attend is connected with the magnetization of transverse degrees of freedom. The constant A is determined by the grid type. The state with the minimun energy is achieved for a volume-centered grid for which A ≅ 1,4. For other types of grids the constant value A will be somewhat less. Let us choose the dependence of the internal energy of equilibrium state in the form (9) satisfying to both asimptotics (7) and (8) in order to simplify further calculations:

$$U_0(n,T_{||}) = \frac{T_{||}}{2} + U_c(T_{||}) = \frac{T}{2}{}_{||} - e^2(\pi n)^{1/3}\sqrt{\frac{e^2(\pi n)^{1/3}}{T_{||} + e^2(\pi n)^{1/3}}} \tag{9}$$

Just after fast acceleration the nonequilibrium correlative energy U_c is small in comparison with $e^2 n^{1/3}$.Really the relative position of electrons practically do not change during the time of acceleration. Therefore the correlative energy expressed in the $e^2 n^{1/3}$ units does not change and equals the correlation energy calculed near a cathode surface where the condition of a weak nonideally of plasma is fulfilled ($T_{||} \gg e^2 n^{1/3}$). For the usual thermocathode's temperature Tk = 0.1 eV near it's surface we have:

$$\frac{T_{||}}{e^2 n^{1/3}} \cong \frac{T_k}{e^2}\left(\frac{e}{j}\sqrt{\frac{T_k}{m}}\right)^{1/3} \cong \frac{T_k^{7/6}(\text{eV})}{j^{1/3}(\text{A/cm}^2)} \cdot 1300$$

and the condition of a weak nonideally of plasma is fulfilled with a good store. In this case we can neglect in equation (6) by values U_c' and $T_k^2/2W$ in comparison with $U_0(n,T_{||}) \cong e^2 n^{1/3}$ (see (9)). As a result if $T_k^2/2W \ll e^2 n^{1/3}$, we have :

$$0 = \frac{T}{2}{}_{||} - e^2(\pi n)^{1/3}\sqrt{\frac{e^2(\pi n)^{1/3}}{T_{||} + e^2(\pi n)^{1/3}}} \tag{10}$$

The solution of this equation is

$$T_{||} \cong 1.9\, e^2 n^{1/3}. \tag{11}$$

We carried out the numerical simulations to get a more accurate dependence of the electron gas internal energy upon temperature. If the magnetic field is sufficiently strong one can neglect the transverse movement of electrons taken into account only the longitudinal movement and not changed the transverse coordinates. At the beginning the electrons are situated inside a uniformly positive charged globe to have the crystal (or random) sites. They are given the longitudinal velocities according to the Maxwell distribution. For the solution of the movement equations we used a second order scheme with changed steps. To calculate a dependence of the electron gas internal energy upon temperature we introduce the friction force proportional to the particle velocity.Owing to the weak fading ($\tau_{fade} = 8\tau_{plasma}$) the system goes through the states with different temperatures and the particles move fit by fit to the places where the system attained a minimun of the energy.The calculated dependence of the electrongas internal energy upon temperature is shown on Fig. 1 for region: $0.1\, e^2 n^{1/3} < T_{||} < 2e^2 n^{1/3}$.The corresponding values of the correlative energy is equal to:

$$U_c (T_{||} = 0) = - (1.33 \pm 0.05)\, e^2 n^{1/3} \tag{12}$$

for the volume centered grid and

$$U_c (T_{||} = 0) = - (1. \pm 0.1)\, e^2 n^{1/3} \tag{13}$$

for the random initial position of electrons. Errors in (12) and (13) are defined full quantity of particles using in calculations (see Fig. 2).

More high value of correlative energy for the random deposition in comparison with the volume-centered grid is explained by "freezing" of the transverse coordinate fluctuations in the longitudinal magnetic field .That gives the sufficient correlation energy increase hence the additional corrections in (9) and (11). In this case after the end of the longitudinal-longitudinal relaxation the electron temperature in the rest frame is:

$$T \cong 1.6\, e^2 n^{1/3} \tag{14}$$

Time of the longitudinal-longitudinal relaxation have the same order as a period of the plasma oscillations. In contradistinction to the transverse-longitudinal relaxation the longitudinal-longitudinal one is not suppressed by a magnetic field, and as a rule it defines the longitudinal temperature. Begun from the electron energy 100 eV and the current density $4\text{mA} / \text{cm}^2$ the second item in (5) is equal to the first one and $T_{||} \cong 1.5 \cdot 10^{-4}$ eV.

4. THE ADIABATIC ACCELERATION.

We introduce the parameter of adiabaticity of acceleration λ as a relative change of the longitudinal temperature due to acceleration during a plasma oscillation period :

$$\lambda = \frac{1}{T_{\parallel}} \frac{dT_{\parallel}}{dt} \frac{1}{\omega_p} \tag{15}$$

During the acceleration of electrons in a gun operating in the "mode 3 / 2" (and more than that in the gun operating in the mode of emission limitation) the adiabaticity criterion $\lambda \ll 1$ is not satisfied. Really for Pearce gun the parameter λ is equal to $2\sqrt{2}$ and does not change during the acceleration (9). In the case of sufficiently slow acceleration $\lambda \ll 1$ the plasma oscillations have enough time to mix the density fluctuations, and the increasing of the longitudinal velocity spread due to the longitudinal-longitudinal relaxation is suppressed by a following acceleration.

Let us find the change of the longitudinal electrontemperature during slow (adiabatic) acceleration. The change of electron internal energy at acceleration on dW is:

$$dE = -T_{\parallel} \frac{dW}{2W} + \frac{1}{3} U \frac{dn}{n} = T_{\parallel} \frac{dn}{n} + \frac{1}{3} U \frac{dn}{n} \tag{16}$$

where U is the potential energy of electrons : $U(T_{\parallel}) = U_c(T_{\parallel}) - U_c(0)$.
Hence, we have a differential equation describing the temperature variation during acceleration:

$$\frac{\partial U_0}{\partial T_{\parallel}} dT_{\parallel} = \left(T_{\parallel} + \frac{1}{3} U\right) \frac{dn}{n} \tag{17}$$

Fig. 3 shows the solution of this equation in dimensionless variables, where $U_0(n, T_{\parallel})$ expressed by (9).
Fig. 4 shows dependencies of the longitudinal temperature (according to (17)) upon the current at beam energy 470 eV, for the initial temperature of electrons emitted from an oxide cathode: Tk = 0.1 eV, and GaAs photochatode: tk = 0.03 eV. At the same figure are shown corresponding dependencies for the usual (fast) acceleration in the Pierce gun. It is necessary to emphasize, that adiabatic acceleration means the usual acceleration in the Pierce gun up to the energy W_{min}, and following wuch slower acceleration up to full energy. W_{min} is defined by conditions of obtaining of the essential current. Therefore the temperature at Fig. 4 is a function of the full current and does not depend on beam radius. Fig. 5 shows dependencies as on Fig. 4, but for parameters of ion storage ring with electron cooling (electron energy is equal 100 keV).

5. DESCRIPTION OF THE INSTALLATION.

The temperature relaxation studying in electron beam was carried out at the modified variant of "Solenoid model" installation (7). A general scheme of the installation is given on Fig. 6, and main parameters are the following:

Electron energy	50 - 800 eV
Electron beam current	0.01 - 10 mA
Radius of electron beam	1 mm
Magnitude of magnetic field	1 - 4 kG
Solenoid length	2.88 m
Accelerating structure length	0.25 m
Length of drift space available for the moving energy analyzer	1.8 m
Vacuum	5.10^{-8} Torr

The electron gun (13), placed in the magnetic field of the solenoid (12). The electron beam moves along the solenoid axis to the energy analyzer (6), that is the electron collector at the same time. The electron gun has a "smooth" optics (10). It means, that the anode aperture size is significantly lager than the beam diameter. In the strong longitudinal magnetic field that allows to get very small transverse velocities of electrons, which appears due to nonideal gun optics. After fast acceleration the electron beam passes through the adiabatic acceleration structure of five pipes with 16mm diameter.

In order to realize the adiabatic acceleration, the potential of the accelerating structure was distribueted along the beam axis in accordance with dependence $\phi \sim Z^{4/3}$. This one corresponds to the constant value of λ- parameter (15). The full lenght of the adiabatic structure was selected to provide more slow acceleration, than in the Pierce gun. It corresponds to condition $\lambda < 1$, which is necessary for damping of the longitudinal-longitudinal relaxation during the acceleration. To simplify the current control, the gun anode and the devider of the adiabatic structure have the same high voltage source. The voltage between cathode and anode determines beam current.

At the fast acceleration regime all structure for the adiabatic acceleration has zero potential, that converts it to usual drift space. As in the previous case, the voltage in between cathode and anode determines beam current, but the acceleration up to full energy takes place immediately in the gap between the anod and the first pipe of structure. It is necessary to note, that acceleration in this case will be faster than in Pierce gun, and parameter λ will be lager, then Pierce one ($\lambda = 2\sqrt{2}$). For example for beam current 6μA and energy 1500 eV $\lambda_{max} \cong 60$.

After adiabatic structure the beam passes through the drift space (5), which consists of five pipes, with 10mm radius.The energy analyzer (6) is placed inside these pipes and can move along their axis. All these pipes usually has zero potential, if it is not emphasaized especially.

The scheme of measurements of the longitudinal temperature is the same as described in (7). This method is based on the analysis of the energy spread in a fineelectron beam cut from the main beam with small hole (0.025mm radius). The analysis is performed with decelerating electric field of the analyzer diaphragm. One can obtain an integral distribution function over energy by varying the diaphragm potential on respect to the cathode potential and simultaneously measuring the current of the collector (9). Fig. 7 shows the integral distribution function for electrons over the energy and its derivation - the distribution function over energy. In calculations of the energy spread the integral distribution function is used. The central part of the integral distribution function from 0.25F to 0.75F is approximated by stright line. The least squares method is used for this approximation. The energy spread ΔW is determened as slope of this line.

There are two the most important factors, that determines quality of the analyzer. The first is connected with defocusing of the fine electron beam by the immersion lens of a small hole. It leads to additional longitudinal energy spread in a fine beam. Actually, the focus distance of that lens is f = 4d, where d is the distance between the hole and the analyzer diaphragm. Taking into account the relation between the decelerating electric field and the electron energy ($W=e_{el} \cdot d$), it is easy to obtain an additional energy spread:

$$\Delta W = W \cdot (r \cdot 4d)^2 \qquad (18)$$

The whole transverse electric field is concentrated near the small hole. The size of this hole is much less than Larmour spiral step (~4mm), as a result an influence of magnetic field is negligible in the process of defocousing.In this case the additional longitudinal energy spread has linear dependence on the beam energy, and determines the energy resolution for the analyzer. Fig.8 shows the measured dependence of the longitudinal energy spread upon the beam energy for the analyzer with different (r / d) values.The most convenient way to increase an analyzer resolution is to decreasse r / d factor. But we have here additional limitations. The decreasing of r leads to decreasing of the signal-noise factor like r^2, and no another hand the increasing of the distance d requires a careful adjusting of the analyzer elements and high homogenity of the magnetic field in the analyzer region.

Another factor which determines the analyzer quality is the influence of the surface oxide films. Noncontrol potentials, which appears under the beam influence on this films strongly distorts the electric field inside energy analyzer. That significantly disfigures the shape of the measured electrons distribution function and leads to additional noncontrol longitudinal energy spread in a cutting beam. Fig.9 shows dependence of the shape of the distribution at the fixed current for the analyzer, made from different materials (copper and tantalum). The strong distortion of the distribution function for copper analyzer can not be explained by an axial-symmetric electric field in the hole. Experiments showed, that these effects for the tantalum analyzer after heating it by the beam are much less. The best way, to prevent any effect like these is to use gold for the small cutting hole, the analyzing diaphrgm and the collector.

6. TRANSVERSE - LONGITUDINAL RELAXATION.

Data obtained in experiments on the NAP-M and the "Solenoid model" has given only indirect information about dependence of the longitudinal temperaature upon coordinate along the beam, because the energy's spread is measured in the single point at the end ofthe drift space. Using of the moving energy analyzer allows to carry out direct observation of the transverse-longitudinal relaxation process in the electron beam. Fig. 10 shows the dependence of the longitudinal energy spread upon the coordinate Z for various magnitudes of magnetic field and beam current. Slope of the curve directly corresponds to the rate of the transverse-longitudinal relaxation. The strong damping of the relaxation by the longitudinal magnetic field is clearly seen. The irregularity of the experimental curves at large currents is connected with the considerable transverse beam displacement due to excitationof ion-electron transverse oscillation (12). The ionization of residual gas leads to the ion accumulation inside the electron beam right up to the equality ion and electron densities.

To check the dependence (4) the numerical steering of the constants C_1, C_2, C_3 to experimental data was fulfilled by the least squares method. So we have:

$$\frac{dT_{\parallel}}{dz} = \frac{\pi e^3 j}{W}\sqrt{\frac{m}{T_{\perp}}}\, C_1 \exp\left\{-\frac{C_2 e^2}{\rho_{\perp}\,(e^2 n^{1/3} + C_3 T_{\parallel}}\right\} \qquad (19)$$

For the adopt we used data from 24 experimental sets of temperature dependence upon coordinate: three values of magnetic field magnitude (1,2,3 kG) and eight values of current (from 0.1to 10mA) for every field magnitude. In every set the energy spread was measured in 17 points along drift space through the 10 cm step. The results show that experimental data is in satisfactory agreement with formula (19) for the following constants:

$$C_1 = 7.4^{+2.6}_{-1.7} \qquad C_2 = 0.35^{+0.18}_{-0.12} \qquad C_3 = 0.35^{+0.18}_{-0.12}$$

The poor accuracy of C_1 - C_3 (up to 50% by χ^2 - criteria) is explained by the weak dependence of formula (19) undersimultaneous varying of the constants. Let us note, that if it is necessary to estimate the relaxation rate out of the accepted region (magnetic field 1-3kGs, the electron density $\sim 10^6$-10^9 cm^{-3}, $T \cong 0.1$eV) formula (19) have to used with the certain care. The considerable difference between C_1-C_3 and previous results (4) (see (9)) is connected with the strong dependence of the expression (19) upon the initial temperature. In the first experiments the temperature was measured only at the end of the drift space and the initial temperature was unknown. For the calculations of initial temperature was used $T_{\parallel init} = 2e^2\, n^{1/3}$ that has led to mistakes in the work (9).

7. LONGITUDINAL-LONGITUDINAL RELAXATION.

For direct observation of the longitudinal-longitudinal relaxation (with damping of the transverse-longitudinal relaxation by strong magnetic field) the electron beam after forming in the electron gun was transported on low energy ($\cong$ 50eV) to the first gap between pipes attainable for the moving analyzer (see Fig.6) where beam was accelerated to the full energy (800eV). Fig.11 shows the dependence of the longitudinal temperature upon the distance from the point of fast acceleration in dimensionless variables. It is clearly seen that temperature relaxation occur for the time the same as the plasma oscillation period. For comparison the results of computer simulation of the longitudinal-longitudinal relaxation are shown on the same figure. The effect of the final temperature decreasing for more slowly acceleration is shown on the Fig.12. The upper curve corresponds to the fast acceleration (between points A and B), the lower one is obtained at two-step acceleration of the beam with the same parameters. At first it is the acceleration up to the intermediate energy 250eV (between points A and B), then the drift (15cm), where relaxation of temperature occurs,after that acceleration up to the full energy (between points C and D) and observation of the following relaxation. Fig.13 shows the dependence of the longitudinal temperature upon distance along the last part of the drift space after fast and slow (adiabatic) acceleration in the electron gun. The difference in initial temperatures is connected with damping of longitudinal-longitunal relaxation by adiabatic acceleration. The regular growth of temperature along coordinate is connected with the transverse-longitudinal relaxation. This figure clearly demonstrates the growth of the longitudinal-longitudinal relaxation rate with the temperature grouth.

Part of the experiments was carried out on the installation prototype with fixed energy analyzer located at a distance of 40 cm from the gun. In this experiment the accelerating structure with the full lenght of 35 cm was used. Fig.14 shows the dependence of the electron longitudinal energy spread in the rest frame upon the full beam energy for adiabatic and fast acceleration at current 100μA and magnetic field 1.2 kG. The solid line is a result of calculationsby theoretical dependence for the fast acceleration:

$$\Delta E = \sqrt{T_k^2 + C\,2\,W\,e^2\,n^{1/3}}$$

where the constant value C = 1.6 (1.9) is given by (11) and (14). For the adiabatic acceleration the curve correspondsd to the model mentioned above (17) and is considerable below than the experimental plots due to the small lenght of drift space (35cm) as that lenght in not sufficient for achievement of small parameters λ i.e. the good adiabaticity of acceleration. In fact for beam current 100μA and energy 600eV the parameter $\lambda \cong 0.5$. The decreasing of λ can be achieved by the current increasing, but this leads to the increasing of the initial energy of the adiabatic acceleration (that decrease itsefficiency for fixed full energy) and increase that transverse-longitudinal relaxation. The growth of the energy spread in the low energy region observed in Fig.14 is connected with the transverse-longitudinal relaxation.

Besides the adiabatic and fast regimes of acceleration the intermediate regime was realized. In this case the electron beam was accelerated in one part of the structure (up to 15 cm), the another part was the drift space, so the adiabatic parameter $\lambda \cong 1$ (remind that is $\cong 0.5$ for the adiabatic acceleration and $\lambda_{max} \cong 60$ for the fast one acceleration). The results for intermediate regime are between experimental curves for adiabatic and fast acceleration (see Fig.14). Note that with current decreasing the curves for the adiabatic and fast regimes begin to diverge at the smaler energies. The reason is that the flight time from cathode to collector is small in comperison with the period of plasma oscillations. In this case there's no one plasma oscillation and the longitudinal-longitudinal relaxation does not occur. Fig.16 shows that temperatures for adiabatic and fast regimes do not differ at small current (< 5μA) as electrons have no enough time for relaxation.

8. PREPARING OF EXTRA COLD BEAMS.

An application of photocathodes, that allow to obtain electrons with extremely low temperatures (13,14), gives some additional opportunities in using the adiabatic acceleration (see Figs 4 and 5). For general thermoemission cathode the temperature of emitted electrons is determined by cathode temperature ($\cong 0.1$eV). For the photocathode, which emits electrons from GaAs surface at liquid nitrogen temperature under the infrared radiation (800nm) the longitudinal energy spread is (13): $\Delta E < 0.03$eV and the transverse one is $\Delta E_{\perp} < 0.001$ eV.

Using of such photocathode do not give global gain in the achievement of minimal temperatures in the cooled beam of the heavy particles as transverse degree of freedom of electrons is excluded from kinetic of cooling, but longitudinal temperature is mainly determined by the electron density and the method of acceleration. However the decreasing of the transverse temperature of electrons leads to decreasing of the internal scattering in magnetized beam and oppurtinity to increase the electron density in cooling region without violation of magnetizing of ion-electron scattering at the fixed magnetude of magnetic field.

Electron beams with a temperature $T_{\parallel} < 0.01\, e^2 n^{1/3}$ are of great interest. At such low temperature the crystal, ordering appears in an electron beam. For obtaining such cool electron beam an adiabatic acceleration up to a high energy is required. But beam current is not higher then 20μA even at acceleration up to $\cong$ 50keV as an the adiabatic acceleration had to begin from very low energy, that bans extration of the large current. Within the current range used in the real experiments on electron cooling it is difficult to attain such a superlow temperature.

REFERENCES.

1 Budker G.I. - Atomic energy, 1967, v. 22, p. 346.

2. Dementyev E.N., et al., Measurement of Thermal Noise of Proton beam in a Storage ring NAP-M. -JTP. 1980, v.50 n.8 p.1717.

3. Parhomchuk V.V., Physic of Fast Electron Cooling, Doctoral Thesis. - Novosibirsk, 1985.

4. G.I.Budker, A.F. Bulushev, N.S. Dikansky, et. al., in: Proc.5-th All - Union Conf. on Charged Particle Accelerators.- Nauka, 1977, Vol.1, p.236.

5. Boersch H. , Experementelle Bestimmung der Energieverteilung in thermich ansgeloster Electronenstrahlen, Zs.Phys., 139, 115 (1954).

6. V.I. Kudelainen, V.A. Lebedev, et. al., Temperature relaxation in a magnetized electron beam. - Sov. Physic. JETP 56 (6), 1982.

7. N.S. Dikansky,V.I. Kokoulin, N.Ch.Kot, et. al., Fast Electron Cooling for a Region of small relative Velocities, in: Proc. XIII Intern. Conf. on High Energy Accel. - Nauka, 1987, v.1, p. 330.

8. L.D.Landau, E.M. Livshits, Statistical Physics, v.V., part 1 - Nauka, 1776.

9. N.S. Dikansky, V.I. Kudelainen, V.A.Lebedev, et. al., Ultimate Possibilities of Electron Cooling - Preprint 88-61 INP, Novosibirsk, 1988.

10. V.A.Lebedev, A.N. Sharapa, Forming of electron beam with small Transverse Velocities in the system with longitudinal Magnetic field. - Sov. Phys., JTP, v. 57, n.5, p. 975.

11. N.S.Dikansky, D.V. Pestricov, An Influence of Odering Effects on Novosibirsk, 1984; - Sov. Phys. JTP, v. 58, n.2, p. 289 and n.3, p. 505.

12. A.V. Burov, V.I.Kudelainen, V.A. Lebedev, et. al., The Experimental Research of Compensated State of the Electron Beam. - Preprint 89-116, INP, Novosibirsk, 1989.

13. Feigerle C.S., Pierce D.T., Sieler A., et. al. - Appl. Phys. Lett. 1986, v. 44, 866.

14. D.T. Pierce, R.J.Cellotta, G.C. Wang, et al. - Rev. Sci. Instr., 51 (4) Apr. 1980, p. 478.

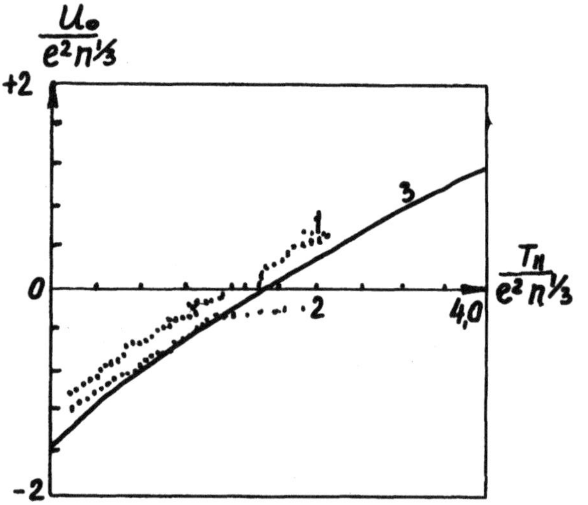

Fig.1.- Dependence of the internal energy of electron gas equilibrium state upon temperature.Points are results of computer simulations (1 - for the random initial disposition, 2- for the volume-centered grid), the smooth line is the result of calculation by formula (9).

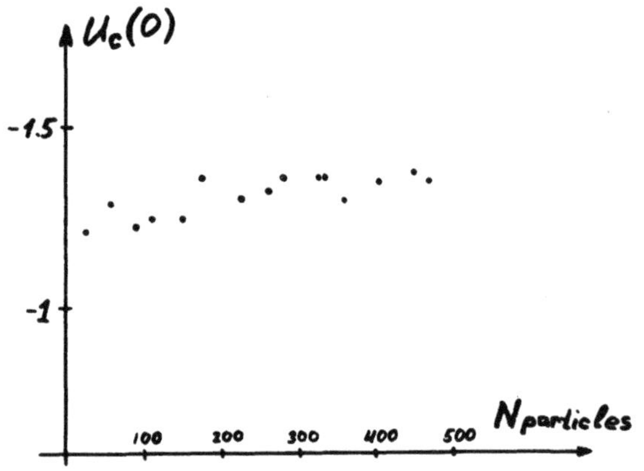

Fig.2 - Dependence of computer simulation results of the correlative energy at $T_{\parallel} = 0$ upon number of particles.

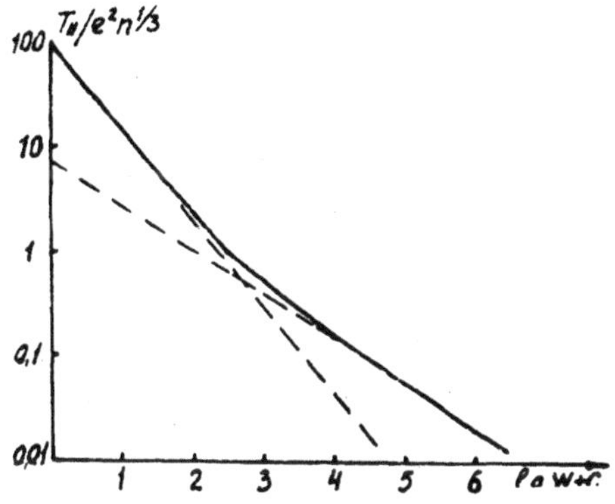

Fig.3 - Variation of the electron beam longitudinal temperature at an adiabatic acceleration as a function of the electron energy variation. C is the constant of integration. The dashedlines show asymptotics in the region of low $T_{\parallel} \propto 1/$ w 7/12 and high $T_{\parallel} \propto 1/w$ temperatures.

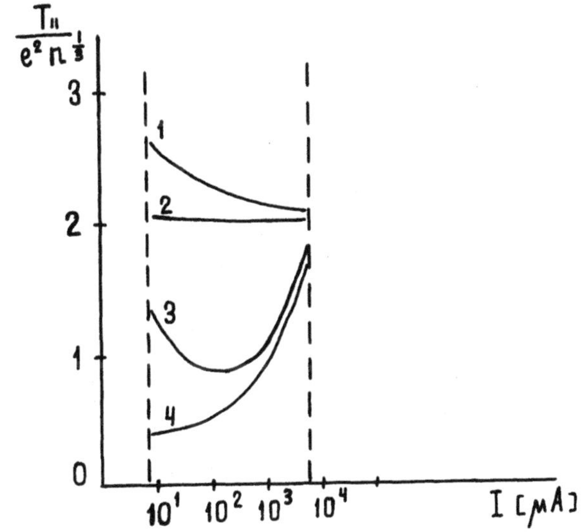

Fig.4 - Calculed dependence of the longitudinal temperature of electron beam upon beam current at W = 470eV for cathods with different temperature of emitted electrons: (1,3 - Tk = 0.1 eV; 2,4 - Tk = 0.03 eV) for fast (1,2) and slow (3,4) acceleration .

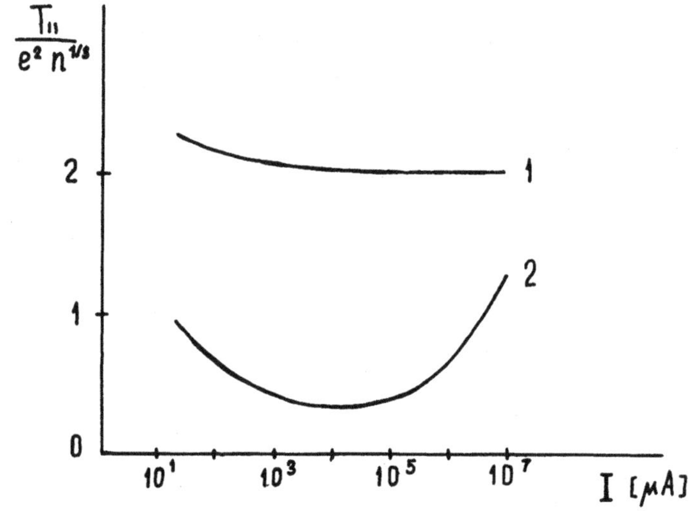

Fig.5 - Calculate dependence of the longitudinal temperature of electron beam upon beam current at tipical parameters of ion storage rings (W = 100keV, Tk = 0.3eV) for fast (1) and slow (2) acceleration.

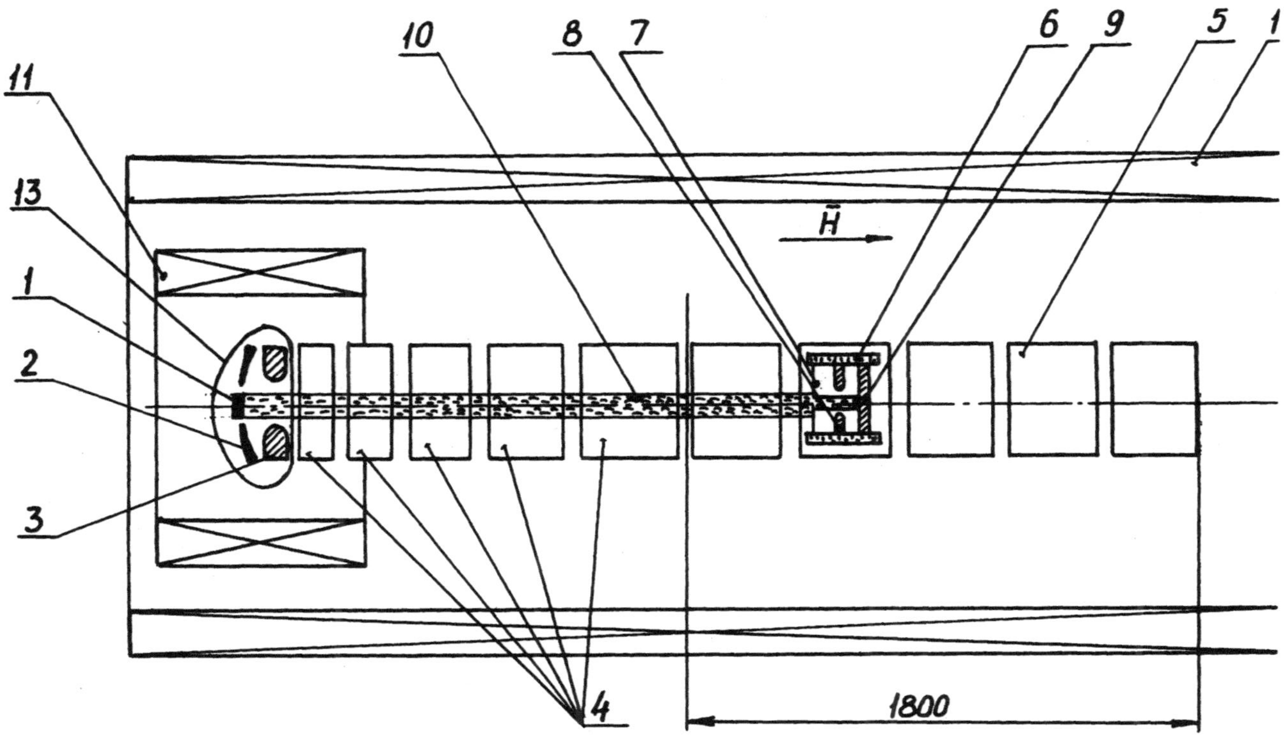

Fig.6 - The scheme of the experimental installation.

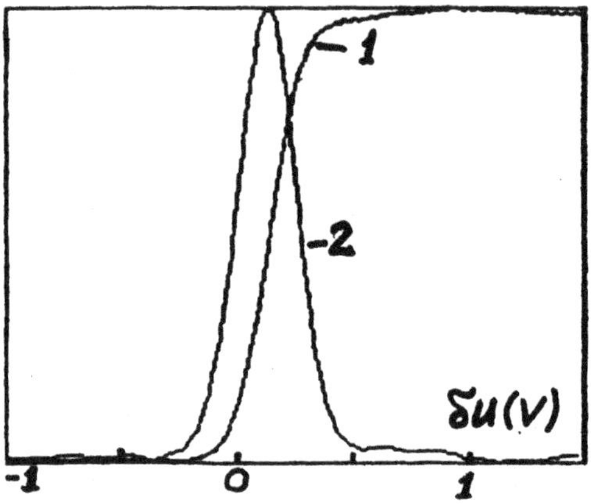

Fig.7 - The measured integral distribution function of electrons over energy (1) and its derivation (2).

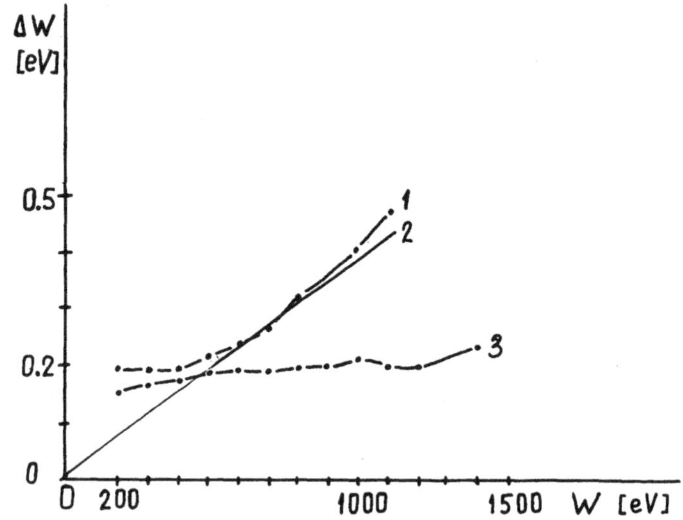

Fig.8 - Dependence of the energy spread upon the beam energy for fast acceleration at different sizes of the cutting hole. $I = 1.5\ \mu A$, $H = 1.2kG$.

1 - $d = 5mm$, $r = 0.15mm$

2 - is the result of calculation by formula (18) at $d = 5mm$, $r = 0.15mm$.

3 - $d = 10mm$, $r = 0.07mm$

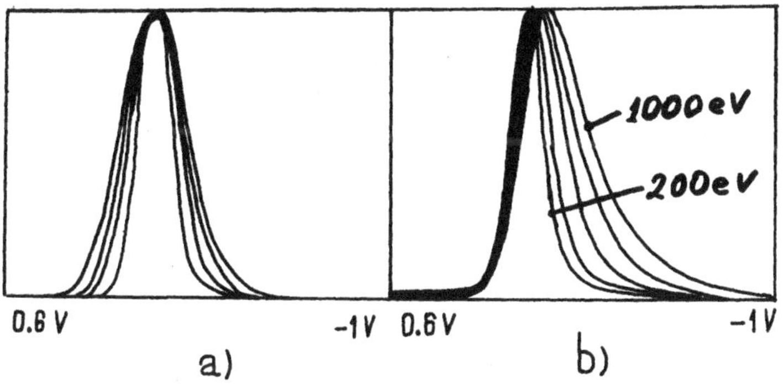

Fig.9 - Dependence of the distribution function of electrons over energy upon everage energy at fixed current for copper () and tantal diaphragms, d=5mm, r=0.15mm, W=200-1000eV, I=20μA.

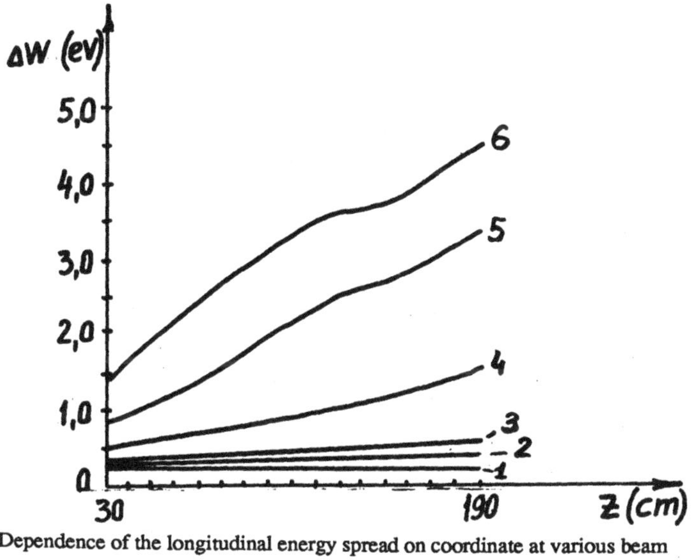

Fig.10a)Dependence of the longitudinal energy spread on coordinate at various beam current:

1- 100μA, 2- 400μA, 3 - 300μA, 4 - 1600μA, 5 - 3200μA, 6 - 6400μA, 7 - 9000μA

for various magnitude of magnitude field: a- H = 1kGs,

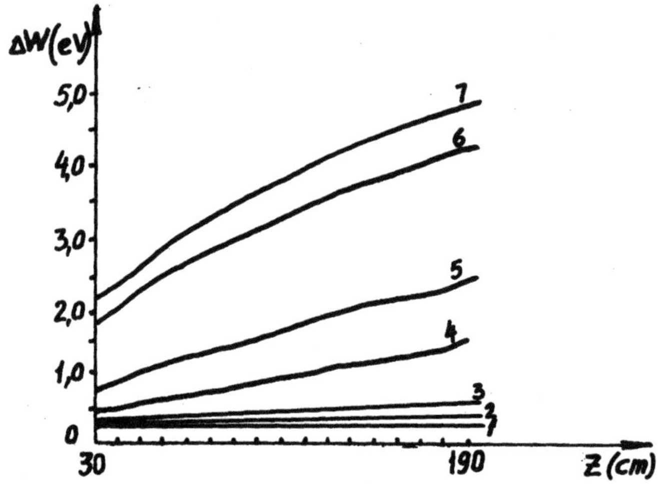

Fig.10b)Dependence of the longitudinal energy spread on coordinate at various beam current:

1- 100μA, 2- 400μA, 3 - 300μA, 4 - 1600μA, 5 - 3200μA, 6 - 6400μA, 7 - 9000μA

for various magnitude of magnitude field: b-H = 2kGs,

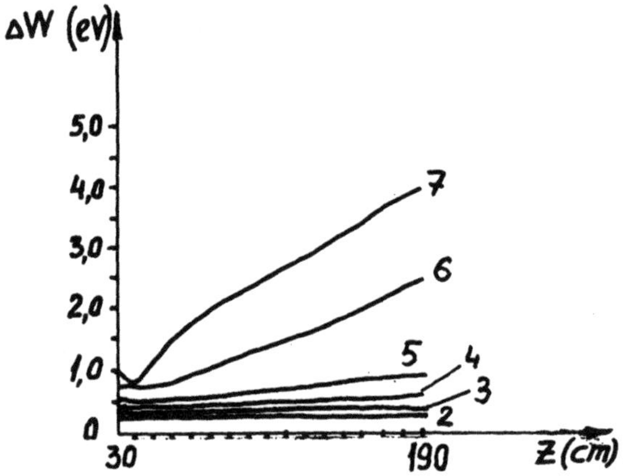

Fig.10c)Dependence of the longitudinal energy spread on coordinate at various beam current:

1- 100μA, 2- 400μA, 3 - 300μA, 4 - 1600μA, 5 - 3200μA, 6 - 6400μA, 7 - 9000μA

for various magnitude of magnitude field: c- H = 3kGs.

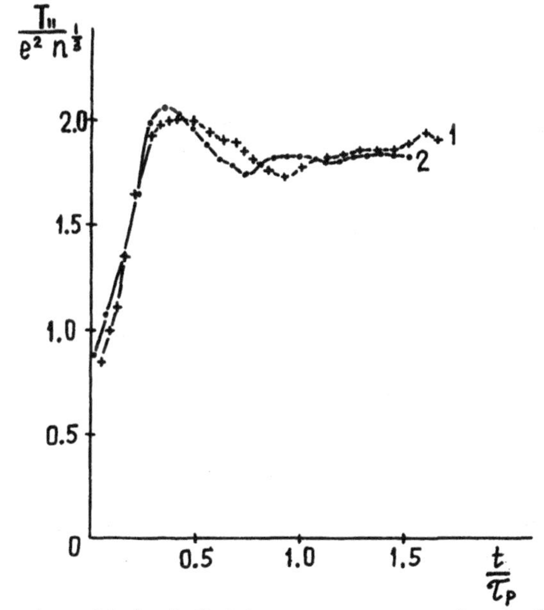

Fig.11-Dependence of the longitudinal electron temperature upon time after fast accelaration.

++++ - is a result of computer simulation.

. . . . - is the experimental plots.

I = 300μA, W = 800eV, n = 3.73 · 107 cm^{-3}, v = 1.6 · 10^9 cm / c.

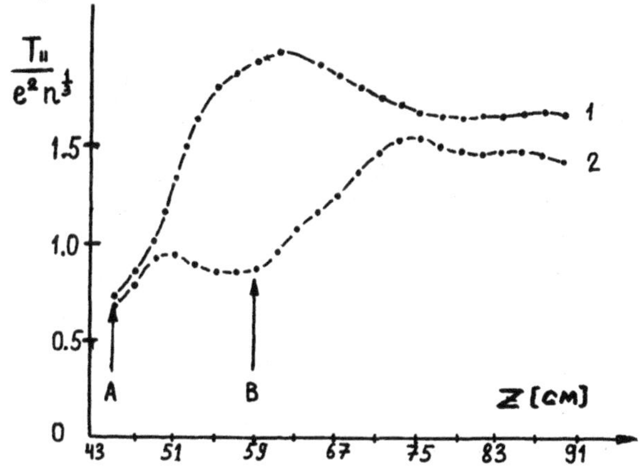

Fig.12-Dependence of the longitudinal electron temperature upon coordinate for fast acceleration (1) and acceleration with drift space (2). I = 100μA, W = 800eV. A- is the first position of the first acceleration gap, B- is the second one.

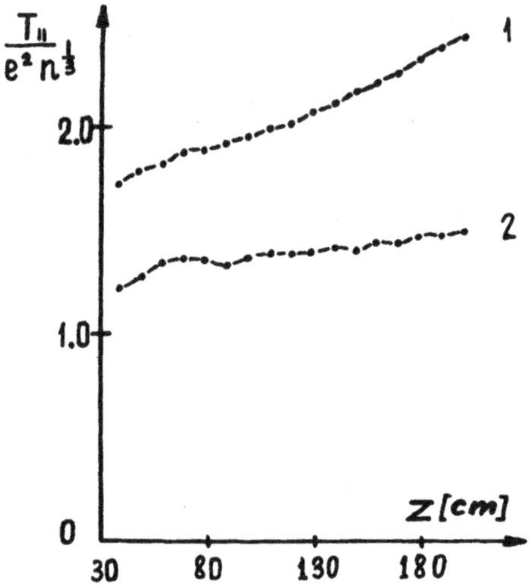

Fig.13- Dependence of the longitudinal electron temperature on coordinate after fast (1) and slow (2) acceleration. I = 200μA, W = 470eV.

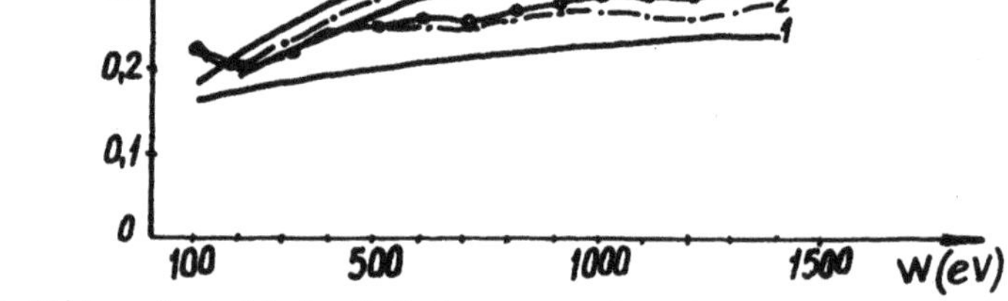

Fig.14aDependence of the longitudinal energy spread upon beam energy at different current for various regims of acceleration. H = 1.2kG. a) -I = 100μA,
Smooth lines are theoretical dependences: 1 - for the adiabatic acceleration, 4.6 for the fast acceleration according to (5) (4 - c = 1.6, 6 - C = 2).
Experimental results: 2 - slow acceleration, 3 - intermediate regim, 5 - fast acceleration.

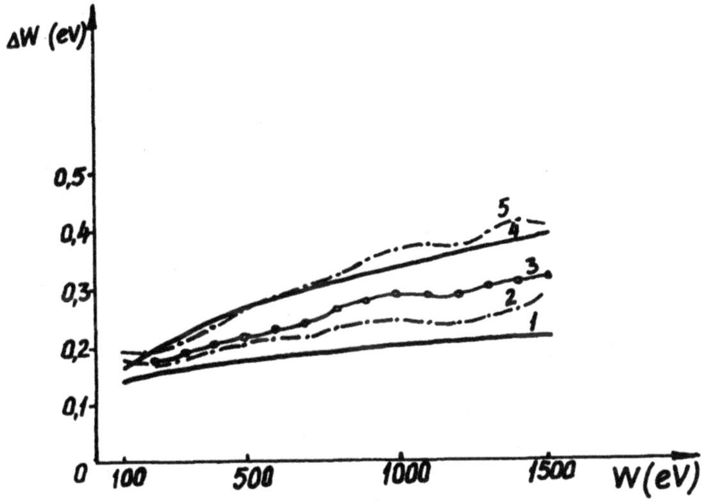

Fig.14b)Dependence of the longitudinal energy spread upon beam energy at different current for various regims of acceleration. H = 1.2kG. b) -30μA.
Smooth lines are theoretical dependences: 1 - for the adiabatic acceleration, 4.6 fo the fast acceleration according to (5) (4 - c = 1.6, 6 - C = 2).
Experimental results: 2 - slow acceleration, 3 - intermediate regim, 5 - fast acceleration.

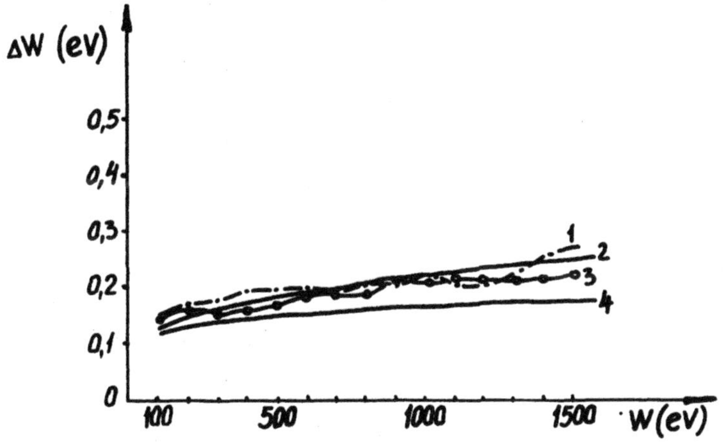

Fig.15- Dependence of the longitudinal energy spread upon beam energy for small current (I = 5μA) 1 - fast acceleration, 3 - slow acceleration.

CRYSTALLINE BEAMS

N. Dikansky - D. Pestrikov
Institute of Nuclear Physics, 630090 Novosibirsk, USSR

ABSTRACT

The report reviews some specific properties of extremely cold beams of heavy particles in storage rings, when the beam can form regular structures along its orbit. Generally, the dimension and symmetry of such crystalline beams depends on its intensity, which can cause the difference in the phase transition for the beams of low and high intensity. The possible diagnostics of regular structures in stored beams will be also discussed.

1. The interest to the ordering phenomena in extremely cold Coulomb systems periodically appears in different fields of physics. Using Monte-Carlo and molecular dynamic (MD) simulations Hansen and others [1,2] showed that the infinite classical one-component ion plasma (OCP) can undergo the phase transition into solid state and form the bcc lattice (so-called Wigner crystal), if its plasma coupling parameter

$$\Gamma = \frac{(Ze)^2}{aT} \tag{1}$$

exceeds the critical value $\Gamma_c = 170$. Here Ze is the charge of the ion,

$$a = \left(\frac{3}{4\pi\rho}\right)^{1/3} \tag{2}$$

is so-called Wigner-Seitz radius, ρ is the density and T is the temperature of ions.

Successful experiments on the electron cooling [3,4,5] demonstrated it as the reliable tool to obtain in storage ring the beams of heavy particles (protons, antiprotons and ions) with extraordinary small temperatures (see in *Table 1*). Due to dominant contribution into the friction force from magnetized electrons [3], after the cooling ion (for short) temperature T can reach the longitudinal temperature of cooling electrons

$$T = T_e \tag{3}$$

rms beam radius $\sigma_\perp$	$5\ 10^{-3}$ cm
$T_\perp$	150 K
$\Delta p/p$	$1.4\ 10^{-6}$
T	1.5 K
longitudinal cooling rate λ	$100\ c^{-1}$
transversal cooling rate $\lambda_\perp$	$12\ c^{-1}$

Table 1. Parameters of the low-intensity cooled proton beam in NAP-M [3].

Experimentally [6] T_e is estimated by $T_e \sim e^2\ \rho_e^{1/3}$. Thus, for beams of the multi charged ions, which have the density of the same order of magnitude as electron beam, this yields

$$\Gamma_i = Z^2 \Gamma_e \tag{4}$$

Even if $\Gamma_e \cong 1$, for ion beams with $Z \geq 13$ one may expect that after the cooling Γ_i will exceed the critical value and the stored beam will undergo the phase transition by forming ordered structures along the orbit with equal spacing between particles. Due to the focusing of particles trajectories towards the given closed orbit, the simmetry of such beams, which now are referred to as crystalline ones, can be very different from that, found for the infinite OCPs. Both from general and particle accelerator physics points of view the study of crystalline beams is among the most fascinating problems for the ion storage rings, which are currently planned, under construction, or already in operation.

2. Originally, the possibility for the crystallline beams in ion storage rings was discussed in [7,8] to explain the anomalies in the dependence of the longitudinal temperature of the cooled proton beam in NAP-M on the beam intensity *N*. These data, obtained by the recalculation from Schottky noise measurement [9] (see fig.1), were in strong disagreement with predictions of the existing IBS theory especially due to the indication of the threshold, which clearly can be seen in fig.1.

At least the qualitative agreement with observations can be obtained [8], if one takes into account two important effects, which usually are ignored in conventional IBS calculations. The first is the separation in the focusing system of particles collisions into fast and adiabatic and the shrinking for cold beams of the region of impact parameters, corresponding to fast collisions. The second is the enhancement with the beam cooling of the contribution into the beam relaxation from the energy exchange between particles and beam coherent fluctuations. This channel is especially important for the relaxation of magnetized particles. It is contributed by coherent mode, which decay during time intervals τ_{fl} comparable with the beam relaxation time

$$\tau_{fl} \sim \tau_{rel} \tag{5}$$

If all modes are stable, this channel can be significant, if the number of fluctuations N_{fl}, satisfying (5), is comparable with the number of particles in the beam *N*. The relaxation stops, when the temperatures of particles and coherent modes become equal $T = T_{fl}$. Once they are fixed by the cooling device at the very low level, this will mean strong ordering and, generally, strong Coulomb interaction of particles inside the beam at the equilibrium. This circumstance reflects the common point that ordered states are caused by the enhancement of the collective phenomena at low temperatures.

The simmetry of the ordered state depends on the beam intensity. In the normal, coasting beam the motion of particles near the given closed orbit is described by the following equations

$$z = a_z \cos\psi_z \ , \quad x = a_x \cos\psi_x + \eta \frac{p}{p_s} \ ,$$

$$\Delta p = p - p_s \ , \quad p_{x,z} = \frac{p_s}{R_0}\frac{d}{d\theta}(x,z) \ , \tag{6}$$

$$\theta = \omega_s t + \varphi \ , \quad \psi_{x,z} = \omega_s \nu_{x,z} \ , \quad \varphi = \alpha_p \omega_s \frac{\Delta p}{p_s} \ ,$$

$$\alpha_p = \gamma^{-2} - \gamma_{tr}^{-2} \ , \quad I_{x,z} = \frac{p_s}{2R_0}(\nu \cdot a^2)_{x,z}$$

Here smooth focusing approximation was used, $p = \gamma m v$, $v = R_0 \omega_s$, ω_s is the beam revolution frequency, s marks the values referred to the closed orbit, $mc^2 \gamma_{tr}$ is the ring transition energy, η is the dispersion function of the ring. Without coherent oscillations the distribution function of such a beam is uniform in betatron phases $\psi_{x,z}$ and φ.

$$f_{st} = f(I_x, I_z, \Delta p) \tag{7}$$

Coherent oscillations of the beam are described by non-uniform additions to f_{st}

$$f = f_{st} + \sum_{m_x m_z} \sum_n f(I_x, I_z, \Delta p, t) \exp(i m_x \psi_x + i m_z \psi_z + i n \varphi) \tag{8}$$

Integer numbers (m_x, m_z,n) describe the simmetry of normal coherent oscillations (betatron (m_x, m_z), longitudinal (n), or combined). If the Coulomb interaction between beam particles predominates, coherent modes decay due to Landau damping, caused by corresponding frequencies spreads or, probably, due to the beam cooling. For non-relativistic ion beam the main contribution into the betatron frequency spread is caused by the Coulomb tune shift

$$\Delta\nu = \frac{Z^2}{A}\,\frac{Nr_p}{2\pi\varepsilon_\perp (v/c)^2} \quad , \quad r_p = e^2/mc^2 \tag{9}$$

Here A is the ion mass number, $\varepsilon_\perp$ is the beam emittance. If $\Delta\nu$ is in the conventional ranges (say, for NAP-M $\Delta\nu \lesssim \Delta\nu_{th} = 0.15$), due to strong dilution of betatron coherent oscillations only one-dimensional (1-d or string) ordering may be expected. Since the lifetime of the longitudinal oscillations is determined by the revolution frequency spread $\Delta\omega$, the level of the ordering in the beam is controlled by the ratio of the average longitudinal particles spacing $a = 2\pi R_0/N$ to the beam Debye radius

$$r_D = R_0 / n_D \;,\; n_D = \frac{2NZ^2e^2R_0}{\pi\sigma_\perp^2 T_\parallel} \quad , \; T_\parallel = \frac{pv}{\alpha_p}\,(\Delta\omega/\omega_s)^2 \tag{10}$$

via parameter

$$\Gamma^{(1)} = \left(\frac{2\pi n_D}{N}\right)^2 = \frac{4Z^2e^2a_\parallel}{\sigma_\perp^2 T_\parallel} \quad , \quad \sigma_\perp^2 = \varepsilon_\perp R_0 / \nu_\perp \tag{11}$$

The calculations for the string beam, consisting of particles spread over the disks with areas $\pi a_x a_z$, showed [8] that in the region $\Gamma^{(1)} \ll 1$ its basic thermodynamic variables are only slightly modified by the Debye-Huckel corrections. On the contrary, in the region $\Gamma^{(1)} \gg 1$ the most favorable *N*-particle distribution function for the beam becomes

$$f_N \,\alpha\, \exp\left\{ -\frac{n_D^2}{2}\sum_{k=1}^{N}\left(\varphi_k - \frac{2\pi k}{N}\right)^2 - \frac{\Delta p^2 \alpha_p}{2m\,T_\parallel} \right\} \tag{12}$$

which corresponds to one-particle density

$$\rho(\varphi) = \frac{n_D}{(2\pi)^{1/2}} \sum_{k=1}^{N} \exp\left\{ -\frac{n_D^2}{2}\left(\varphi - \frac{2\pi k}{N}\right)^2 \right\} \tag{13}$$

with equal spacing $a_\parallel$ between particles, executing longitudinal oscillations with rms amplitude

$$\sqrt{<\varphi^2>} = 1/n_D << 2\pi \ / N.$$

Respectevely, the internal energy and specific heat of the beam get the values

$$E = NT_{\|} , \ C = N \ ,$$

which are characteristic for 1-d crystal chain. Without the calculation of thermodynamic variables for the liquid state, the critical value for $\Gamma^{(1)}$ can be estimated using the Lindemann melting criterion [1]

$$\gamma_L^2 = < \sum_{k=1}^{N} \left(\frac{N\varphi_k}{2\pi} - k \right)^2 > / N \ , \quad (\gamma_L)_c = 0.17 \qquad (14)$$

Since for a chain $\Gamma^{(1)} = 1/\gamma_L^2$ this yields $\Gamma_c^1 \cong 3$ and therefore, for data from *Table 1*

$$N_c = 1.5 \ 10^5, \ a_{\|} = 3 \ 10^{-2} \ \mathrm{cm} \gg \sigma_\perp \ .$$

These values are in general agreement with criterion for string ordering

$$\lambda = N a / \Pi \leq 0.7,$$

which was obtained in [10] using MD simulations, but are out of the range of measurements at NAP-M (see fig.1).

In fact, the real longitudinal temperature of the beam in NAP-M might differ from the value given in Table 1, which was recalculated from Schottky noise powers using formula in Debye-Huckel approximation ($2\pi n_D \ll N$). To exclude this we can estimate from data of fig.1 and using [11]

$$\frac{1}{T}\left(\frac{\partial P}{\partial \rho}\right)_T = \lim_{n \to 0} \frac{N}{< |\rho_n|^2>} \qquad (15)$$

the isothermal compressibility $(\partial P/\partial \rho)_T$ of the beam (here $\delta\rho$ is the fluctuation of ρ, P is the pressure). Fig.2 shows that in the range of measurement $(\partial P/\partial \rho)_T$ strongly exceeds the value specific for the ideal gas $(\partial P/ \partial \rho)_T = 1$. Thus we may conclude that in these experiments $\Gamma^{(1)}$ (or Γ) was high enough. To estimate $T_{\|}$ we may adopt, for istance, $\Gamma^{(1)} \geq 1$ for $N \leq 3 \ 10^7$ (see figs 1,2). Then eq. (11) yields $T \leq 2.6 \ 10^{-6}$ eV. This value rather corresponds to $T_e \cong T_c^2 / 2W_e$ (We is the electron kinetic energy, T_c is the cathode temperature). Nevertheless, these data give no indication was the beam in solid or liquid state.

For heavy ion beams the string ordering can be expected for

$$N \ll N_c = \frac{Z^2}{A} \frac{4r_p\Pi}{\Gamma_c\, \alpha_p\, \sigma_\perp^2\, (\Delta p / p)^2\, (v/c)^2} \tag{16}$$

This threshold increases, when the beam energy is tuned to the transition energy of the ring. One more limitation on *N* is implied by the requirement $\Delta\nu \le \Delta\nu_{th}$,which reads

$$a_\parallel\, \sigma_\perp^2 \ge \frac{Z^2}{A} \frac{r_p\, R_0^2}{\nu\Delta\nu\, (v/c)^2}$$

For parameters of NAP-M this yields $N \le 3\ 10^7$, which is very close to the point, where the curve in fig.2 changes the slope. The violation of the particles magnetization in the string beam can be expected, when [8]

$$N \le N_c = \frac{2\pi\nu_{x,z}}{\alpha_p\, (\Delta p / p)}$$

3. For 3-d crystallization the damping of betatron coherent oscillations must be suppressed. For non-relativistic energies of ions it may occur only in beams of high intensity, where due to Coulomb repulsion coherent oscillations of normal beams become unstable and therefore $\Delta\nu$ is out of conventional range. If the normal beam is injected in the ring, before getting the transverse ordering during the cooling it must cross the critical point (the threshold of instability) with associated blow-up of the beam noise [8] and enhancement of IBS [11]. This is the main difference with the case of string ordering, when such an instability is not necessary. Qualitativelly, the beginning of such processes in the beam may indicate the data in fig. 3. On the right slope one has $(\partial P/\partial \rho)_T / T \propto \rho^{-5}$, which is specific for the vicinity of the liquid-to-gas critical point.

General properties of the instability must be very close to that caused by the beam-beam interaction [12]. It may be expected, when the cooling tunes towards nonlinear resonances

$$\Delta_{m,n} = m_x\nu_x + m_z\nu_z - 2n \approx 0 \tag{17}$$

from the above. Typically such resonances have buckets (see in fig.4), where particles can be stored, if cooling is not too strong. For particular resonances (the modest beam density) the helix (or shell) structure

$$m_x\Delta\nu_x\, (\varepsilon_x, \varepsilon_z) + m_z\Delta\nu_z\, (\varepsilon_x, \varepsilon_z) = \Delta_{mn}$$

can be expected.

Originally such shell structures for ion beams have been found in MD simulations [13] (see in fig.5). It was found also that on the shell surface particles are arranged in hexagonal patterns. All these phenomena appear, if Γ exceeds at least 100 [10,13]. These simulations and the calculation of the beam energy minima [10] showed that as far as Γ is large, the beam radial structure is mainly controlled by the dimensionless density λ. For instance [10], one has: the smallest distance between particles 1.77 a, the radius of the beam 1.15 $\lambda^{1/2}$, the difference of radii between shells 1.54 a, and the total number of shells. 75 $\lambda^{1/2}$. In the agreement with more general expectations [11], when N increases, new shells appear gradually [13] without crossing new critical temperatures. Recently shell structures have been detected in ion traps with use of the laser cooling [14].

The influence of the alternating gradient focusing and bending of particles trajectories on the beam ordering as a heating sources was studied in [13] and more systematically in [15].

This is very delicate problem since the ring lattice has to accept beams both below and well above space charge limit. It has been found that without bending the Coulomb crystal is elastic enough to keep ordering with deformations due to periodic focusing. The calculations [15] indicate also that high symmetry lattices and small betatron tunes are favorable.

One more obstacle for 3-d ordering is expected [13] from specific for storage rings modulation of particle azimuth by the radial betatron oscillations - so-called shear force. With normal beams this modulation is responsible for the radial-longitudinal coupling. But in ion beam with $\Gamma \gg 1$ conventional normal coordinates (x, z, θ) must be strongly coupled by Coulomb interaction. In fact, simulations [13] indicate in the solid state helix particles trajectories. This can campensate shear provided the step of the helix is not too large. Once in the crystalline state the beam can be replaced by the ideal gas of normal oscillations, a particle feel the bending moving either in a magnet or in a straight section of the ring. Though, simulations show [15] that shear force can be important at least in the region $\Gamma \leq 100$, where the long-range order may not exist. It can be suppressed in rings with large bending radii and strong focusing [13]. Simulations [13,16,17] also shows that in the liquid flows with radial velocity gradient shear can lead to new ordering into strings, which in the beam cross section are arranged in hexagonal patterns.

4. Since the transition of the real beam into the solid state is still too hard problem for both theoretical and numerical analysis, experimental observation of crystalline beams is of primary importance. This can be done using more or less standard techniques

- scattering experiments with co-moving beams and measurements of the beam structure factor

$$(S (n) = < |\delta\rho_n|^2 > /N) \quad ;$$

the measurement of the beam response functions and spectra; detectin of the coherent light of probe electron beam in the field of the crystalline one. Complementary possibilities appear in the diagnostic of unfully stripped ion beams using lasers [17].

In fact, the measurement of the friction force presents the simplest experiment with scattering. No anomalies, which could be referred to the beam structure, have been detected here up to date. Schottky noise measurements at NAP-M seem to indicate the anomaly in IBS. Though, the interpretation of those results needs additional data.

The study of coherent oscillations in the beam can yield frequency spectra and especially Q-values of normal modes. As it was discussed before, useful data concerning thermodynamic state of the beam can be obtained from (longitudinal and transversal) Schottky noise measurements, even if performed in the usual frequency range. For these purposes the detection of multipole coherent oscillations (quadrupole, sextupole, etc.) can be very informative.

REFERENCES.

[1] Hansen J.P., Phys.Rev. A8, 3096, 1973; Pollock E.L., Hansen J.P. Phys.Rev. A8, 3110, 1973.

[2] Ichimaru S. Rev.Mod.Phys., 54, 1017, 1982.

[3] Budker G.I. Collection of papers. M. Nakua 1982; Budker G.I., Skrinsky A.N. U.Ph.N., 124, 561, 1978; Parkhomchuk V.V., Skrinsky A.N. Phys. Elem. Part.Atomic Nucl., 12, 557, 1981.

[4] Bell M., et al. NIM, 190, 237, 1981; Poth H., et al. Proc. of the Workshop on Crystalline Beams, GSI-89-10, 96, 1989.

[5] Forster R., et al. IEEE Trans. on Nucl. Sci., NS-28, 2386, 1981.

[6] Kudelainen V.I., et al. JTEPh, 82, 2056, 1982.

[7] Parkhomchuk V.V. Proc of ECOOL-84, KfK 3846, 71, 1984.

[8] Dikansky N.S., Pestrikov D.V. ibid, p.275, 1984; Same authors. Physics of Intense Beams in Storage Rings. Novosibirsk. Nauka, 1989.

[9] Dementiev E.N., et al. Sov. Phys. Tech. Phys., 25, 1001, 1981.

[10] Hasse R.W., Schiffer J.P. CSI-89-10, 33, 1989.

[11] Landau L.D., Livshiz E.M. Statistical Physics. M.Nauka, 1964.

[12] Dikansky N.S., Pestrikov D.V. Part. Acc., 12, 27, 1982; Same authors. Preprint INP 90-14, Novosibirsk, 1990.

[13] Rahman A., Schiffer J.P. Phys.Rev.Lett., 57, 1133, 1986;

Schiffer J.P. GSI-89-10, 2, 1989.
[14] Bollinger J.J., et al. GSI-89-10, 231, 1989.
[15] Hofmann I., Struckmeier J. GSI-89-10, 140, 1989.
[16] Habs D. GSI-89-10, 43, 1989.
[17] Hess S. GSI-89-10, 321, 1989.

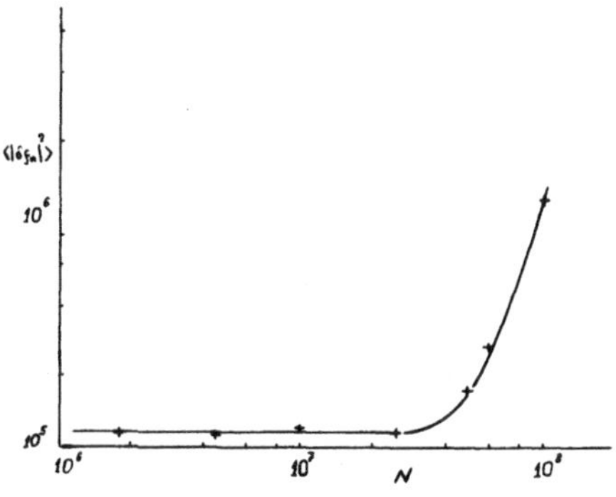

Fig. 1 - The Schottky noise of a cooled beam in NAP-M vs N.

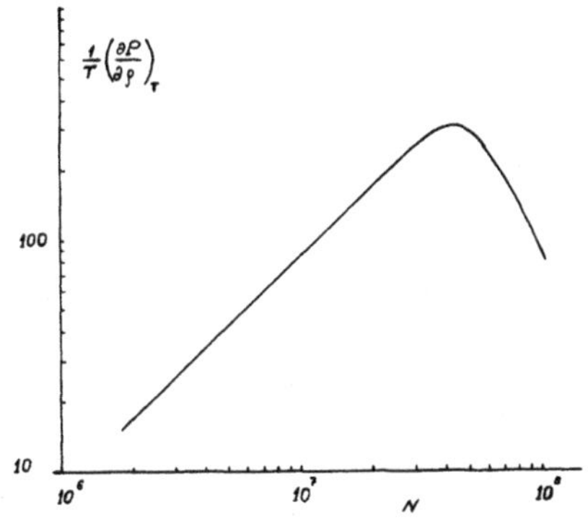

Fig. 2 - The isothermal compressibility of the beam vs N recalculated from fig.1.

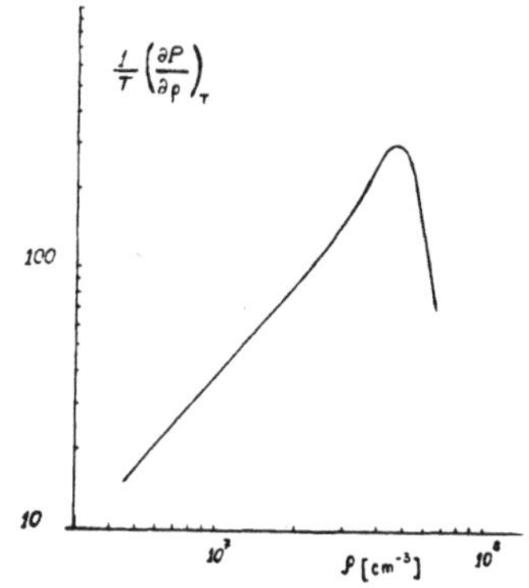

Fig. 3 - The same as in fig.2 vs the beam density ρ.

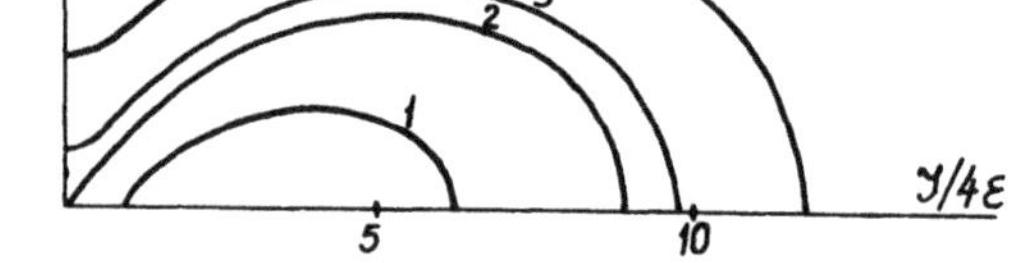

Fig. 4 - The upper right quarter of phase trajectories of non-linear coherent oscillation driven by the Coulomb repulsion near resonance $\Delta = \nu - n << \Delta\nu$; $\Delta/\Delta\nu = 0.1$, trajectory 2 is the separatrix; $\sqrt{2JK_0/\nu}$ is the amplitude of coherent oscillations.

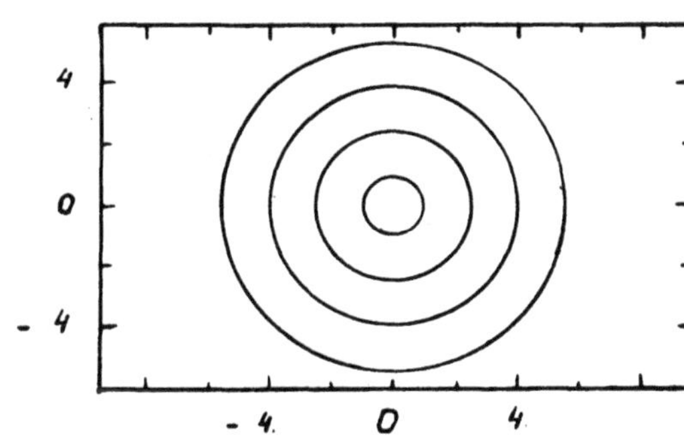

Fig. 5 - Schematic position of shells in the beam cross section from MD simulations [13]; $\Gamma = 2\ 10^4$, $\lambda = 26$.

MEASUREMENTS OF THE LONGITUDINAL ENERGY DISTRIBUTION OF A COLD ELECTRON BEAM FROM GaAS

A.V. Aleksandrov, N. S. Dikansky, N. Ch. Kot, V. I. Kudelainen, V. A. Lebedev, P.V. Logacev, B. M. Smirnov
(Institut of Nuclear Physics So An USSR, Novosibirk)

R. Calabrese, S. Bigoni, G. Bonora, V. Carassiti
(Dipartimento di Fisica Università di Ferrara and INFN Ferrara, Italy)

L. Tecchio, G. Abbrugiati
(Dipartimento di Fisica Sperimentale Università di Torino and INFN Torino, Italy).

1. INTRODUCTION.

The use of cold Gallium Arsenide photocathodes as emitters of electrons [10], allows to obtain, for the first time, low temperature electron beams in comparison with normal hot cathodes. Nevertheless the experimental results of the measurements carried out with this beam [2], have showed a quick increase of the temperature whenever electron current and energy are increased. The attempt of this work is to study this effect. The observed increase in the temperature has been showed to be connected with a deep intensity modulation of the laser radiation on the band of hundreds megahertz.

2. EXPERIMENTAL SET UP.

The measurements of the longitudinal energy distribution of the electron beam obtained from the Gallium Arsenide photocathode have been carried out on a modified "Solenoid Type" equipment. The scheme of the cathodic group is represented in Figure 1; the fundamental parameters of the equipment are reported in the following Table.

Electron energy	up to 800 eV
Solenoid magnetic field	0.5 - 4 KGs
Range distance for the longitudinal temperature detector from the cathode surface	0.3 ÷ 1.8 m
Lenght of accelerating structure	0.25 m
Photocathode diameter	2 mm
Vacuum	5.10^{-11} Torr

Table. Parameters of the experimental equipment.

The electron beam, generated from an electron gun immersed in the solenoid magnetic field, is carried along the magnetic field to the collector. The electron gun has been designed according to the standard Pierce optics. Directly from the first anode of the gun starts the acceleration section; the inner diameter is 16 mm and the beam, along the whole distance from the first anode to the measurements collector, travels in a constant section channel. All tubes are isolated from each other and from the zero potential, too; this allows to get the required distribution of potential along the beam. Afterwards, if not expressely specified, the tubes are supposed to be connected with earth. The measurements of the energy distribution of the electrons is carried out with the retarding potential method and is described in detail elsewhere [3]. For measurements, the collector can be moved along the beam within the limits 0.3 to 1.8 m. from the cathode surface. In order to diminish the optical distortion of the gun, cesium evaporators are installed in the external part of the anode electrode and through its slits are directed on the photocathode surface. Oxygen is passed into the vacuum channel near the photocathode through a long tube. Oxygen introduction is controlled by a leak valve. The laser beam is introduced through a window into the vacuum chamber and is oriented on the crystal surface through a mirror. In ref. [3] a Helium-Neon laser with 0.63 µm wavelenght and 7mW power has been used.

The high vacuum is carried out by ion pumps and non evaporable getters (N.E.G.). The ion pumps have a total capacity of 800 l / s. and are installed on the two outputs of the solenoid vacuum chambers, while the N.E.G. (four modules WP 1250) are installed in the middle of it. The vacuum chamber has been heated to 250°C.

3. PHOTOCATHODE ACTIVATION.

The GaAs crystal surface has been cleaned with pure acetone before its installation in the vacuum chamber. The complementary cleaning of the photocathode surface, before the activation, is possible by heating the GaAs crystal at 640°C by a tantalum heater (fig.1). During the measurements, the photocathode can be cooled to the temperature of the liquid nitrogen. The electric insulation of the cathodic group from the liquid nitrogen container is obtained through a ceramics insulating tube. The cathode temperature is controlled by a thermocouple fixed on the cathode electrode. The procedure of cleaning and activation of the photocathode is described in the ref. [4]. The activation is obtained by evaporating cesium on the GaAs crystal. The photoelectric current appears about 15 min. after the starting of the cesium evaporators; it reaches a peak value and then decreases. At this moment oxygen is introduced into the cathode region causing in this way an additional current increase. The current peak value is obtained by a subsequent evaporation of cesium and oxygen (Fig. 2). The activation procedure ends with the oxygen introduction and afterwards the current still goes on increasing for a while. The maximun current emission has been of 50 μA obtained illuminating the photocathode with Helium - Neon laser of 7mW power. This corresponds, without considering laser intensity loss due to reflections and lasses in the diaphragm (according to the valutations they can reach more than 50%), to a quantum efficiency of about 1.4% . The life time of the photocathode is 48 hours for a $5 \cdot 10^{-11}$ torr. vacuum and 40 μA current.

4. EXPERIMENTAL RESULTS.

In order to control the quality of the gun optics and the regularity of the photocathode illumination and emission, the distribution of the current density on the beam section is measured. This is obtained by a transversal scanning (by magnetic corrections) of the electron beam in the small aperture at the entrance of the collector. Fig.3 puts in evidence the subsequent variations in the distribution of the current density with the passage from the regime "3 / 2" (a) to the regime of limited emission (d), when the voltage on the gun anode changes, and Fig.4 puts in evidence the relative variations of the integral function concerning the distribution of the electrons along the longitudinal energy :

$$F(\delta E_{\parallel}) = \int f(\delta E'_{\parallel})\, d\delta E'_{\parallel} \qquad (1)$$

measured in the middle of the beam. Here $dE_{\parallel}$ is the shift of the electron energy from the energy which corresponds to the tension on the cathode. Both the figures show the considerable dependence of the beam properties on the gun working regime. If the distortion in the distribution of the current density seems to be understandable and is connected with the non homogeneity of the laser emission and radiation, the sudden increase, on the contrary, in the energy dispersion when passing from the regime "3 / 2" to the regime of the limited emission seems to be quite incomprehensible. As a matter of fact, the observed energy dispersion (more than 2eV) exceeds both the photon energy (1.96 eV) and its space charge of the electron beam,

$$\phi = \frac{2I}{V_0}\left(\ln\left(\frac{b}{a}\right) + \frac{1}{2}\right) \cong 0.7\ \mathrm{V} \tag{2}$$

when the electron current is 40μ A and the cathode tension is 40V. Here V_0 is the electron velocity, a and b are respectively the beam and vacuum channel rays.

In order to explain the increase in the energy dispersion, we shall consider a simple model. Let's suppose that the laser radiation intensity is modulated in high frequency:

$$I(t) = I_0 + \delta I \cos\omega_0 t \tag{3}$$

In the rEGIME of limited emission, the electron current and density will be modulated in the same way in the drift region. The presence of the density space modulations in the beam, leads to an excitation of the complementary longitudinal movement and consequently to an increase in the scattered energy. Let's suppose that the magnetic field is strong enough to neglect the electron transversal movement. The linear equations in the particle rest frame system will be:

$$\begin{aligned} & m\frac{\partial V_z}{\partial t} = eE_z \\ & \frac{\partial n}{\partial t} + n_0\frac{\partial V_z}{\partial z} = 0 \\ & \Delta\phi = -4\pi ne \end{aligned} \tag{4}$$

where e and m are respectively the electron charge and mass, n_0 is the non excited electron density. For a longitudinally unlimited beam, it is possible to find a solution of the type $e^{i(kz-\omega t)}$, where $k = \omega_0 / V_0$ is the primary state of the density space modulation. If the beam radius is small in comparison with the vacuum chamber radius, the dependence of the beam features on the radial coordinate can be neglected. In this case the last equation of (4) can be written as:

$$(k^2 + k_\perp^2)\,\phi = 4\pi ne\ , \quad \text{where } k_\perp = \frac{1}{a}\sqrt{\frac{2}{\ln(b/a)}}$$

After simple computations, a relationship between the velocity harmonics and the density perturbations is obtained:

$$V_z = \frac{\omega_p}{\sqrt{k^2 + k_\perp^2}} \frac{n}{n_0}$$

In considering again the system of laboratory coordinates, longitudinal energy perturbation in function of the time and of the longitudinal coordinate is obtained :

$$\frac{\delta E\,(z,t)}{E_o} = \frac{2\omega_p}{V_o \sqrt{k^2 + k_\perp^2}} \frac{n}{n_o} \sin\left(\frac{\omega_p\, z\, k}{V_o \sqrt{k^2 + k_\perp^2}}\right) \cos\,(\omega_o\, t) \qquad (5)$$

where

$$\omega_p = \sqrt{4\pi n_o\, e^2 / m}$$

is electron plasma frequency.

For a beam having an energy E_0=40eV, a density $n_o = 6.10^7\ cm^{-3}$ and a density modulation $n / n_0 = 0.5$, we obtain $\delta E_{max} = 4eV$. Such a high value of the longitudinal dispersion is connected with its increase when passing from the system of the beam coordinates to the laboratory system:

$$\delta E = \sqrt{2Eo\ \delta E'} \qquad (6)$$

where $\delta E'$ is the energy dispersion in the beam system.

The relaxation time corresponds in magnitude to the period of the plasma oscillations and, according to the example above mentioned, the lenght, along which the increase in the energy dispersion for the considered parameters occurs, is equal to:

$$l_{rel} = \frac{\pi}{2} \frac{V_o \sqrt{k^2 + k_\perp^2}}{\omega_p\, k} \cong 3.4\ cm$$

For the normal thermocathodes the modulation of the cathode current density is determined only by thermic fluctuations and these non-omogeneities are negligible, beacuase the longitudinal energy is [5] :

$$T_{\|} = \frac{T_k^2}{2E_0} + 1.6\, e^2 n_0^{1/3}\ , \quad \delta E = \sqrt{2\ Eo\ T_{\|}} \qquad (7)$$

where T_k is the cathode temperature.

In Fig. 5 are represented comparatively the relationships concerning the longitudinal energy dispersion in the particle rest frame system in unit $e^2\, n^{1/3}$ (a), the transit time of electron from the gun to the collector in periods of plasma oscillations (b), the electron current I and the microperveance $I / W^{3/2}$ (c,d) with respect to the difference of voltage between cathode and anode. The energy dispersion is calculated in the same way as in ref. [5], measuring the slope of the integral function of the distribution among the values (0.3 - 0.7) of the maximum. For an electron current lower than the so-called critical current, the gun always works in the regime "3 / 2" and the current density modulation in the gun is strongly depressed. In comparison with the étotal beam current, we obtain an evaluation of the modulation depth of the laser radiation $\Delta I / I_0 = 0.5$. We can observe that during the electron passage from the gun to the collector several dozens plasma oscillations are carried out.This, also with small beam parameters variations, leads to consistent phase variations containing the coordinate z in the expression (5). Because of the homogeneity in this phase, the dependence of the longitudinal energy distribution on the longitudinal coordinate during the measurements disappears. To the additional homogeneity corresponds also the current heterogeneity in the beam section and the corrisponding strong dependence of the plasma frequency on the transversal coordinate. But also for the gun working regime of "3 / 2", when the energy distribution is minimun, its value exceeds twice $e^2n^{1/3}$.This is probably connected with weak fluctuation of the electron density for the harmonics higher than the laser radiation fluctuations (f = 1GHZ), when the oscillation period can be compared with the transit time from the cathode to the anode, (1 - 2) 10^{-9} sec.

In order to become convinced that the observed increase in the width of the energy distribution is dependent on the beam transport system and that it is not a photocathode propriety, we have carried out the following experiments.

To establish the relaxation time of the electron beam density modulations, we have used the following measurement scheme. Between the cathode and the anode we have applied a high voltage (about 300 V), so that the gun can work in a regime of limited current. Immediately after the acceleration the beam was slown down to 4 eV energy and was moved with this energy to a fixed distance. Afterwards the beam was accelerated to a total energy of 300eV and moved to the collector. If during the transport the density non-homogeneities are responsible of the increase in the longitudinal energy distribution, this one was contained by the subsequent acceleration. Neverthless if this relaxation was unsuccessful, the density modulations passed mostly in the accelerated beam, and this caused after the relaxation a much higher value of the width of the energy distribution. By reducing the transport sector lenght of the low energy, we observe a jump which increase the longitudinal energy distribution in the beam. In Fig. 6 we have reported the dependence existing between this distribution and thetransport sector lenght. As we can see in the figure, the distance along which the reaction occurs is about 20 cm, which correspond to about 10 plasma oscillations.

The fact that the electric field at the cathode surface does not influence the energy distribution has been verified measuring the correlation between the longitudinal temperature and the cathode voltage, when the beam energy has been fixed. In this case when the difference of tension between cathode and anode was increased, the gun entered in the "3 / 2" regime (and there was a corresponding absence of a correlation between the current and the voltage) and the distribution function did not change during the anode voltage variations.

In order to make sure of the truthfulness of what has been above described, we have directly measured the fluctuation spectrum of the laser radiation power in the range 0 - 1.5 GHz. The measurements have been carried out by a photoelectric receiver with wide spectrum semiconductors and the results are reported in the Figure 7. The spectrum consists of equidistant peaks connected with the interference of the different longitudinal modes of the laser radiation. The distance between close modes is defined by the resonators lenght and is equal to $\Delta f = c / 2L = 251$ MHz.The calculated potential modulation depth is about 50%, which well coincides with the evaluation obtained by studying the longitudinal relaxation procedure above mentioned.

The represented spectrums are not stable enough. The relative intensity of the peaks changes in a discontinuous way, and this evidently changes the character of the relaxation in the electron beam and considerably impedes the realization of the experiments and the interpretation of the results, too.

According to the known spectrum of the power fluctuations of the laser radiation in linear approximation, the shape of the distribution integral has been calculated. By summing up along the laser radiation power harmonics from (5) we obtain:

$$\frac{\delta E}{Eo} \cong \frac{4\,\omega_p}{\pi\, Vo} \sum_n \frac{In}{Io} \frac{\cos(n\omega_o t + \varphi_n)}{\sqrt{k_n^2 + k_\perp^2}}\;, \quad k_n = n\omega_o / Vo$$

The energy distribution function will be proportional to:

$$f(\delta E) \approx < 1 / \left| d(\delta E)/dt \right| >$$

where the average is calculated on the phases φ_n.

Figure 8 shows the integral function of the energy distribution obtained by numerical calculations for two modes with an intensity ratio of 4 / 3. We can see the conformity between the shape of the integral function of the enrgy distribution obtained by numerical calculation and the experimental one (see Fig. 4).

The experiments carried out well show that the observed increase in the energy distribution, when the current and the energy of the beam obtained from the photocathode are increased, is connected with the laser radiation power modulation because of the interference of the close modes which derive from laser. It has been showed that, in order to obtain a low temperature, it is necessary excite the photocathode with a single mode laser source.

REFERENCES.

1) D.T. Pierce, R.J. Celotta, G.C. Wang, W.N. A.Galegjs, C.E. Kuyat S.R. Mielczarek. GaAs spin polarized electron source. Rev. Sci. Instrum. 51 (1980), 478.
2) C.S. Feigerle, D.T. Pierce, A. Stiler, R.J. Celotta, Appl. Phys. Lett. 44 (1984) 866.
3) R. Calabrese, F. Ciccacci, V.A. Lebedev, V.V. Parkhomchuk, A.A. Sery and L. Tecchio. Il Nuovo Cimento 101A (1989) 333.
N.S. Dikansky, V.I. Kudelainen, V.A. Lebedev, I.N. Meshkov, V.V. Parkhomchuk, A.A. Sery, A.N. Skrinsky and B.N. Sukhina; Ultimate Possibilities of electron cooling; INP, Preprint 88-61, Novosibirsk, URSS.
4) R. Calabrese, V. Kudelainen, A. Novokhaski and L. Tecchio. A high intensity electron source from GaAs.
5) A.V. Aleksandrov, N.S. Dikansky, N.C. Kot, V.I. Kudelainen, V.A. Lebedev, P.V. Logachov. Temperature relaxation and adiabatic acceleration of magnetized electron flux; these proceedings.

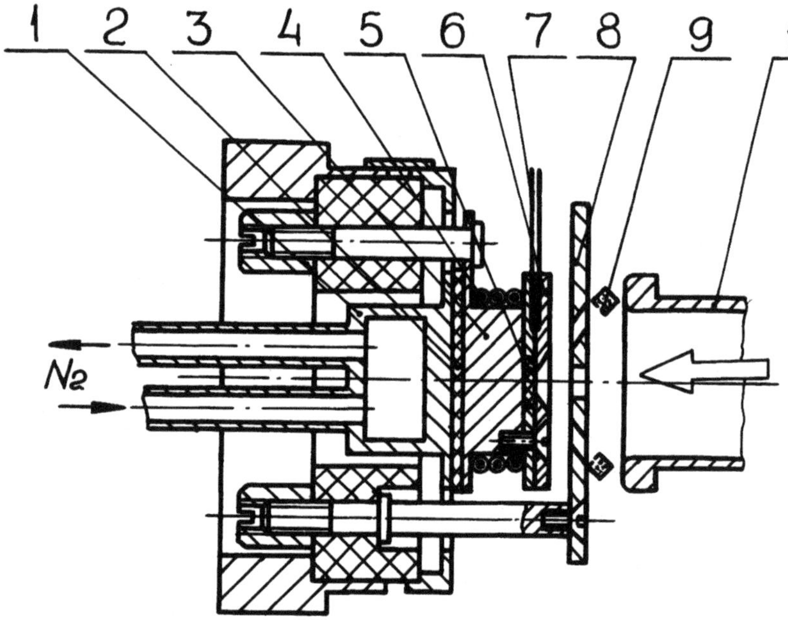

Fig. 1 - The cathode assembly: 1 - cold vase of liquid nitrogen, 2 - ceramics disk, 3 - molibdenum support for the GaAs crystal, 4 - heater, 5 - GaAs crystal, 6 - Pierce electrode (Mo), 7 - thermocouple, 8 - anode, 9 - cesium dispenser, 10 - first electrode of the accelerating structure.

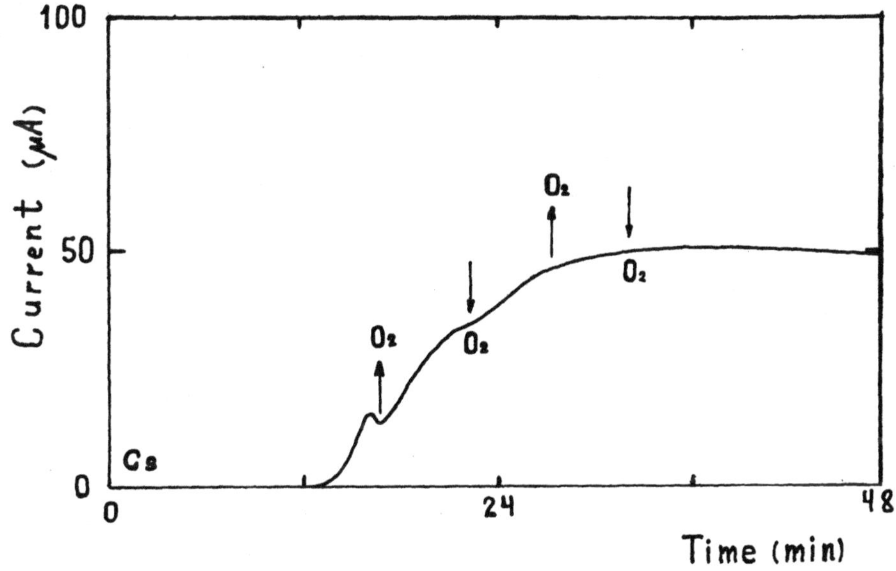

Fig. 2 - Dependence of the photoelectric current on the time in the activation process.

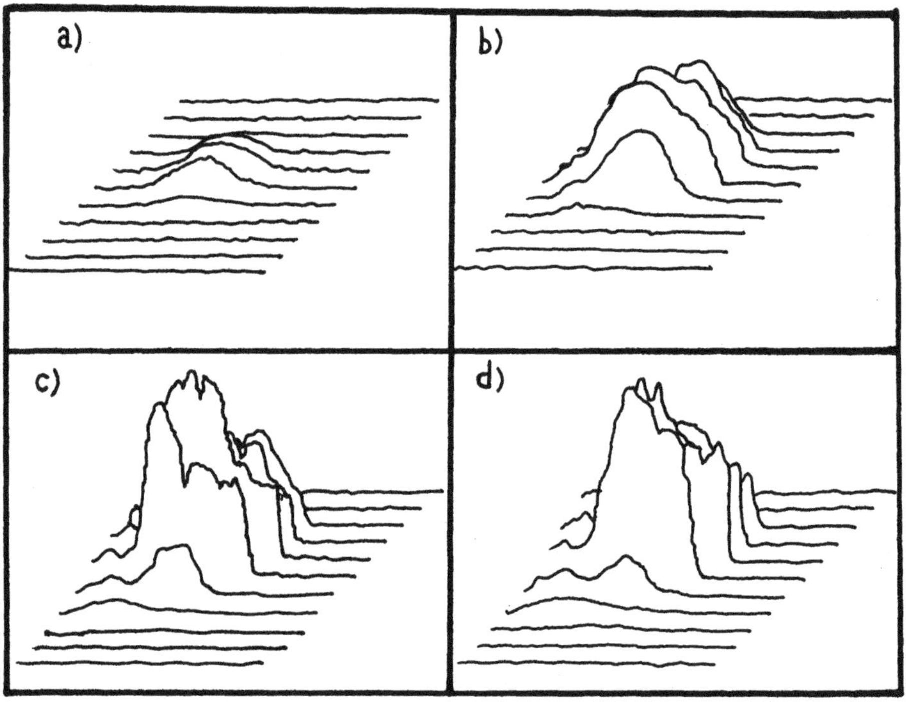

Fig. 3 - Current density distribution at different conditions:
a) W = 7eV, I = 0.8μA b) W = 18eV, I = 13μA
c) W = 30eV, I = 20μA d) W = 40eV, I = 21μA

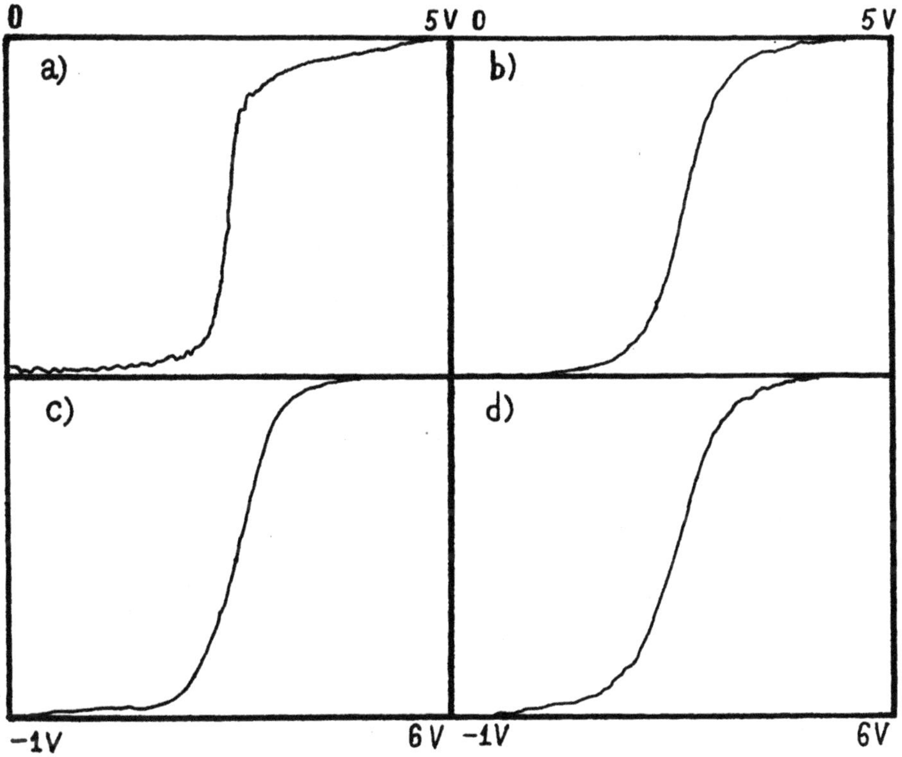

Fig. 4 - Integral function of electron distribution versus the longitudinal energy. The gun parameters in the cases a),b),c),d), Fig. 3 are given in.

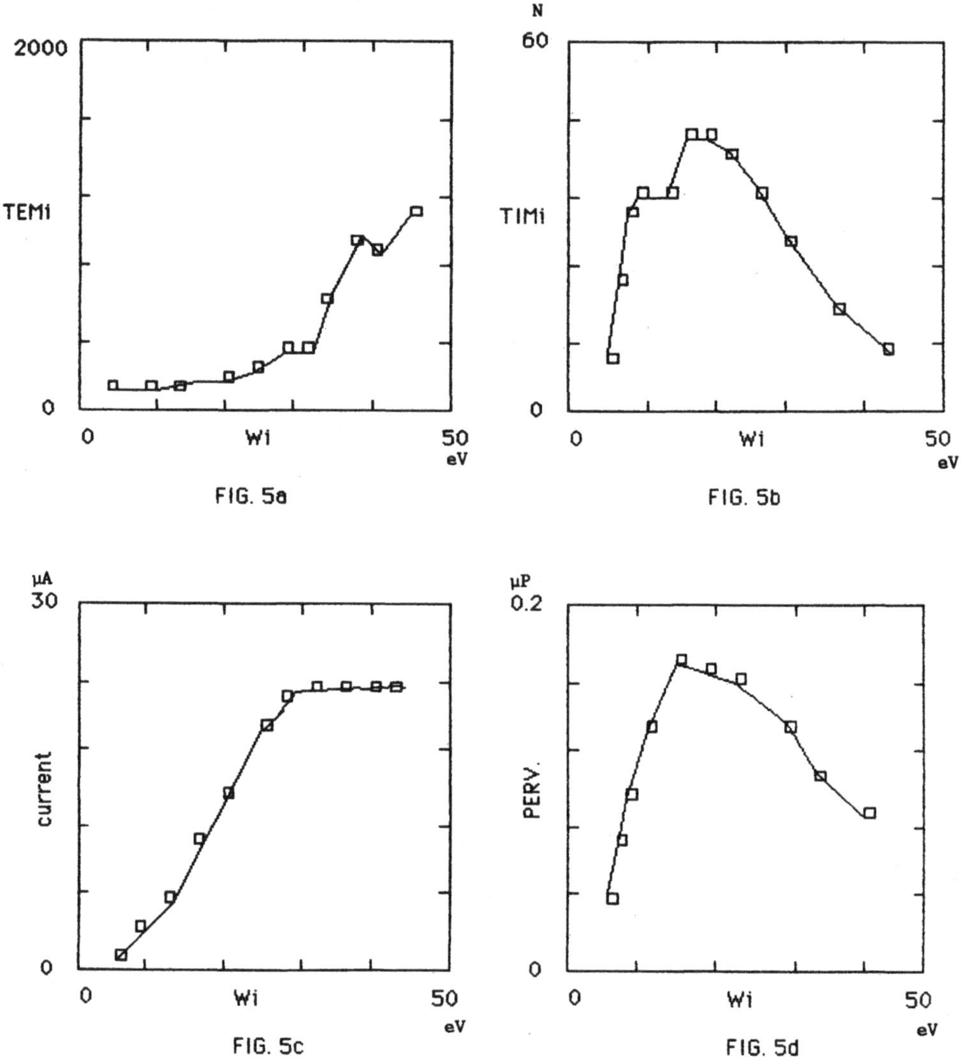

FIG. 5a FIG. 5b FIG. 5c FIG. 5d

Fig. 5 - a) Longitudinal energy dispersion in function of the beam energy, in $e^2n^{1/3}$ units;

b) Number of plasma oscillations along the drift region, at different beam energies;

c) Electron current at different energies;

d) Perveance at different beam energies.

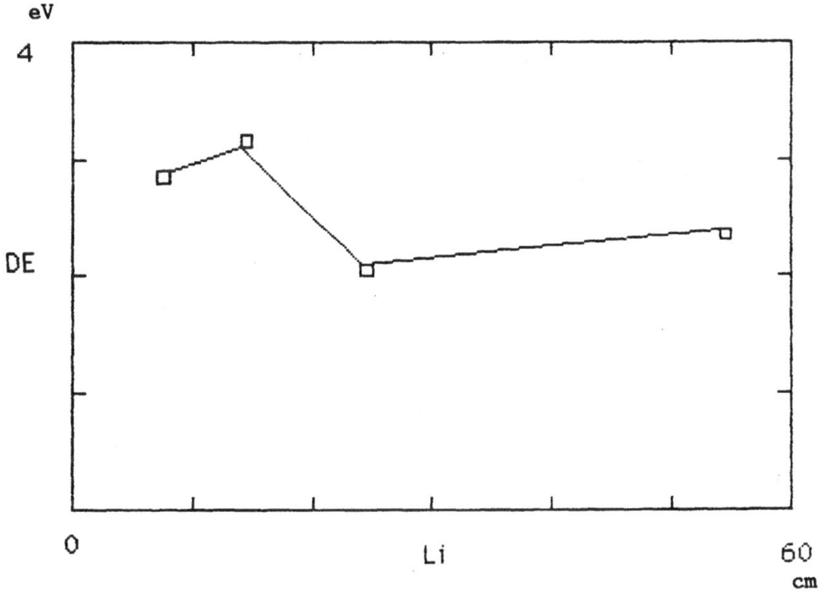

FIG. 6

Fig. 6 - Longitudinal energy dispersion at low beam energy (4eV), along the drift region.

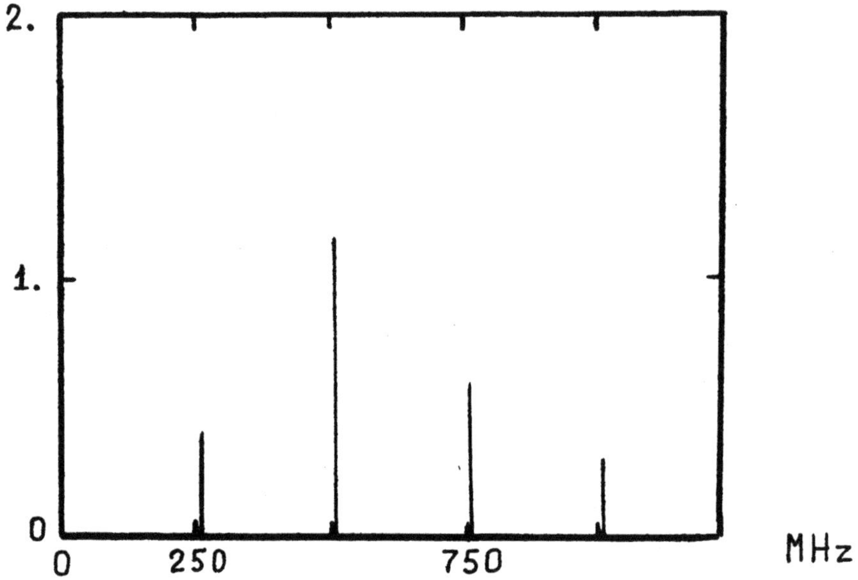

Fig. 7 - Spectrum of the power fluctuation of the He-Ne laser (0.63μm,7mW).

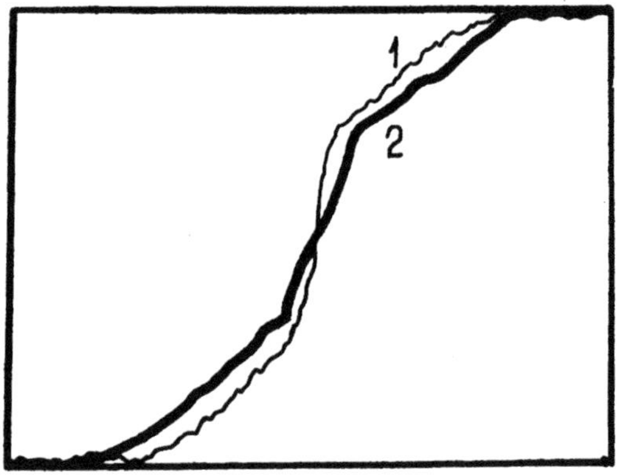

Fig. 8 - Integral function of the energy distribution. The beam parameters are W = 100eV and I = 2μA and the horizontal scale is: curve 1→20eV curve 2→7.2eV

LIST OF PARTICIPANTS

E. ADAMIDES, INFN - Laboratori Nazionali di Legnaro, Legnaro, Padova
A. ARAGONA, INFN - Laboratori Nazionali di Frascati , Frascati, Roma
G. P. BEZZON, INFN - Laboratori Nazionali di Legnaro, Legnaro, Padova
M. BIANCONI , PROEL Tecnologie S.p.A., Firenze
G. BISOFFI, INFN - Laboratori Nazionali di Legnaro, Legnaro, Padova
I. BOSCOLO, Dipartimento di Fisica dell'Università, Milano
J. BOSSER, C.E.R.N.- PS/AR,Geneve, Switzerland
R. CALABRESE, Dipartimento di Fisica dell'Università, Ferrara
G. CARUGNO, INFN - Laboratori Nazionali di Legnaro
M. CASTELLANO, INFN - Laboratori Nazionali di Frascati, Frascati, Roma
N. CAVALLO, INFN - Dipartimento di Scienze Fisiche, Napoli
M. CAVENAGO, INFN - Laboratori Nazionali di Legnaro, Legnaro, Padova
F. CERVELLERA, INFN - Laboratori Nazionali di Legnaro, Legnaro, Padova
F. CEVENINI, INFN - Dipartimento Scienze Fisiche, Napoli
V. CHIMENTI, INFN - Laboratori Nazionali di Frascati, Frascati, Roma
A. CLOZZA, INFN - Laboratori Nazionali di Frascati, Frascati, Roma
M. CONTE, INFN - Dipartimento di Fisica dell'Università, Genova
A. DAINELLI, INFN - Laboratori Nazionali di Legnaro, Legnaro, Padova
P. DALPIAZ, INFN - Laboratori Nazionali di Legnaro, Legnaro, Padova
K. DANARED, Manne Siegbahn Institute of Physics, Stockholm, Sweden
G. DI MASSA, INFN -Dipartimento di Scienze Fisiche, Napoli
N. DIKANSKY, INP, Novosibirsk, USSR
T. ELLISON, Indiana University Cyclotron Fac.,Bloomington, IN, USA
P. FAVARON, INFN - Laboratori Nazionali di Legnaro, Legnaro, Padova
G. FORTUNA, INFN - Laboratori Nazionali di Legnaro, Legnaro, Padova
U. GASTALDI, INFN - Laboratori Nazionali di Legnaro, Legnaro, Padova
A. GHIGO, INFN - Laboratori Nazionali di Frascati, Frascati, Roma
V. GUIDI, INFN - Laboratori Nazionali di Legnaro, Legnaro, Padova
D. HABS, MPI Fur Kernphysik, Heidelberg, Germany
J. HANGST,Fermi Lab., Batavia, USA
L. HERMANSSON, The Svedberg Laboratory, Uppsala, Sweden
K. KILIAN, Institut fur Kernphysik, Julich, Germany
M. KLENOID, Institut fur Angewandte Physik, Frankfurt, Germany
P. KRAUSE, MPI fur Kernphysik, Heidelberg, Germany
J. M. LAGNIEL, CEN, LNS, Saclay, France
V. LEBEDEV, Nuclear Physics Institute, Novosibirsk, USSR

R. LEY, C.E.R.N. PS/AR, Geneve, Switzerland
E. LUPPI, INFN - Dipartimento di Fisica dell'Università, Ferrara
M. R. MASULLO INFN - Dipartimento di Scienze Fisiche, Napoli
H. O. MEYER, Department of Physics, Indiana University, Bloomington,IN, USA
M. MUSIC, MPI Fur Kernphysik, Heildeberg, Germany
S. PAVLOV, Institute for Nuclear Research, Kiev, USSR
R. PENGO, INFN - Laboratori Nazionali di Legnaro, Legnaro, Padova
W. PETRICH, MPI Fur Kernphysik, Heidelberg, Germany
D. PESTRIKOV, Institut of Nuclear Physics, Novosibirsk, USSR
F. PETRUCCI, Dipartimento di Fisica dell'Università, Ferrara
R. POLLOCK, Indiana University Cyclotron Facility, Bloomington, IN, USA
M. PUSTERLA, Dipartimento di Fisica dell'Università di Padova, Padova
K. G. RENSFELT, Manne Siegbahn Institute of Physics, Stockholm, Sweden
A. RUGGIERO, Brookhaven National Laboratory, Upton, NY, USA
G. F. SERAFINI, INFN - Laboratori Nazionali di Frascati, Frascati, Roma
A. SHARAPA, Institut of Nuclear Physic, Novosibirsk, USSR
V. STAGNO, Dipartimento di Fisica dell'Università, Bari
M. STECK, MPI Fur Kernphysik, Heidelberg, Germany
H. J. STEIN, KFA - IKP, Julich, Germany
B. SUKHINA, Institut of Nuclear Physic, Novosibirsk, USSR
T. TANABE, Institute for Nuclear Study, University of Tokyo, Japan
L. TECCHIO, Dipartimento di Fisica dell'Università, Torino
B. TIVERON, INFN - Dipartimento di Fisica dell'Università, Padova
G. TRANQUILLE, C.E.R.N.- PS Division, Geneva, Switzerland
V. VARIALE, INFN - Dipartimento di Fisica dell'Università, Bari
A. WOLF, MPI Fur Kernphysik, Heidelberg, Germany
A. ZHMENDAK, Institute for Nuclear Research, Kiev, USSR
S. ZWICKLER, MPI Fur Kernphysik, Heidelberg, Germany

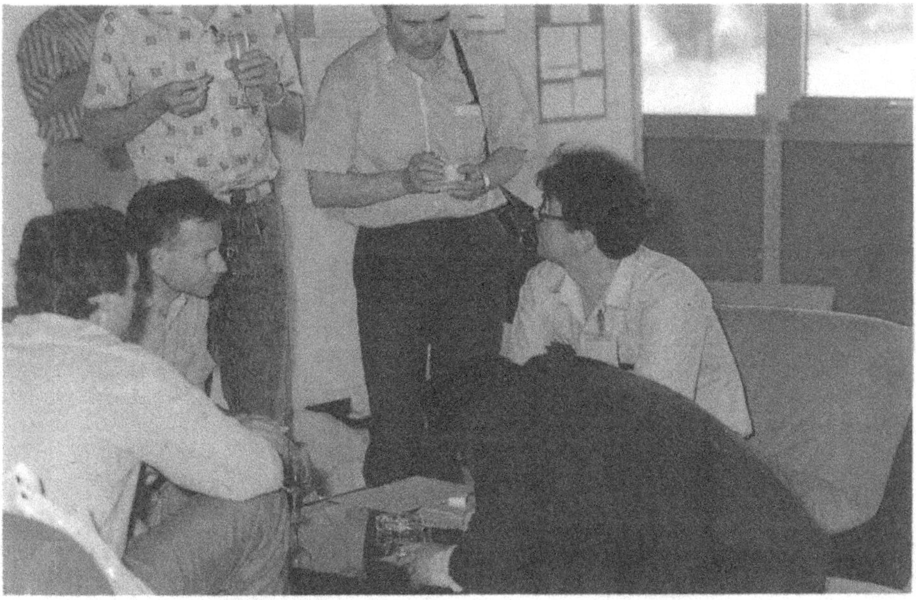

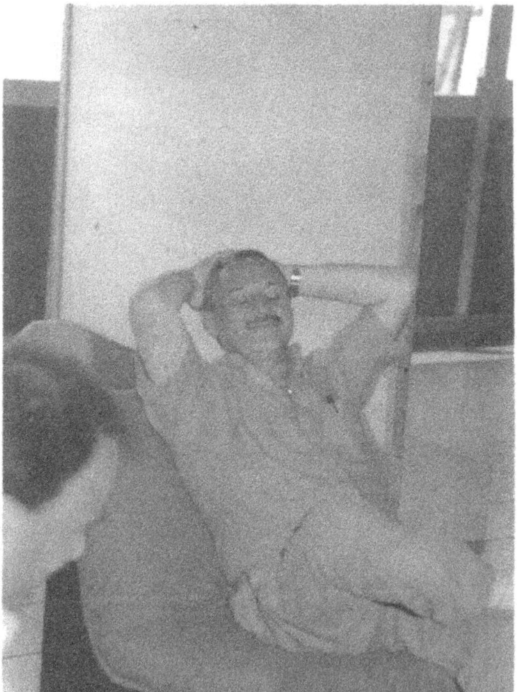

www.ingramcontent.com/pod-product-compliance
Lightning Source LLC
LaVergne TN
LVHW082255241026
850861LV00008B/53
* 9 7 8 9 8 1 0 2 0 5 1 1 9 *